Handbook of Porphyrin Science

with Applications to Chemistry, Physics, Materials Science, Engineering, Biology and Medicine

HANDBOOK OF PORPHYRIN SCIENCE
with Applications to Chemistry, Physics, Materials Science, Engineering, Biology and Medicine

ISSN: 1793-9518

Vol. 1 Supramolecular Chemistry
Vol. 2 Synthesis and Coordination Chemistry
Vol. 3 Synthetic Methodology
Vol. 4 Phototherapy, Radioimmunotherapy and Imaging
Vol. 5 Heme Proteins
Vol. 6 NMR and EPR Techniques
Vol. 7 Physicochemical Characterization
Vol. 8 Open-Chain Oligopyrrole Systems
Vol. 9 Electronic Absorption Spectra — Phthalocyanines
Vol. 10 Catalysis and Bio-Inspired Systems — Part I
Vol. 11 Catalysis and Bio-Inspired Systems — Part II
Vol. 12 Applications
Vol. 13 Synthesis and Structural Studies
Vol. 14 Theoretical, Electron Transfer and Physical Studies
Vol. 15 Biochemistry of Tetrapyrroles
Vol. 16 Synthetic Developments — Part I
Vol. 17 Synthetic Developments — Part II
Vol. 18 Applications and Materials
Vol. 19 Biochemistry of Tetrapyrroles — Part II
Vol. 20 Chlorophylls and Related Systems
Vol. 21 Catalysis
Vol. 22 Biophysical and Physicochemical Studies of Tetrapyrroles
Vol. 23 Synthesis
Vol. 24 Coordination Chemistry and Materials
Vol. 25 Vitamin B12 and Related Systems
Vol. 26 Heme Biochemistry

Vol. 27 Erythropoiesis, Heme and Applications to Biomedicine
Vol. 28 Chlorophyll, Photosynthesis and Bio-inspired Energy
Vol. 29 Porphyrias and Sideroblastic Anemias
Vol. 30 Heme Proteins — Part II
Vol. 31 Synthesis II — Part II
Vol. 32 Materials
Vol. 33 Applications II — Part II
Vol. 34 Harnessing Solar Energy
Vol. 35 Cummulative Index for Volumes 1–34
Vol. 36 BODIPYs and Chlorins: Powerful Related Porphyrin Fluorophores
Vol. 37 Synthesis and Reactivity of Exotic Porphyrinoids
Vol. 38 Towards Green Chemistry
Vol. 39 Towards Diagnostics and Treatment of Cancer
Vol. 40 Nanoorganization of Porphyrinoids
Vol. 41 Novel Porphyrinoid Precursors
Vol. 42 Towards Tuned Properties of Porphyrinoids
Vol. 43 Design of Precursors for Sustainable Chemistry
Vol. 44 Bio-Inspired Porphyrin Scaffolds for Synthesis and Catalysis
Vol. 45 Phthalocyanine Synthesis and Computational Design of Functional Tetrapyrroles

Handbook of Porphyrin Science

with Applications to Chemistry, Physics, Materials Science, Engineering, Biology and Medicine

Volume 45

Phthalocyanine Synthesis and Computational Design of Functional Tetrapyrroles

Editors

Karl M. Kadish
University of Houston, USA

Kevin M. Smith
Louisiana State University, USA

Roger Guilard
Université de Bourgogne, France

World Scientific

NEW JERSEY • LONDON • SINGAPORE • BEIJING • SHANGHAI • HONG KONG • TAIPEI • CHENNAI • TOKYO

Published by

World Scientific Publishing Co. Pte. Ltd.
5 Toh Tuck Link, Singapore 596224
USA office: 27 Warren Street, Suite 401-402, Hackensack, NJ 07601
UK office: 57 Shelton Street, Covent Garden, London WC2H 9HE

Library of Congress Control Number: 2011290361

British Library Cataloguing-in-Publication Data
A catalogue record for this book is available from the British Library.

HANDBOOK OF PORPHYRIN SCIENCE
With Applications to Chemistry, Physics, Materials Science, Engineering, Biology and Medicine — Volume 45: Phthalocyanine Synthesis and Computational Design of Functional Tetrapyrroles

ISBN 978-981-120-180-6

For any available supplementary material, please visit
https://www.worldscientific.com/worldscibooks/10.1142/11320#t=suppl

Typeset by Stallion Press
Email: enquiries@stallionpress.com

Printed in Singapore

Contents

Preface for Volume 45

We had planned to periodically release five-volume sets of the Handbook, but because of various unavoidable delays and our desire to release a set to celebrate ICPP-9 in 2016, our previous set (Volumes 41–44, May 2016) was aesthetically incomplete. We have rectified this by releasing the present Volume 45. In this Volume our authors have comprehensively addressed the science of phthalocyanine syntheses, as well as computational methodologies for the design of novel tetrapyrroles, and structural and electronic properties of chiral porphyrin systems. Yet again our contributors have exceeded our expectations in both the quality of writing and in the content of their individual chapters. We asked them to write "a review for the ages," and we anticipate that what you now see will remain relevant for at least the next 15 years, and possibly longer.

The three editors once again wish to thank many individuals — our authors, our copy editors, our production team and also World Scientific Publishing Company, for continuing to support us in the illustration of new developments with porphyrins, phthalocyanines, related tetrapyrroles, and their applications. We hope our readers will enjoy browsing these pages as much as we, the editors, enjoyed organizing the process that produced them.

Karl M. Kadish (Houston, USA)
Kevin M. Smith (Baton Rouge, USA)
Roger Guilard (Dijon, France)
February 15, 2019

Preface for Volumes 1–10

Although the porphyrin and tetrapyrrole research area was regarded as "fully matured" during the 20th century, as evidenced for example by the awarding of numerous Nobel Prizes to its principal researchers, new advances and accomplishments in the field still amaze us as editors. The area continues to blossom and to expand into new areas of science and applications that would probably never have occurred to our 20th century heroes. An earlier *Porphyrin Handbook* assembled the large amount of factual data that had been accumulated during the 20th century. Our new venture, the *Handbook of Porphyrin Science* takes a completely new look at our research area and comprehensively details the contemporary science now appearing in the scientific literature that would indeed have been hard to predict even 10 years ago. In particular, fundamentally new methodologies and potential commercial applications of the beautiful compounds that we all love are exemplified, fully recognizing the subtitle of the series — "with applications to chemistry, physics, materials science, engineering, biology and medicine".

The three of us have complementary expertise in physical chemistry, synthetic and bioorganic chemistry, and in synthetic and mechanistic organometallic chemistry; this has enabled us to cover the whole field of porphyrin science and applications, and to devise comprehensive volume and author content. As of the date of writing, between the three of us, we have published more than 1600 tetrapyrrole research articles, and hold 31 patents related to commercial applications of porphyrin science. So we do know our field, and this has enabled us to assemble a first-rate group of experts who have written comprehensive up-to-date chapters with accuracy and authority; we thank our authors for their cooperation and willingness to go along with our highly ambitious schedule for production of these volumes.

We look forward to comments from our readers, and to suggestions that might enable us to expand our basic interests and scientific coverage even further. Meanwhile, we hope that porphyrin researchers, old, new and of the future, will enjoy reading these volumes just as much as we enjoyed planning and, with the help of World Scientific Publishing Company, producing them from manuscript to published article, in a timely manner.

Karl M. Kadish (Houston, Texas, USA)
Kevin M. Smith (Baton Rouge, Louisiana, USA)
Roger Guilard (Dijon, Bourgogne, France)
January, 2010

Contributing Authors for Volumes 45*

Milan Balaz
Yonsei University
Seoul, 03722, Republic of Korea
Chapter 217

Nina Berova
Columbia University,
New York, NY 10027, USA
Chapter 217

Semyon V. Dudkin
Nesmeyanov Institute of
Organoelement Compounds of
Russian Academy of Sciences
(INEOS RAS),
Moscow, 119991, Russia
Chapter 215

Alessandro D'Urso
Università di Catania
95125 Catania, Italy
adurso@unict.it
Chapter 217

Yoonbin A. Joh
Yonsei University
Seoul, 03722, Republic of Korea

and

University of New Hampshire
Durham, NH 03824, USA
Chapter 217

Mutsumi Kimura
Shinshu University,
Ueda 386-8567, Japan
Chapter 215

Nagao Kobayashi
Shinshu University
Ueda 386-8567, Japan
Chapter 215

Roberto Purrello
Università di Catania
95125 Catania, Italy
Chapter 217

Martin J. Stillman
University of Western Ontario
London, Ontario N6A 5B7, Canada
stillman@uwo.ca
Chapter 216

Krisztina Varga
University of New Hampshire
Durham, NH 03824, USA
Chapter 217

Satoshi Yamamoto
University of Washington
Seattle, WA, 98195-2600, USA
Chapter 215

Angel Zhang
The University of Western Ontario
London, Ontario, Canada N6A 5B7
Chapter 216

*Full contact information for authors can be found on the title page of each chapter. For a complete list of Contributing Authors for Volumes 1-45, please vist https://www.worldscientific.com/series/porsci.

Contents of Volumes 45*

Volume 45

*For a complete list of Contents for Volumes 1-45, please vist https://www.worldscientific.com/series/porsci.

215 Phthalocyanine Synthesis

Satoshi Yamamoto,* Semyon V. Dudkin,† Mutsumi Kimura‡ and Nagao Kobayashi‡

*Department of Mechanical Engineering, University of Washington, Box 353600, Seattle, WA, 98195-2600, USA

†Department for Aliphatic Organoboron Compounds, N.A. Nesmeyanov Institute of Organoelement Compounds of Russian Academy of Sciences (INEOS RAS), Moscow, 119991, Russia

‡Division of Chemistry and Materials, Faculty of Textile Science and Technology, Shinshu University, Ueda 386-8567, Japan

List of Abbreviations

abs	absolute
acac	acetylacetonate
Ad	adamantyl
Ar	argon
Bu	butyl
Bn	benzyl
CBA	*o*-cyanobenzamide
1-ClN	1-chloronaphthalene
3-ClP	3-chloropyridine
DABCO	1,4-diazabicyclo[2.2.2]octane
DBN	1,5-diazabicyclo[4.3.0]non-5-ene
DBU	1,8-diazabicyclo(5.4.0)undec-7-ene
DI	1,3-diiminoisoindoline
DMAE	*N,N*-dimethylethanolamine
DMF	*N,N*-dimethylformamide
Et	ethyl
Et_3N	triethylamine
EtOH	ethanol
HMDS	hexamethyldisilazane
i	*iso-*
L	ligand (in general case)

MOA	ammonium molybdate
Me	methyl
MeOLi	lithium methoxyde
MeOH	methanol
Mes	mesityl
MW	microwave irradiation
n.r.	not reported
OAc	acetyl
P	pressure
PA	phthalic anhydride
PI	phthalimide
$PhNO_2$	4-nitrophenol
PN	phthalonitrile
Pr	propyl
Pr^i	*iso*propyl
Py	pyridine
Pyz	pyrazine
RT	room temperature
SuperPc	superphthalocyanine
TBA	tetra-*n*-butylammonium bromide
TCB	1,2,4-trichlorobenzene
TCNB	1,2,4,5-tetracyanobenzene
TEBAC	triethylbenzylammonium chloride
tert-Bu	*tert*-butyl
THF	tetrahydrofuran
TMGT	1,1,3,3-*N,N,N′,N′*-tetramethylguanidinium trifluoroacetate
p-TsOH	*p*-toluenesulfonic acid monohydrate
Δ	melting

I. Introduction

Phthalocyanines (Pcs) and metallophthalocyanines (MPcs) are well-known functional molecules due to their unique optical and electrochemical properties, as well as their ability to coordinate a wide range of metal ions.[1,2] Phthalocyanines were accidentally discovered as a byproduct in 1907,[3] and Linstead characterized the phthalocyanine structure in 1934.[4] The structure of Pcs is similar to a porphyrin ring, being an 18π aromatic macrocycle in which *meso*-carbons of benzoporphyrin are replaced with nitrogen atoms. Figure 1 shows the structures of metal-free phthalocyanine (H_2Pc) and MPc, as well as the numbering scheme traditionally used for their naming convention. The internal and external positions of the fused benzene ring are also commonly known as the α and β positions, respectively. Due to their remarkable photophysical properties and extreme chemical, thermal, and photostability, Pcs and MPcs have been used as dyestuffs and

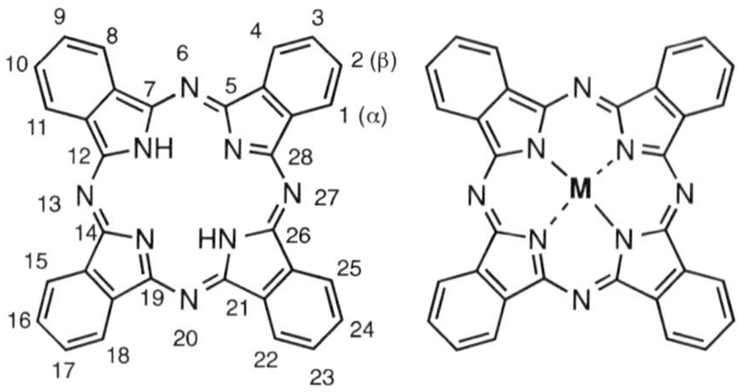

Figure 1. Chemical structure of metal-free phthalocyanine (left, with nomenclature) and metallophthalocyanine (right).

pigments for many years,[5–7] with more than 5×10^5 g of Pcs and MPcs now being synthesized per year.[8] For the last few decades, they have found other potential applications as promising materials in numerous fields such as catalysis,[9] organic thin-film transistors,[10] optical discs,[11] sensors,[12] nonlinear optical materials,[13] photovoltaic devices,[14] probes for bio-imaging,[15] and photodynamic therapy.[16]

Pcs and MPcs are usually synthesized by cyclo-tetramerization of phthalonitrile,[17,18] 1,3-diiminoisoindoline[19,20] or phthalic acid derivatives (anhydrides, imides, amides and nitriles)[21–24] at high temperature (about 160–240 °C), and often reaction times of *ca.* 30 min to several hours are required. Metal ions act as templates and can dramatically increase the reaction yields. The reaction conditions used for the synthesis of Pcs depend on the inserting metal ion as well as the substituents around the macrocyclic ligand. Studies have been made to lower the temperature required for the reaction.[25] In addition, a synthetic method using microwave[26] or UV light[27] has been reported.

This chapter provides an overview of the synthesis strategy for H_2Pcs and MPcs. In Section II, we will describe the various methods used for the synthesis of H_2Pcs and MPcs. Sections III and IV summarize the synthesis methods of H_2Pcs and MPcs. respectively. As described above, since the reaction conditions of MPcs are strongly dependent on the central metal, a wide variety of synthesis conditions is introduced. Since the synthetic method for water-soluble Pc is different from Pcs which are soluble in general organic solvents, due to their substituents such as carboxylic acid and sulfonic acid, this is summarized separately in Section V.

II. Representative Synthesis Methods

H_2Pcs and MPcs are formed by cyclo-tetramerization of suitably substituted or unsubstituted derivatives of phthalic anhydride,[21,22,24] phthalamide,[21,22,28] phthalimide,[21] phthalonitrile,[29,30] and 1,3-diiminoisoindoline.[19,20] Regarding the reaction mechanism of Pc formation, several routes have been proposed. Baumann *et al.*[33] and Borodkin[34] studied and reported the formation of Pc with an alkoxide. The accepted mechanism for the cyclo-tetramerization involves a nucleophilic attack

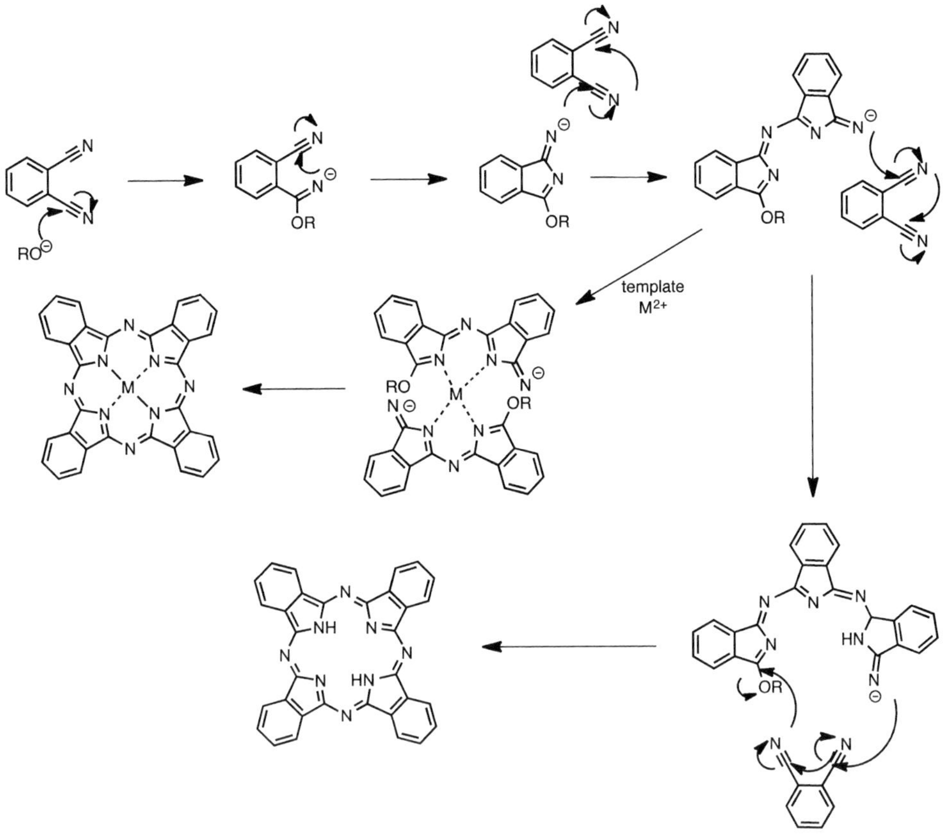

Figure 2. Reaction mechanism of metal-free phthalocyanines and metallophthalocyanines.

on the cyano group of phthalonitrile by the alkoxide anion, leading to the formation of a 1-alkoxy-3-iminoisoindolenine intermediate. The next step in the reaction sequence is the nucleophilic attack of this intermediate upon the cyano group of another phthalonitrile. This forms a dimer of phthalonitrile. The dimer reacts with another phthalonitrile unit in the same way, to form a trimer. The trimer reacts with the 4th phthalonitrile, to give Pc[35,36] (Figure 2). In addition, when a metal salt is added to the reaction mixture, the metal cation acts as a template to which phthalonitrile reacts during cyclo-tetramerization. Therefore, MPc synthesis generally gives a higher yield compared to H_2Pcs. The synthesis of Pcs can be classified into three types (Wyler, phthalonitrile, and isoindoline methods) based on the starting materials. These are summarized in Figure 3 and the following sections.

A. Phthalic Anhydride, Phthalimide, and Phthalamide (Wyler) Methods

Phthalic anhydrides, phthalimides, and phthalic acid are transitional precursors for not only MPcs but also of H_2Pcs (Schemes 1 and 2).[21,22] This method is also known

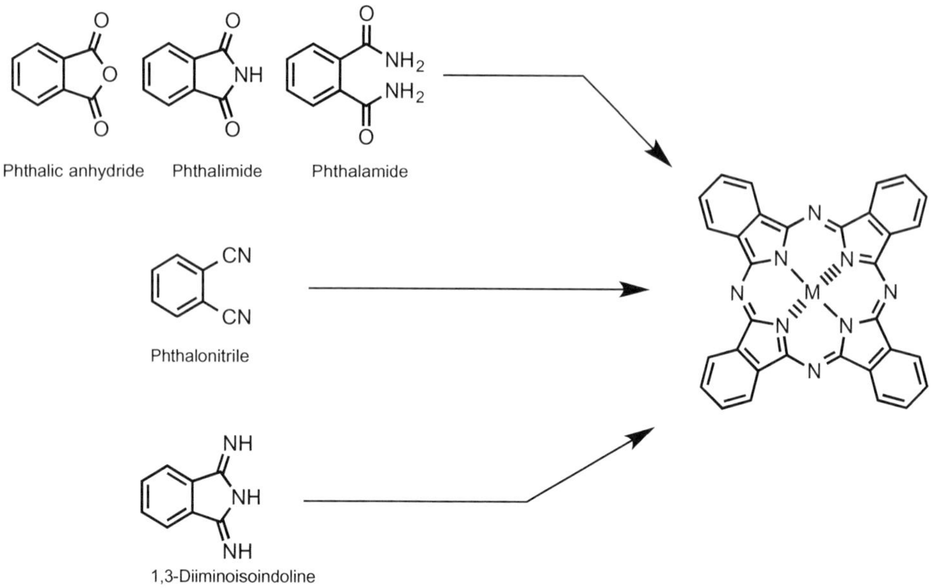

Figure 3. Precursors of synthesis of phthalocyanine.

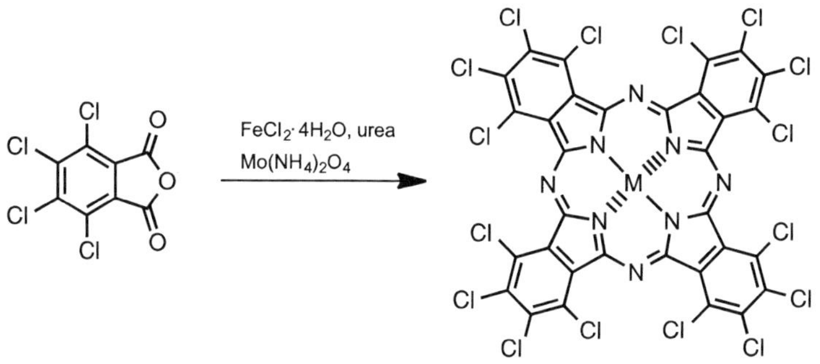

Scheme 1. Synthesis of FePc from phthalic anhydrides.[24]

as the "Wyler method."[23] Most commercial processes use these compounds, in particular phthalic anhydride, as starting materials rather than the more expensive phthalonitriles (Scheme 1). However, compared with the phthalonitrile and isoindoline methods, the number of metal ions that can be used is limited.

General Procedure: Synthesis of FePc from Phthalic Anhydride (Ref. 24)

A suspension of urea (35 g, 0.50 mol), tetrarchlorophthalic anhydride (48 g, 0.17 mol, $FeCl_2{\cdot}4H_2O$ (8.5 g, 0.04 mol), and $(NH_4)_2MoO_4$ (0.2 g, 1 mmol) in $PhNO_2$ (150 ml) was heated to 180–190 °C for 4 h until the evolution of NH_3 stopped. The

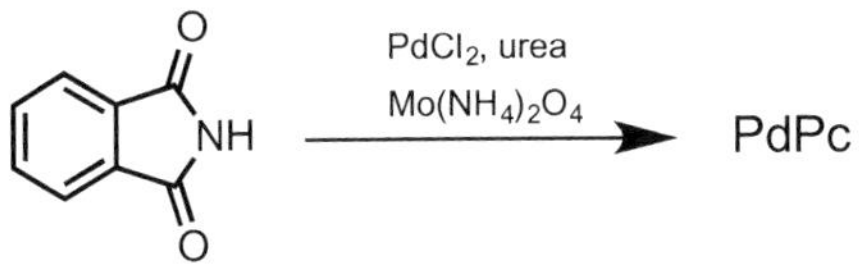

Scheme 2. Synthesis of PdPc from phthalimide[28]

viscous olive-green suspension was cooled to 90 °C, diluted with EtOH (150 ml), and filtered. The residue was washed with boiling water until the washings were neutral, and then treated on a sintered glass disk with acetone and Et_2O, and filtered. Finally, it was extracted for 3 days with PhCl. Yield: (61%).

General Procedure: Synthesis of PdPc from Phthalimide (Ref. 28)

A nitrobenzene solution of palladium(II) chloride ($PdCl_2$), phthalimide, and urea (molar ratio 1:4:14) was refluxed for 4 h in the presence of ammonium molybdate (2% mass of phthalimide) as a catalyst. The resulting mixture was purified by washing in succession with isopropanol, 2–3% NaOH, and 2–3% HCl. The PdPc was recrystallized four times from concentrated H_2SO_4. Yield: (68%).

B. Phthalonitrile Method

Phthalonitriles (1,2-dicyanobenzenes) are a common precursor of H_2Pcs and MPcs (Schemes 3 and 4). A large number of Pc compounds can be synthesized by this method (Table 1).[29,30] The reactivity of phthalonitrile is much higher than that of phthalic acid derivatives. For the synthesis of H_2Pcs, the general procedure is to reflux phthalonitrile with a base such as NH_3, DBU, or DBN in a solvent. In the case of the synthesis of MPcs, phthalonitriles are reacted in high-boiling point solvents (DMF, DMAE, quinoline, 1-chloronaphthalene) in the presence of a metal or metal salt. MPcs can be prepared by a bulk reaction (often without solvents). Heating (typically at more than 200 °C) phthalonitrile with a metal or metal salt provides the corresponding MPc. It should be noted that some metal halides ($FeCl_3$, $CuCl_2$ and $ZnCl_2$) cause mono halogenation of the Pc in a bulk reaction.[17]

General Procedure: Synthesis of Metal-free and Metal Phthalocyanines from Phthalonitrile (Ref. 18)

DBU (0.6 mol) was added dropwise to a solution of phthalonitrile (0.6 mol) in 1-hexanol at 100 °C. After 18 h of heating, the reaction mixture was filtered, to collect the precipitate. The precipitate was washed with water, and purified by extraction with acetone. Yield: (48%).

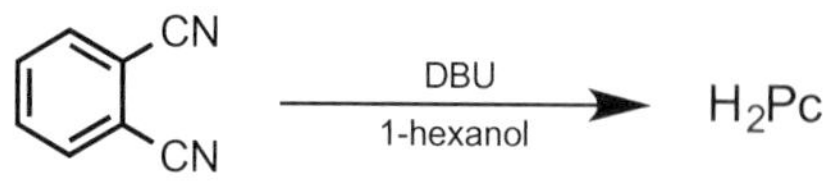

Scheme 3. Synthesis of H_2Pc from phthalonitrile.[18]

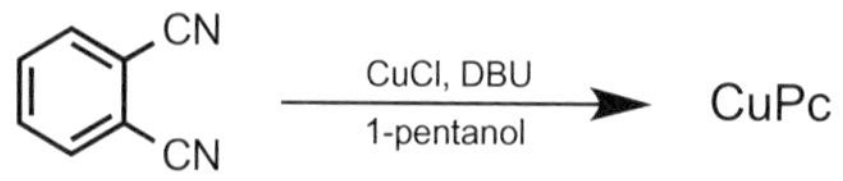

Scheme 4. Synthesis of CuPc from phthalonitrile.[29]

General Procedure: Synthesis of CuPc from Phthalonitrile (Ref. 29)

A reaction mixture of phthalonitrile (10 mmol) and copper(I) chloride (CuCl) (2.5 mmol) in the presence of DBU (10 mmol) was refluxed in 1-pentanol. As the reaction proceeded, a blue or greenish blue powder gradually precipitated. The reaction mixture was filtered, to collect the precipitate. The precipitate was washed with 3% HCl solution, water, and then ethanol, and purified by an extraction with chloroform. Yield: (85%).

C. 1,3-Diiminoisoindoline Method

1,3-Diiminoisoindoline (DI) is also a popular precursor for Pc synthesis (Table 2). DI is almost exclusively formed from the corresponding phthalonitrile by reacting it with ammonia gas in the presence of sodium methoxide in methanol.[30–32] DI derivatives undergo self-condensation without any reagent at high temperature. Refluxing of DI in DMAE in the presence of ammonia provides H_2Pcs (Scheme 5). MPcs can be obtained by adding a suitable metal salt to the cyclo-tetramerization reaction of DI (Scheme 6). For the synthesis of Pcs with some central metals such as silicon, phthalonitrile is not reactive enough as the starting material. Here, it may be advantageous to use DI, which shows higher reactivity. In the case of phthalonitrile having bulky substituents at the 3- and 6-positions, the desired Pcs may not be obtained due to steric hindrance.

General Procedure: Synthesis of Metal-free Phthalocyanine from 1,3-Diiminoisoindoline (Ref. 19)

A mixture of 1,3-diiminoisoindoline (1.138 mol) and DMAE (100 ml) was refluxed for 7 h, during which ammonia was evolved. The reaction mixture was filtered hot, and the obtained purple phthalocyanine crystals were washed thoroughly with ethanol and then acetone. The product was vacuum dried at 80 °C for 3 h. Yield: (86%).

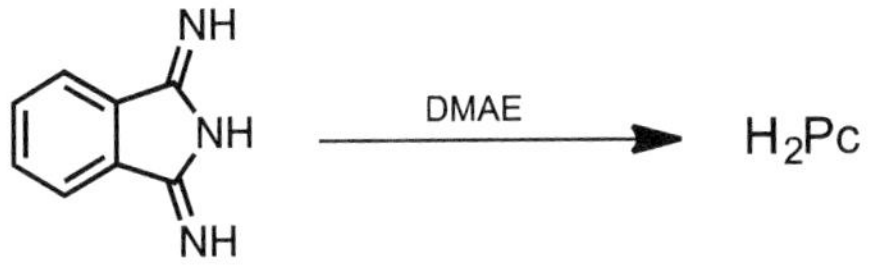

Scheme 5. Synthesis of H_2Pc from DI.[19]

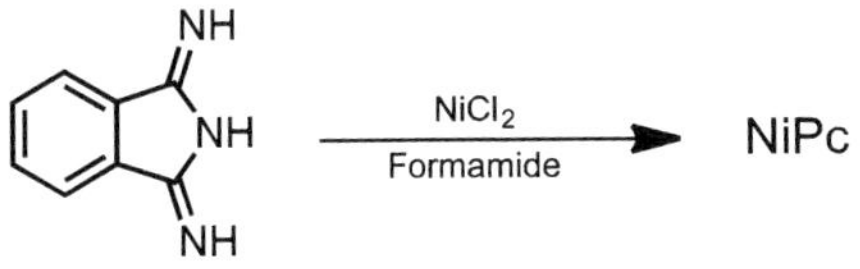

Scheme 6. Synthesis of NiPc from DI.[20]

General Procedure: Synthesis of NiPc from 1,3-Diiminoisoindoline (Ref. 20)

1,3-Diiminoisoindoline (0.2 g) and nickel chloride (0.2 g) were dissolved separately in hot formamide (2.5 and 12.5 ml, respectively), and the hot solutions mixed. The pigment separated almost immediately. Next day, the nickel phthalocyanine was collected, washed with ethanol, and extractively crystallized from chloronaphthalene. Yield: 0.19 g (96%).

III. Synthesis of Metal-Free Phthalocyanines

A large number of metal-free phthalocyanines have been obtained from cyclocondensation of diverse precursors such as phthalonitriles,[18,19,37, 42,43,45, 49,50] 1,3-diiminoisoindolines,[19,20,38,39,43,46–48] phthalic anhydrides and phthalimides[44] under basic conditions in high-boiling anhydrous solvents such as pentanol, *N,N*-dimethylformamide (DMF) and *N,N*-dimethylaminoethanol (DMAE). For the synthesis of H_2Pcs using highly pure starting materials and solvents, the use of quartz or Teflon vessels is important in order to avoid contamination of the macrocycle by metal ions.[41] Usually, synthesis of H_2Pcs gives lower yields compared to MPcs because of the absence of the template effect of metals. High-yield syntheses of H_2Pcs by electrochemical,[43] UV irradiation,[26] and microwave irradiation[27] methods have been reported.

A. Phthalonitrile Methods

There are two main methods for synthesizing H_2Pcs from phthalonitrile. The first is by the addition of a non-nucleophilic base, such as 1,8-diazabicyclo[5.4.0]undec-7-ene (DBU) or 1,5-diazabicyclo[4.3.0]non-5-ene (DBN), and reflux in primary alcohol (1-pentanol).[18,45] The second is by refluxing with a basic solvent such as *N,N*-dimethyl-2-aminoethanol (DMAE).[37] In general, these are mild, clean and direct methods.[41]

Table 1. Synthesis of H_2Pc from phthalonitrile.

S. M.	Regents	Solvent	Conditions	Yield (%)	Ref.
PN	DBU	MeOH	Reflux, 18 h	50	18
PN	DBU	2-Buthoxyethanol	100 °C, 18 h	55	18
PN	DBU	*i*-Octanol	100 °C, 18 h	60	18
PN	DBN (The concentrate is 25 mol%)		200 °C, 4 h	90	37
PN	DBU (The concentrate is 25 mol%)		200 °C, 4 h	82	37
PN	DMAE (The concentrate is 25 mol%)		200 °C, 4 h	24	37
PN	DBU	Ethanol	Reflux, 24 h	*ca.* 20	45
PN	DBN	Ethanol	Reflux, 24 h	*ca.* 43	45
PN	$CeCl_3$	1-Penthanol	Reflux, 24 h	45	49
PN	Ionic liquid,	DBU	140 °C, 10 min	47	50
PN	Ammonia gas	DMAE	Reflux, 7 h	90	19
PN		MeOH	10 kbar, 90 °C, 70 h	55	42
PN	TBA	*i*-BuOH	25 mA, 150 V, 1 h, 90 °C	98	43

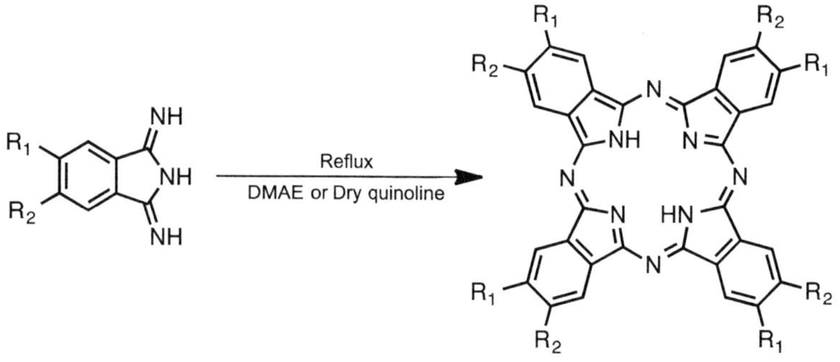

Scheme 7. Synthesis of metal-free Pc from DI.

Refluxing phthalonitriles with sodium or lithium *n*-pentoxide in *n*-pentanol or other alcohols results in alkali-containing Pcs, which can be demetalated with acid and water. Synthesis methods using high pressure[42] and electrochemistry[43] have also been reported.

B. 1,3-Diiminoisoindoline Method

1,3-Diiminoisoindoline (DI) is a useful starting material for H_2Pc synthesis (Scheme 7). The preparation of H_2Pc from DI is generally achieved in a refluxing solution of DMAE.[19,37,46,48] Electrochemical reduction of DI seems to increase the yield, especially when a polar aprotic solvent, such as DMF or DMSO, is used.[43]

Table 2. Synthesis of H_2Pc from 1,3-diiminoisoindoline.

S. M.	Substitutes (R_1 and R_2)	Regent, solvent	Conditions	Yield (%)	Ref.
DI	R_1 and R_2 = Et	DMAE	Reflux, 7 h	23	38
DI	R_1 and R_2 = H	DMAE	Reflux, 7 h	90	19
DI	R_1 and R_2 = H	Dry quinoline	145–150°C, 6 h	55	39
DI	R_1 = Vinyl, R_2 = H	DMAE	Reflux, 24 h	95	46
DI	R_1 = 2-(*N*-2′-cyanoethyl) amino ethylsulfanyl, R_2 = H	dry quinoline	145–150 °C, 6 h	44	47
DI	R_1 = 2-trimethylsilylethyl, R_2 = H	DMAE	150–160 °C, 60 h	28	48
DI	R_1 and R_2 = H	DMSO	40 mA, 42 V, 1 h, 189 °C	98	43
DI	R_1 and R_2 = H	Succinonitrile, Butanol	Boiling, 3.5 h	34	20

C. Demetalation

Some metal ions such as Li(I), Na(I), K(I), Mg(II), Be(II), Ag(II), Cd(II), Pb(II), and Sb(II) are easily removed from the Pc skeleton by acids such as acetic acid.[40] As mentioned above, when lithium or sodium alkoxide is used as a base, an alkali metal-containing Pc is formed in each case, but H_2Pcs can be easily obtained from these by treating with an acid and water.[26] For example, lithium can be removed simply by a work-up with acid.[39] On the other hand, the demetalation reaction of strongly-coordinating ions such as Mn(II), Fe(II), Co(II), Cu(II), Ni(II), Cu(II), and Zn(II) is considered difficult or even impossible without destroying the Pc itself. Luedtke and coworkers have reported a new demetalation reaction of Zn ions using pyridine-HCl.[44]

IV. Synthesis of Metallophthalocyanines

A. Organic Solvent-Soluble Metallophthalocyanines

1. *Group 1: Li_2Pc, Na_2Pc, K_2Pc, Rb_2Pc, and Cs_2Pc*

The alkali metal Pcs have two metal cations in the central cavity because alkali metals are monovalent and small. The *dilithium*-(Li_2Pc, **Li-1**), *disodium*-(Na_2Pc, **Na-1**), and *dipotassium* (K_2Pc, **K-1**) phthalocyanines were obtained by the template condensation (Linstead's method) of phthalonitriles in the presence of the corresponding metals (Scheme 8, Table 3).[21,52] Li_2Pc is sensitive to water and acids;[53] Na_2Pc is more sensitive to moisture and even alcohol compared to Li_2Pc, and is readily demetalated.[21] Li_2Pc and Na_2Pc are useful as intermediates for the synthesis of

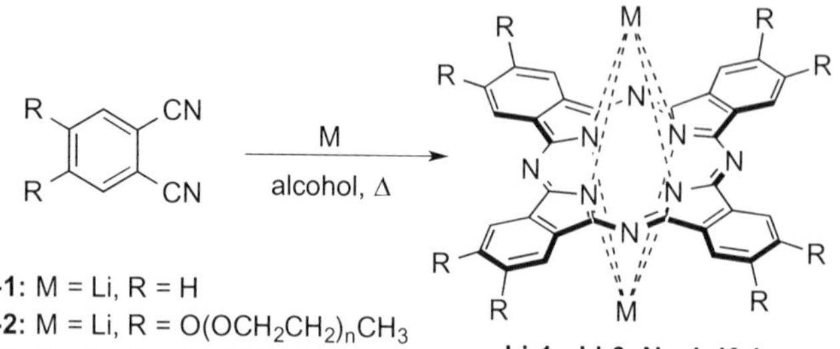

Li-1: M = Li, R = H
Li-2: M = Li, R = $O(OCH_2CH_2)_nCH_3$
Li-3: M = Li, R = $O(CH_2)_{11}CH_3$
Na-1: M = Na, R = H
K-1: M = K, R = H

Li-1 - Li-3, Na-1, K-1

Scheme 8. Syntheses of Li_2Pc, Na_2Pc and K_2Pc.[21,52–54]

Table 3. Preparation of Li_2Pc, Na_2Pc, K_2Pc, Rb_2Pc and Cs_2Pc.

	Conditions	Yield (%)	Ref.
Li-1	Li metal, *n*-pentanol, refluxing, 30 min	52	52
Li-1	Li metal, *n*-pentanol, refluxing, 30 min	75	53
Li-2	Li metal, methanol, refluxing, 3 h	80	54
Li-3	Li metal, methanol, refluxing, 3 h	85	54
Na-1	Na metal, *n*-pentanol, refluxing, 10 min	70	21
K-1	K metal, *n*-pentanol, refluxing, 60 min	n.r.	21
Rb-1	Rb metal, TCBN, 300–350 °C, 10 h	n.r.	55
Cs-1	Cs metal, TCBN, 300–350 °C, 10 h	n.r.	55

H_2Pcs (part or section 3.3) and other MPcs (parts or sections 4.1.2, 4.1.4, 4.1.10 – 4.1.12). Both Li_2Pc and Na_2Pc are now commercially available from different vendors. The procedure for the preparation of Li_2Pcs with four hydrophilic (**Li-2**) or four hydrophobic (**Li-3**) substituents has also been published (Scheme 8, Table 3).[54]

The preparation of the *dirubidium*- (Rb_2Pc, **Rb-1**) and *dicesium* (Cs_2Pc, **Cs-1**) Pcs were described later.[55] **Rb-1** and **Cs-1** were obtained as metal complexes of octacyano-Pc by chemical-vapor deposition of 1,2,4,5-tetracyanobenzene and the corresponding metal chloride (Table 3).

Synthesis of Li_2Pc (Ref. 52)

Phthalonitrile (60 g) was added to a solution of lithium metal (4 g) in 300 mL dry amyl alcohol. On heating, an exothermic reaction occurred and the color changed to a deep blue. The solution was refluxed for a further 30 min. On cooling, 1 L of benzene was added and the mixture left for 3 h. The resulting solid was collected by filtration and extracted in a Soxhlet apparatus using dry acetone. The solvent was removed under reduced pressure, to give the title compound as a crystalline deposit with a purple luster. Yield: 30 g (50%).

2. *Group 2: BePc, MgPc, CaPc, and BaPc*

There are only a few reports of synthesis of *beryllium* Pcs (BePc, **Be-1**) due to the highly toxic nature of beryllium. **Be-1** can be prepared by melting phthalonitrile and beryllium metal.[21] The surface of beryllium was first activated with acid.[21] **Be-1** is unstable under air, and transforms into a hydrated form. In 2006, a slightly modified procedure for the preparation of **Be-1** was published (Table 4).[56]

Magnesium metal is one of the most popular template for the synthesis of MPcs, and numerous syntheses of unsubstituted *magnesium* Pcs (MgPcs), as well as peripherally-substituted MgPcs have been reported (Tables 4–8).[29,57–77] All publications for the preparation of unsubstituted MPc before 1963, including patents, were summarized in Moser and Thomas' monograph.[57] The first procedure for the

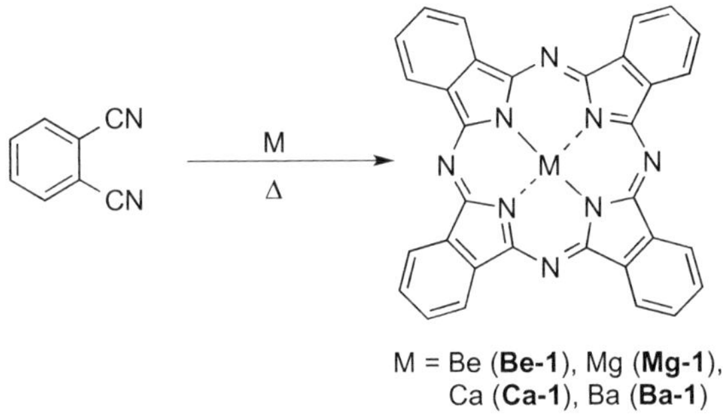

Scheme 9. The general approach of synthesis of unsubstituted BePc, MgPc, CaPc, and BaPc.[21,29,56,58, 59,78]

Table 4. Synthesis of unsubstituted **Be-1**, **Mg-1**, **Ca-1** and **Ba-1**.

	Conditions	Yield (%)	Ref.
Be-1	Be metal, melting, 1.5 h	n.r.	21
Be-1	Be metal, 270 °C, sealed tube, 24 h	n.r.	56
Mg-1	Mg turnings, CBA, 230–240 °C, 2 h	38	58
Mg-1	Mg turnings, H_2Pc, 230 °C, 2 h	n.r.	58
Mg-1	$MgCl_2$, DBU, *n*-pentanol, refluxing, 6 h	53	29
Mg-1	Mg (powder), methanol, RT, 25 d	12	59
Ca-1	Ca metal, *n*-pentanol, refluxing, 5 h	20	58
Ca-1	CaO, 275–280 °C	70	21
Ca-1	$CaCl_2$, Li_2Pc, abs. alcohol, refluxing, 3 h	n.r.	78
Ba-1	BaO, 290 °C, 2 h	50	21

preparation of unsubstituted MgPc was publishted in 1934 by Linstead and Lowe (Scheme 9).[58] Later, a fast, simple procedure was published by Tomoda *et al.*[29] Using this protocol, unsubstituted MgPc was obtained by heating phthalonitrile with $MgCl_2$ in *n*-pentanol in the presence of DBU as a strong organic base (Table 4).[29] Another procedure for **Mg-1** synthesis was published by Leznoff *et al.* in 2000.[59] Indeed, **Mg-1** has been prepared at ambient temperature using phthalonitrile and magnesium powder as starting materials (Table 4).[59]

Nowadays, the preparation of MgPcs focuses on two routes: *i*) synthesizing MgPcs by condensation of substituted phthalonitriles in the presence of Mg(II) ions as a template (Scheme 10, Table 4),[59–75] and *ii*) synthesis of MgPcs by metalation of H_2Pcs or metal exchange of Li_2Pc (Scheme 15, Table 4).[76–78] Thus, magnesium 1,8(11),15(18),22(25)-tetra-substituted Pcs (**Mg-2** and **Mg-3**) were prepared by template condensation of the corresponding 3-substituted phthalonitriles[60,61] with magnesium metal[60] or magnesium(II) acetate[61] in high-boiling alcohols (Scheme 10, Table 5).

In addition, magnesium 2,9(10),16(17),23(24)-tetra-substituted Pcs (**Mg-4** – **Mg-10**) have been obtained by melting the corresponding 4-substituted phthalonitriles with activated magnesium metal[62–64] or magnesium(II) chlorides[65] as well as by treatment of titled nitriles with magnesium(II) chlorides in alcohol (Scheme 11, Table 5).[66] The procedure for preparing **Mg-10** at ambient temperature has been described.[59]

Magnesium 1,4,8,11,15,18,22,25-octa-substituted Pc (**Mg-11**) was prepared by refluxing a propanol solution of 3,6-di(*p-tert*-butylbenzyloxy)phthalonitrile in the presence of magnesium turnings (Scheme 12, Table 6).[67]

Magnesium 2,3,9,10,16,17,23,24-octa-substituted Pcs (**Mg-12** – **Mg-20**) were obtained *via* template condensation of the corresponding 4,5-disubstituted nitriles

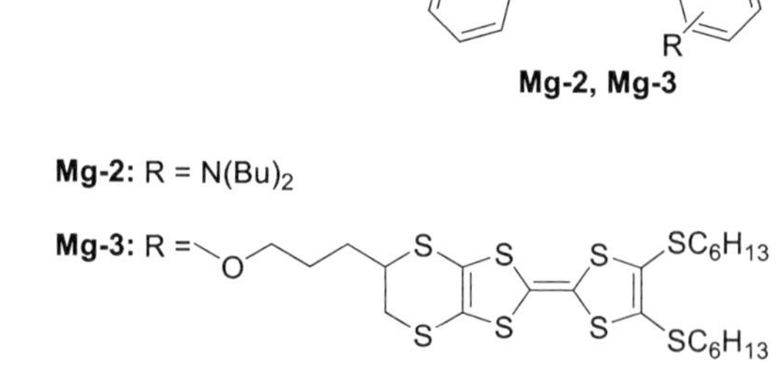

Scheme 10. Preparation of **Mg-2** and **Mg-3**.[60,61]

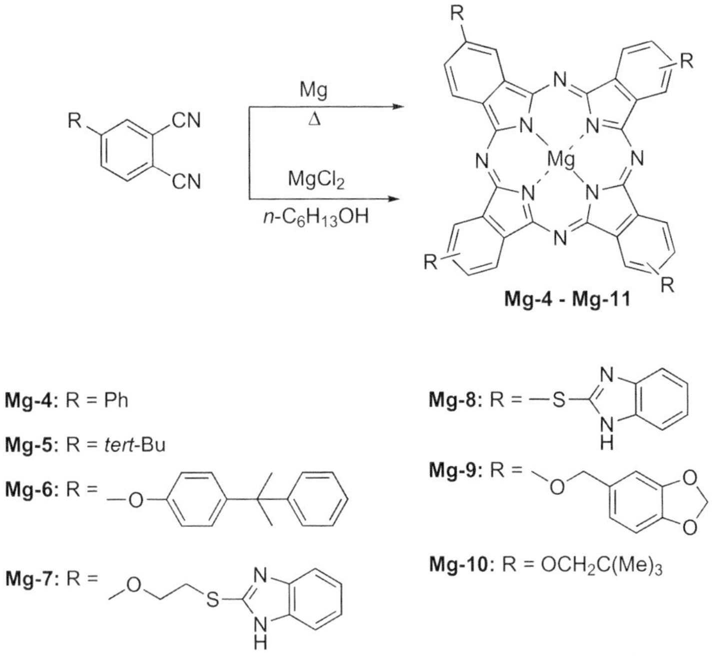

Scheme 11. Preparation of magnesium 2,9(10),16(17),23(24)-tetra-substituted Pcs **Mg-4 – Mg-10**.[59,62–66]

Table 5. Preparation of tetra-substituted MgPcs.[59–66]

	Conditions	Yield (%)	Ref.
Mg-2	Mg (turnings), I_2, *n*-pentanol, refluxing, 12 h	44[a]	60
Mg-3	$Mg(OAc)_2$, *n*-pentanol, refluxing, 12 h	25	61
Mg-4	Mg, MOA, melting, 230–250 °C, 7 h	72	62
Mg-5	Mg, MOA, melting, 210–215 °C, 0.5 h	57.5	63
Mg-6	Mg, sealed tube, melting, 280 °C, 70 h	14	64
Mg-7	$MgCl_2$, DBU, melting, 190 °C, 5 h	27.5	65
Mg-8	$MgCl_2$, DBU, melting, 190 °C, 2 h	36.3	65
Mg-9	$MgCl_2$, DBU, *n*-hexanol, 160 °C, 24 h	44.2	66
Mg-10	Mg (powder), methanol, RT, 16 d	20	59

[a]Mixture of isomers.

using magnesium(II) butoxide or magnesium(II) acetate as a template (Scheme 13, Table 6).[68–72]

On the other hand, highly-hindered magnesium 1,2,3,4,8,9,10,11,15,16, 17,18,22,23,24,25-hexadeca-substituted Pcs **Mg-21** and **Mg-22** were prepared by

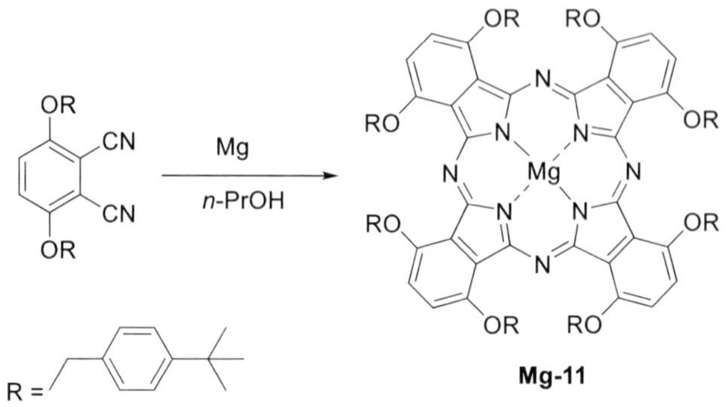

Scheme 12. Preparation of magnesium 1,4,8,11,15,18,22,25-octa-substituted Ps **Mg-11**.[67]

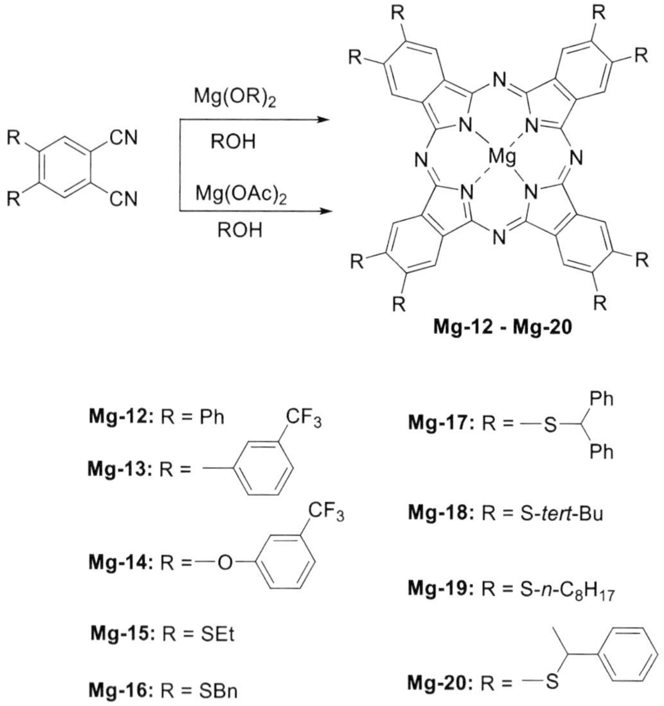

Scheme 13. Preparation of magnesium 2,3,9,10,16,17,23,24-octa-substituted Pcs **Mg-12** – **Mg-20**.[68–72]

treatment of the corresponding phthalonitriles with magnesium(II) octanoate prepared *in situ* from phenylmagnesium bromide (Scheme 14).[73] The same synthetic pathway was used for the preparation of highly-hindered adamantylsubstituted **Mg-23** – **Mg-27** complexes (Scheme 14).[74] Hexadecachloro-substituted **Mg-28** was prepared by template condensation of tetrachlorophthalonitrile with magnesium(II) acetate (Table 7).[75]

Table 6. Preparation of octa-substituted MgPcs.

	Conditions	Yield (%)	Ref.
Mg-11	Mg (turnings), I_2, *n*-propanol, refluxing, 24 h	48	67
Mg-12	$Mg(OAc)_2$, MeOLi, isopentanol, 5 h	65	68
Mg-13	Mg turnings, *n*-octanol:*n*-pentanol mixture (1:1), 170 °C, 1 h	76	69
Mg-14	Mg turnings, *n*-octanol:*n*-pentanol mixture (1:1), 170 °C, 1.5 h	46	69
Mg-15	Mg, I_2, anhydrous *n*-butanol, refluxing, 6 h	49.6	70
Mg-16	Mg, I_2, anhydrous *n*-butanol, refluxing, 6 h	57.8	70
Mg-17	Mg, I_2, anhydrous *n*-butanol, refluxing, 6 h	52.8	70
Mg-18	Mg, I_2, anhydrous *n*-butanol, refluxing, 9 h	31	71
Mg-19	Mg, I_2, anhydrous *n*-butanol, refluxing, 9 h	41	71
Mg-20	Mg, I_2, anhydrous *n*-butanol, refluxing, 6 h	44.4 (R), 48.3 (S), 39.5 (rac)	72

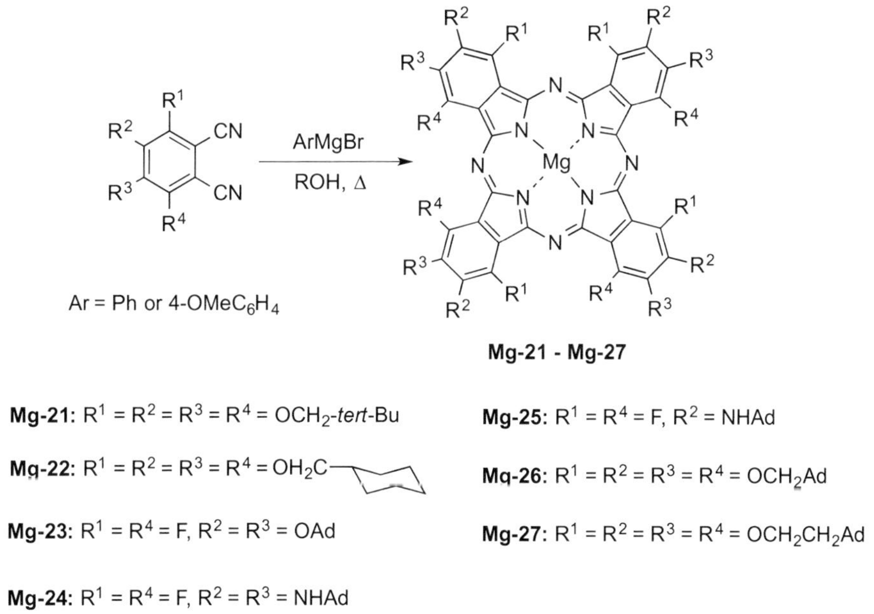

Scheme 14. Preparation of highly hindered complexes **Mg-21** – **Mg-27**.[73,74]

Meanwhile, methods for preparation of MgPcs (**Mg-29** – **Mg-31**) by metalation of peripherally substituted H_2Pcs or metal exchange of Li_2Pc with Mg(II) salts (chloride or acetate) have been published (Scheme 15, Table 8).[76–78]

Calcium Pc (CaPc) can be synthesized by the reaction of calcium metal,[58] calcium oxide[21] or calcium ethoxide[21] with phthalonitrile, as well as by the reaction

Table 7. Preparation of hexadeca-substituted MgPcs.

	Conditions	Yield (%)	Ref.
Mg-21	PhMgBr, *n*-octanol, 120 °C, 4 d	n.r.	73
Mg-22	PhMgBr, *n*-octanol, 120 °C, 4 d	51	73
Mg-23	4-OMePhMgBr, *n*-octanol, 120 °C, 4 d	34	74
Mg-24	4-OMePhMgBr, *n*-octanol, 120 °C, 4 d	17	74
Mg-25	4-OMePhMgBr, *n*-octanol, 120 °C, 4 d	21	74
Mg-26	4-OMePhMgBr, *n*-octanol, 120 °C, 4 d	15	74
Mg-27	4-OMePhMgBr, *n*-octanol, 120 °C, 4 d	20	74
Mg-28	$Mg(OAc)_2$, DBU, isopentanol, 2 h	74	75

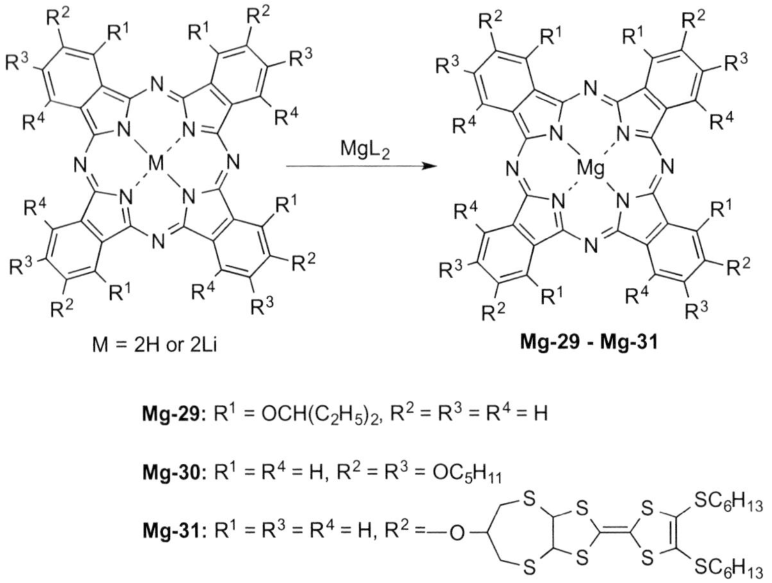

Scheme 15. Preparation of MgPcs *via* metalation of metal-free Pcs or Li_2Pc.[76–78]

Table 8. Preparation of substituted MgPcs.

	Conditions	Yield (%)	Ref.
Mg-29	$MgCl_2$, DMF, refluxing, 4 h	43	76
Mg-30	$MgCl_2$, THF, refluxing, 12 h	89	77
Mg-31	$Mg(OAc)_2$, *n*-pentanol, refluxing, 12 h	12	78

of Li_2Pc and calcium chloride in ethanol.[52] *Barium* Pc (BaPc) can be prepared by the nitrile method using barium oxide as the metal source.[1]

Synthesis of Mg(II) Pc(tert-Bu)$_4^{\beta}$ (Ref. 63)

A mixture of 4-*tert*-butylphthalonitrile (0.27 g), activated magnesium metal (0.04 g) and ammonium molybdate (0.001 g) was heated with stirring for 45 min to 200 °C, then for 15 min at 200–210 °C, and then for 30 min at 210–215 °C, until the melt solidified. Next, the solid was ground and heated with a 3% solution of hydrochloric acid at 50 °C, followed by chromatographic separation on an alumina column, using acetone as eluent, to yield 160 mg (57.5%) of the target compound.

Synthesis of Mg(II) Pc(tert-SBu)$_8^{\beta\beta}$ (Ref. 71)

Magnesium metal (39 mg) and a small crystal of iodine were refluxed in anhydrous butanol (8 mL) for 4 h, then 4,5-bis-*tert*-butylsulfanylphthalonitrile (500 mg) added, and the mixture heated under reflux for another 5 h. The precipitate was filtered, washed with methanol and acetone, dissolved in chloroform, and purified by chromatography on silica with chloroform and later with hexane/ethyl acetate (1:1, 1:2) as eluent, to yield 156 mg (31%) of a dark green solid.

Synthesis of Mg(II) Pc(cyclohexylmethyloxy)$_{16}$ (Ref. 73)

To approximately 2 mL of *n*-octanol, was added 1.0 mL of 1.0 M phenylmagnesium bromide solution in THF. After stirring for 15 min to ensure the complete reaction of the Grignard reagent to form the magnesium alkoxide, tetra(cyclohexylmethyloxy)phthalonitrile (0.25 g) was added. The reaction mixture was heated to 120 °C for 24 h. The resulting material was transferred to test tubes and precipitated out of solution using acidified 95% ethanol: water. After centrifuging and removing the ethanol, the precipitate was washed twice with water to give 130 mg (51%) of a green solid.

Synthesis of M(II) PcPh$_8^{\beta\beta}$ (Ref. 68)

A mixture of 4,5-diphenylphthalonitrile (200 mg), $Mg(OAc)_2{\cdot}4H_2O$ (38 mg) and MeOLi (17 mg) was refluxed in isopentanol (2 mL) for 5 h. The reaction mixture was cooled to room temperature, and a $MeOH$-H_2O (10: 1, v/v) mixture added. The precipitate was filtered and washed with water and methanol. Yield 130 mg (65%).

Synthesis of Mg(II) $PcCl_{16}$ (Ref. 75)

Tetrachlorophthalonitrile (426 mg), $Mg(OAc)_2 \cdot 4H_2O$ (172 mg) and DBU (238 μL) were stirred in 6 mL of boiling isopentanol for 4 h. The reaction mixture was cooled to room temperature, and a MeOH: H_2O (4:1, v/v) mixture added. The precipitate was filtered, washed with a MeOH: H_2O (4:1, v/v) mixture, and dried at 70 °C. Yield: 321 mg (74%).

3. *Group 3: ScPc, $ScPc_2$, YPc, YPc_2, and YPc_3*

Scandium is an early rare-earth element with the smallest radius in Group 3. Similarly to other rare-earth elements of the Periodic Table, scandium can afford both single-decker (ScPcs) and double-decker ($ScPcs_2$) complexes with phthalocyanines. Unfortunately, only a few articles have been published on the preparation of ScPc,[79–81] including the synthesis of $ScPcs_2$[81,82] (Table 9). Unsubstituted *scandium* Pc (Sc(Cl)Pc, **Sc-1**) can be obtained by refluxing $ScCl_3$ and phthalonitrile in 1-chloronaphthalene.[79] An adequate procedure for Sc(Cl)Pc synthesis in a total yield of 64% has been reported (Scheme 16, Table 9).[80]

Tetra-(15-crown-5)-substituted **Sc-2** can be obtained *via* metal insertion of the corresponding H_2Pc using $Sc(acac)_3$ (Scheme 17, Table 9).[81] Only two examples of scandium double-decker phthalocyanine $ScPc_2$ were described.[3,4] Unsubstituted $ScPc_2$ (**Sc-3**) was prepared by Kirin and Moskalev in 1975 (Table 9).[82] The reaction was carried out by melting phthalonitrile with scandium(III) acetate (Scheme 17).[82] Bis(tetra-crown-substituted) $ScPc_2$ (**Sc-4**) can be prepared by metalation of the initial tetra-crown-substituted H_2Pc (Scheme 17, Table 9).[81]

Yttrium Pcs (Y(L)Pc) can be prepared by melting phthalonitrile with yttrium(III) chloride (**Y-1**)[83] or acetate (**Y-2**)[84] as well as by refluxing phthalonitrile in 1-chloronaphthalene in the presence of yttrium(III) chloride[79,85] as the metal source (Scheme 16, Table 10). Yttrium can form double-decker ($YPcs_2$) and triple-decker (Y_2Pcs_3) sandwich-like complexes with Pcs (Scheme 18).[83–94]

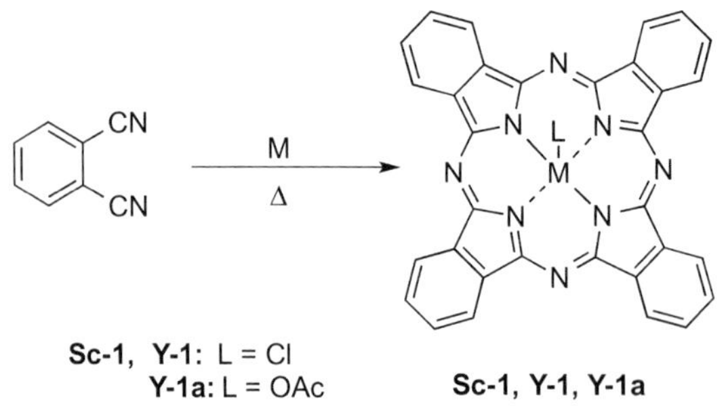

Scheme 16. Syntheses of ScPc and YPsc.[79,80,83–85]

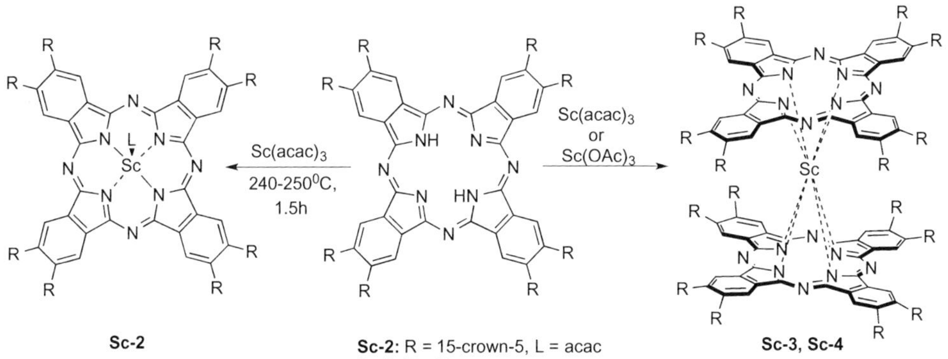

Sc-2

Sc-2: R = 15-crown-5, L = acac
Sc-3: R = H
Sc-4: R = 15-crown-5

Sc-3, Sc-4

Scheme 17. Syntheses **Sc-2** – **Sc-4**.[81,82]

Table 9. Preparation of ScPcs and $Sc(Pcs)_2$.

	Conditions	Yield (%)	Ref.
Sc-1	$ScCl_3$, 1-ClN, refluxing	n.r.	79
Sc-1	$ScCl_3$, 1-ClN, refluxing, 260 °C, 2.5 h	64	80
Sc-2	$Sc(acac)_3$, imidazole:TCB (or 1-ClN: *n*-octanol (7:1)), Ar, refluxing, 1.5 h	97 (97)	81
Sc-3	$Sc(OAc)_3$, melting, 300 °C, 1.5–2 h	17	82
Sc-4	$Sc(acac)_3$, TCB:*n*-octanol (7:1) Ar, refluxing, 1.5 h	70	81

Homoleptic $YPcs_2$ (**Y-3** – **Y-7**) can be prepared by treatment of substituted or unsubstituted phthalonitriles with yttrium salts (acetate or acetylacetonate) in boiling alcohols or 1-cloronaphthalene (Table 10).[85–89] Homoleptic **Y-3** can also be prepared using Li_2Pc as a starting material.[86] Heteroleptic $YPcs_2$ (**Y-8** – **Y-12**) can be prepared by treating phthalonitriles with single-decker YPcs (Scheme 18. Table 10).[87,90] The syntheses of homoleptic (**Y-13**)[91] (Scheme 18) and heteroleptic (**Y-14** – **Y-18**)[92–94] yttrium triple-decker complexes ($Y(Pcs)_3$) has also been reported (Figures 4, 5, Table 10). **Y-13** has been obtained as a by-product in the synthesis of double-decker complex **Y-3**.[91] Homo-dinuclear heteroleptic Y_2Pcs_3 complexes have been known since 2000.[92] Indeed, complexes **Y-14** and **Y-15** (Figure 4) were obtained by condensation of double-decker **Y-6**, dilithium phthalocyanine and ytterbium(III) acetylacetonate in boiling 1,2,4-trichlorobenzene.[92]

Another homo-dinuclear complex **Y-16** (20%) has been obtained by condensation of $Li_2Pc(OC_6H_{13})$, **Y-3** and yttrium(III) acetylacetonate in boiling 1-chloronaphthalene (Figure 4).[93] Hetero-dinuclear heteroleptic triple-decker complexes **Y-17** and **Y-18** were prepared by Tsivadze and Gorbunova in 2011 (Figure 5).[94]

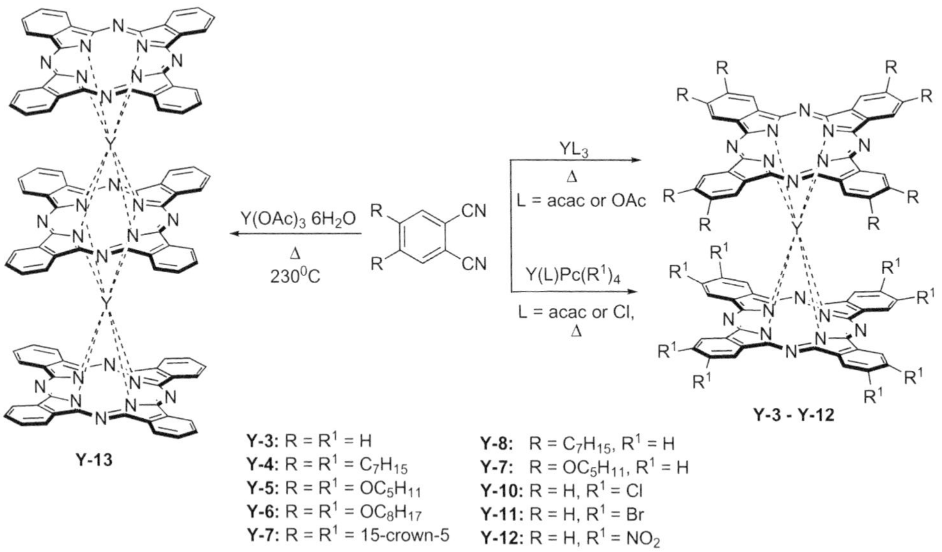

Scheme 18. Syntheses of $YPcs_2$[86–90] and Y_2Pc_3.[91]

Table 10. Preparation of YPcs, $Y(Pcs)_2$ and $Y(Pcs)_3$.

	Conditions	Yield (%)	Ref.
Y-1	YCl_3, melting, 270–280 °C, 1 h	n.r.	83
Y-1	YCl_3, 1-ClN, refluxing	n.r.	79, 85
Y-2	$Y(OAc)_3$, melting, 300 °C	n.r.	84
Y-3	YCl_3, 1-ClN, refluxing	n.r.	85
Y-3	$Y(OAc)_3$, hexanol, DBU, refluxing, 3 h	6.5	86
Y-3	Li_2Pc, $Y(acac)_3$, TCB, refluxing, 10 h	30	87
Y-4	$Y(acac)_3$, *n*-pentanol, DBU, refluxing, 12 h	45	87
Y-5	$Y(acac)_3$, *n*-pentanol, DBU, refluxing, 12 h	75	87
Y-6	$Y(acac)_3$, *n*-pentanol, DBU, refluxing, 12 h	55	88
Y-7	$Y(OAc)_3$, 1-ClN, DBU, 260 °C	20	89
Y-8	Y(acac)Pc, *n*-pentanol, DBU, refluxing, 12 h	10	87
Y-9	Y(acac)Pc, *n*-pentanol, DBU, refluxing, 12 h	19	87
Y-10	$Y(Cl)PcCl_4$, 300 °C, melting, 2 h	28	90
Y-11	$Y(Cl)PcBr_4$, 300 °C, melting, 2 h	55	90
Y-12	$Y(Cl)Pc(NO_2)_4$, 300 °C, melting, 2 h	41	90
Y-13	$Y(OAc)_3$, melting, 230 °C	n.r.	91
Y-14	Li_2Pc, **Y-6**, $Y(acac)_3$, TCB, refluxing, 12 h	81	92
Y-15	Li_2Pc, **Y-6**, $Y(acac)_3$, TCB, refluxing, 12 h	4	92
Y-16	$Li_2Pc(OC_6H_{13})$, **Y-3**, $Y(acac)_3$, 1-ClN, refluxing, 3 h	20	93
Y-17	La_2Pc, **Y-5**, $Yb(acac)_3 \times nH_2O$, 1-ClN, 10 min	75	94
Y-18	La_2Pc, $Yb[(15\text{-crown-}5)_4Pc]$,$Y(acac)_3 \times nH_2O$, 1-ClN, 10 min	61	94

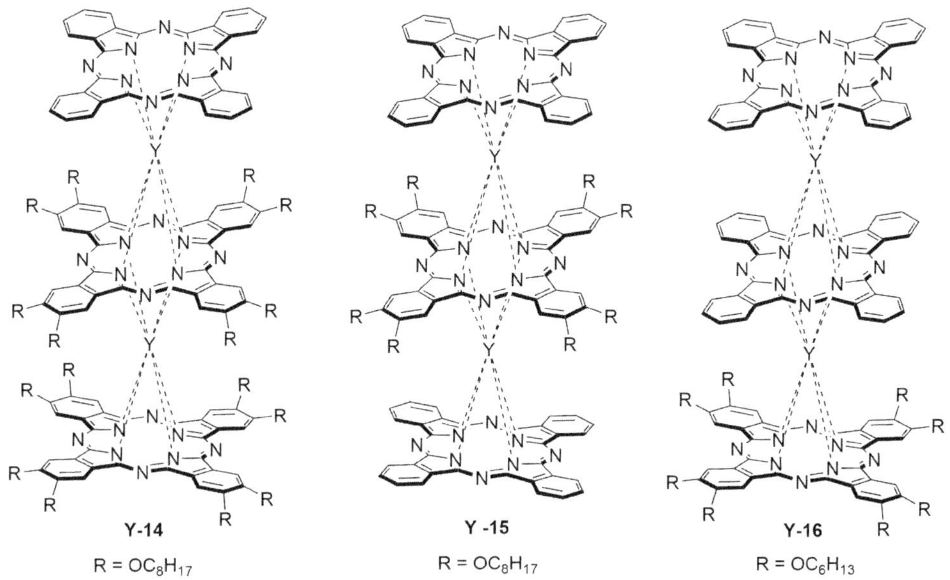

Figure 4. Chemical structures of heteroleptic homo-nuclear yttrium triple-decker complexes.[92,93]

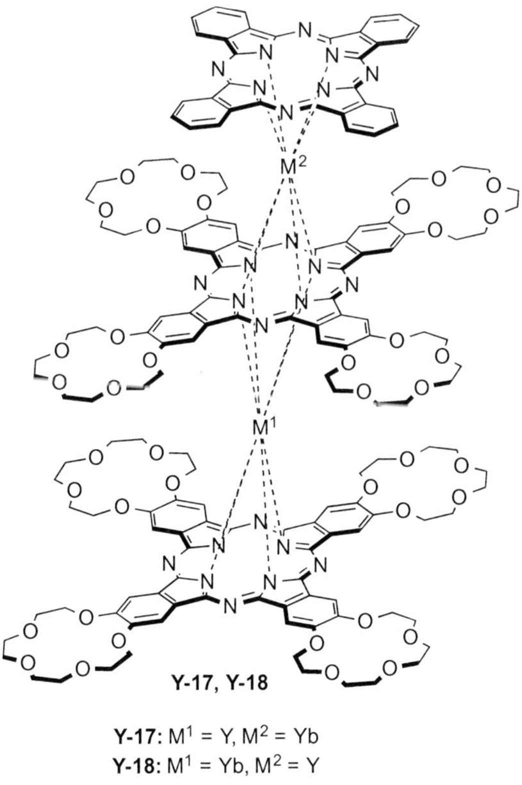

Figure 5. Chemical structures of heteroleptic hetero-nuclear yttrium/ytterbium triple-decker complexes.[94]

For the preparation of **Y-17**, the reaction was carried out between unsubstituted $LaPc_2$, **Y-5** and ytterbium(III) acetylacetonate, and for the preparation of **Y-18**, the reaction was carried out between unsubstituted $LaPc_2$, ytterbium bis(tetra-15-crown-5)Pc and yttrium(III) acetylacetonate, (Table 10).[94]

Synthesis of Sc(Cl)Pc (Ref. 80)

A mixture of phthalonitrile (3.08 g) and 1-chloronaphthalene (6 mL) was heated to 180 °C, and $ScCl_3$ (0.454 g) was added to the beige solution. The reaction mixture was heated to 260 °C and maintained at this temperature for 6 h, during which time it turned deep red, purple and finally dark blue. The reaction mixture was removed from the oil bath, opened to air and briefly allowed to cool. Additional 1-chloronaphthalene (10 mL, undried) was added to the flask, and then the mixture was hot-filtered through a glass frit. The obtained dark blue powder was rinsed with dichloromethane. Sublimation at 200 °C for 2 h afforded 1.14 g of a blue powder of predominantly Sc(Cl)Pc (plus some $ScCl_3$) in 64% total yield.

Synthesis of YPc_2 (Ref. 86)

A mixture of Li_2Pc (53 mg) and $Y(acac)_3 \times H_2O$ (150 mg) in anhydrous TCB (10 mL) was refluxed for 10 h under nitrogen. The resulting dark blue solution was cooled and evaporated under reduced pressure, to give a residue, which was chromatographed on a silica gel column with chloroform as eluent. A pale blue band containing a trace amount of H_2Pc was developed, followed by a greenish-blue band which was collected and concentrated. The crude product was purified by recrystallization from a mixture of chloroform and methanol. Yield: 17 mg (30%).

4. *Complexes of the Lanthanide and Actinide Elements*

The first experimental articles focusing on the synthesis of Pcs with lanthanoids (LnPcs) was published by Plyushchev and Shklover in 1964.[95,96] Later, Kirin, Moskalev and Makashev published the procedure of the preparation of so-called double-decker Pc complexes.[97] This heralded the beginning of a new era of the chemistry of phthalocyanines. Since 2000, numerous research papers, as well as reviews and book chapters, focusing on the synthesis and applications of rare-earth Pcs have been published.[98–111] A comprehensive review on the synthesis of rare-earth tetrapyrolic complexes was published by Pushkarev, Tomilova and Nemykin.[112] In view of this, in this subsection we will only briefly discuss the preparation of the LnPcs.

There are two approaches for the synthesis of the lanthanoid *single-decker* Pcs — (i) by direct template condensation of so-called phthalogenes, namely (un)substituted phthalonitriles or 1,3-diiminoisoindoline, in the presence of the corresponding lanthanide metal salt, and (ii) metal insertion into H_2Pcs. The simplest method of the direct template condensation of phthalogenes is a melting of the initial compounds at 190–300 °C.[90,113,114] Indeed, this method has been used to prepare the tetra-substituted LnPcs **Ln-1** – **Ln-4** (Scheme 19, Table 11).

Another direct approach for the synthesis of the LnPcs is to use alcohols (typically *n*-pentanol ~ *n*-octanol) as reaction media in the presence of a strong organic base (typically DBU). The series of unsubstituted (**Ln-5**),[115,116] tetra- (**Ln-6** – **Ln-8** and **Ln-9** – **Ln-13**),[117–121] as well as hexadeca-substituted (**Ln-14**)[75] LnPcs have been prepared by this approach (Schemes 20 and 21. Table 11).

The unsubstituted **Ln-5** was prepared using 1,3-diiminoisoindoline as a phthalogene (Scheme 20).[122]

Single-decker LnPcs have been prepared by Tomilova and coworkers in 1985 *via* metal exchange of Li_2Pc using DMSO as a solvent (Scheme 22).[123] Furthermore, later on this procedure has been used by Lomova and coworkers.[124,125] Later, a method for preparation of the *single-decker* LnPcs *via* metal insertion into a metal-free ligand in *o*-dichlorobenzene was described (Scheme 22).[68,126–129] The series of unsubstituted (**Ln-5**), tetra-substituted (**Ln-15** – **Ln-18**), as well as hexadeca-substituted LnPcs (**Ln-14**)[75] have been prepared using both of these methods (Table 11).

For the synthesis of the lanthanide *homoleptic double-decker* Pcs ($Ln(Pc)_2$), two approaches can be used — (i) by direct template condensation of (un)substituted phthalonitriles in the presence of a corresponding lanthanide metal salt, and

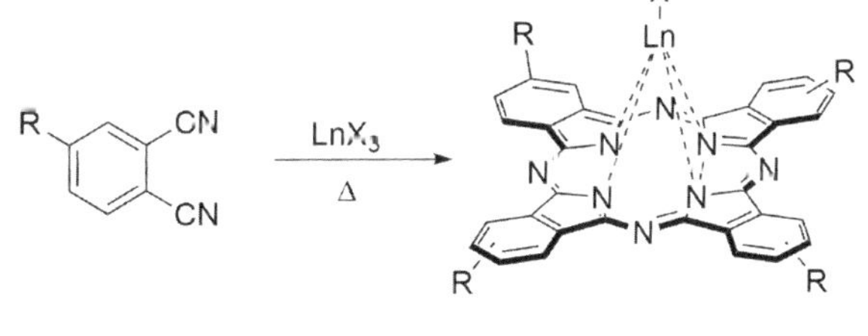

Ln-1: R^1 = NO_2, Ln = Nd, Sm, Ho, Er, Lu, X = Cl
Ln-2: R^1 = Cl, Ln = Sm, Ho, Er, Lu, X = Cl
Ln-3: R^1 = Br, Ln = Er, X = Cl
Ln-4: R^1 = I, Ln = Lu, X = OAc

Scheme 19. Synthesis of LnPcs by melting of initial compounds.[90,113,114]

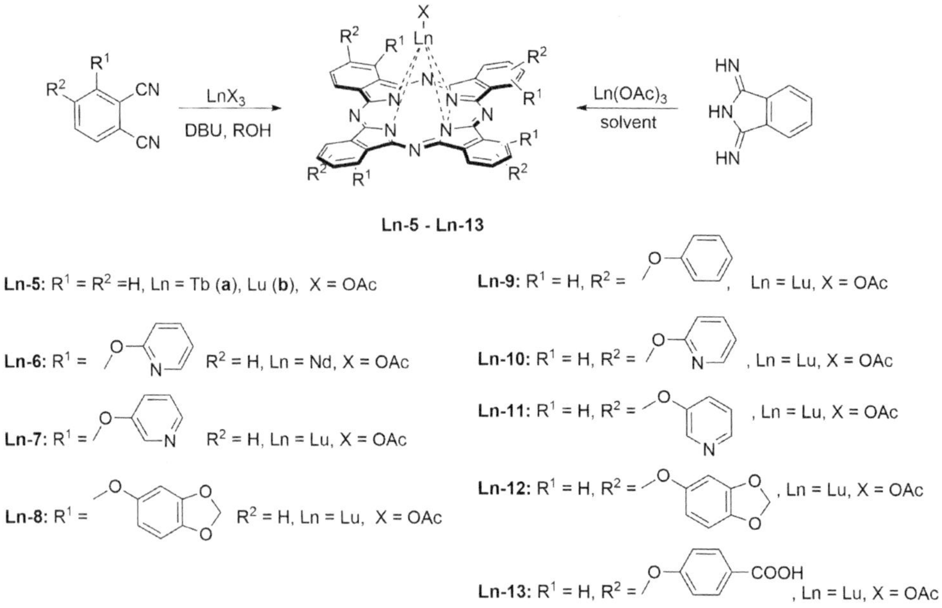

Scheme 20. Syntheses of unsubstituted (**Ln-5**) and tetra-substituted LnPcs (**Ln-6** – **Ln-13**).[114–121]

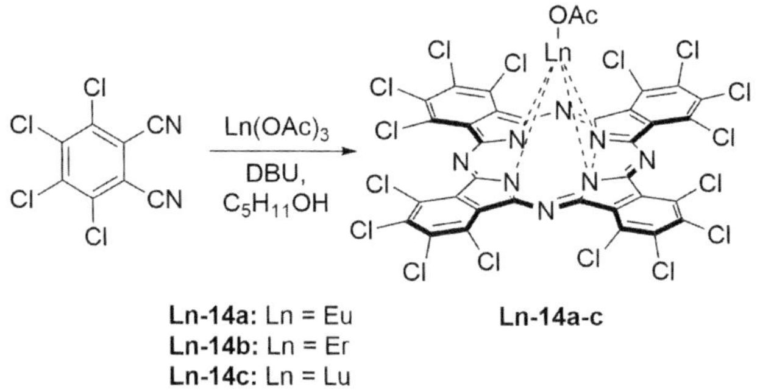

Scheme 21. Synthesis of dodeca-substituted LnPcs (**Ln-14**).[75]

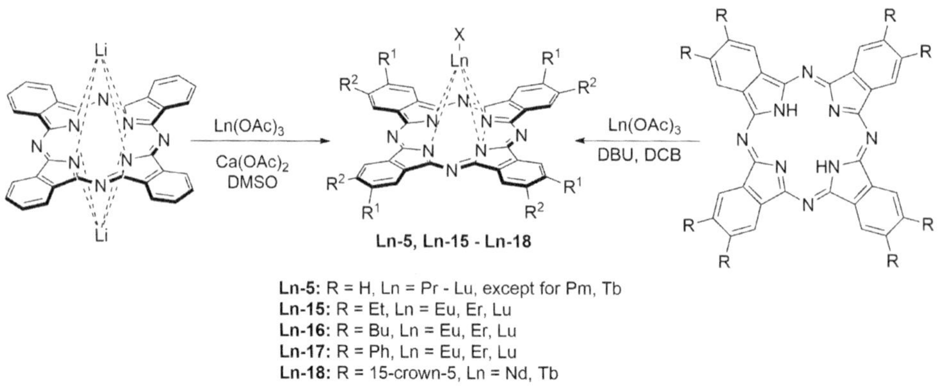

Scheme 22. Syntheses of LnPcs by metal exchange reaction of Li_2Pc or *via* metal insertion into metal-free Pcs

Table 11. Preparation of *single-decker* LnPcs.

	Conditions	Yield (%)	Ref.
Ln-1	$LnCl_3$, melting, 250–270 °C, 2 h	55–74	90, 113
Ln-2	$LnCl_3$, melting, 270–280 °C, 2 h	34–76	90, 113
Ln-3	$ErCl_3$, melting, 270–280 °C, 2 h	31	90
Ln-4	$Lu(OAc)_3$, melting, 370 °C, 5 min, sealed tube	n.r.	114
Ln-5	PN, $Ln(OAc)_3$ *n*-hexanol, DBU, refluxing, 1.5 h	46–75	115, 116
Ln-5	Li_2Pc, $Ln(OAc)_3$, $Ca(OAc)_2$, DMSO, refluxing, 1 min	90–93	123–125
Ln-6	$Nd(OAc)_3xH_2O$, *n*-pentanol, DBU, refluxing, 21 h	23	117
Ln-7	$Lu(OAc)_3$, *n*-pentanol, DBU, 135 °C, 5 h	61	118
Ln-8	$Lu(OAc)_3$, *n*-pentanol, DBU, 170 °C, 17 h	26.9	119
Ln-9	$Lu(OAc)_3$, *n*-pentanol, DBU, refluxed, 7 h	25	120
Ln-10	$Lu(OAc)_3$, *n*-pentanol, DBU, refluxed, 7 h	23	120
Ln-11	$Lu(OAc)_3$, *n*-pentanol, DBU, 135 °C, 5 h	67	118
Ln-12	$Lu(OAc)_3$, *n*-pentanol, DBU, 170 °C, 17 h	16.3	119
Ln-13	$Lu(OAc)_3$, *n*-pentanol, DBU, refluxed, 7 h	10	121
Ln-14	PN, $Ln(OAc)_3xnH_2O$, isopentanol, DBU, refluxing, 4 h	44–78	75
Ln-14	H_2PcR_{16}, $Ln(OAc)_3xnH_2O$, DCB, DBU, refluxing, 4 h	68–81	75
Ln-15	$Ln(OAc)_3xnH_2O$, DCB, DBU, 180 °C, 3–4 h	89–96	126
Ln-16	$Ln(OAc)_3xnH_2O$, DCB, DBU, 180 °C, 3–4 h	95–97	126
Ln-17	$Ln(OAc)_3xnH_2O$, DBU, DCB, refluxing, 2 h	60–94	68
Ln-18	$Nd(OAc)_3xnH_2O$, DBU, DCB:DMF (1:1) 1 h	quant.	127

(ii) metal insertion into metal-free Pcs. Higher temperatures, longer reaction times and larger radii of the lanthanide ions favor the formation of the *double-decker* LnPcs, in contrast to the *single-decker* LnPcs.

The preparation of the *double decker* LnPcs *via* direct template synthesis of phthalogenes can be made under different conditions: (i) by melting of the starting materials at 250–300 °C; (ii) by using high-boiling solvents (*e.g.* pentanol, isopentanol, hexanol or sulfolane) as the reaction medium.

The first example of $Ln(Pc)_2$ was obtained *via* fusion of the starting materials at high temperature by Kirin, Moskalev and Makashev in 1965 and 1967.[97,128] Later, the series of unsubstituted (**Ln-19**),[129–132] octa-substituted (**Ln-20** – **Ln-23**)[129,130,135] as well as hexadeca-substituted (**Ln-24** – **Ln-26**)[130–134] $Ln(Pc)_2$ were obtained by this method (Scheme 23, Table 12). On the other hand, $Ln(Pc)_2$ can be prepared by direct template condensation of the phthalogenes by the use of high-boiling solvents as the reaction medium, and a series of the hexadeca-substituted (**Ln-27** – **Ln-39**) $Ln(Pc)_2$ has been obtained similarly (Scheme 23, Table 12).[88,136–144]

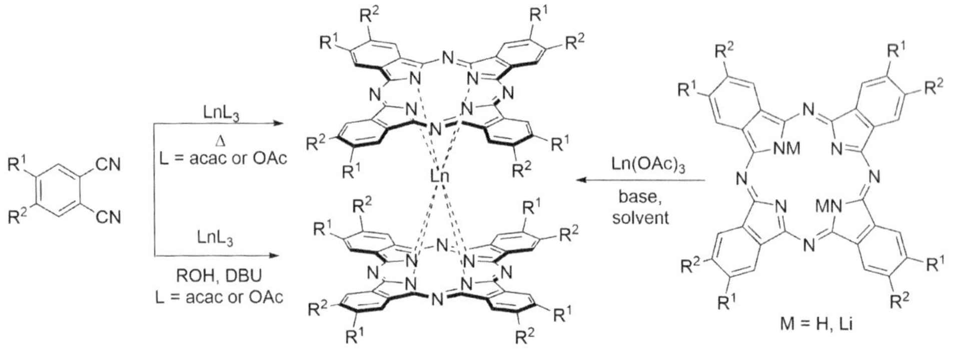

Ln-19, Ln-20 - Ln-36

Ln-19: $R^1 = R^2$ = H, Ln = La, Pr, Nd, Sm-Lu
Ln-20: R^1 = *tert*-Bu, R^2 = H, Ln = La, Pr, Nd, Sm, Gd, Er, Lu
Ln-21: R^1 = OPh, R^2 = H, Ln = Lu
Ln-22: R^1 = Ph, R^2 = H, Ln = Lu
Ln-23: RR = 15-crown-5, Ln = Ng, Gd, Yb, Lu
Ln-24: R^1 = Me, R^2 = H, Ln = Tb, Lu
Ln-25: R^1 = Bu, R^2 = H, Ln = Eu, Dy, Er, Lu
Ln-26: R^1 = OMe, R^2 = H, Ln = Lu
Ln-27: R^1 = Et, R^2 = H, Ln = Eu, Dy, Er, Lu
Ln-28: R^1 = C_7H_{15}, R^2 =H, Ln = Eu, Gd
Ln-29: $R^1 = R^2$ = OPr, Ln = Sm, Tm, Lu
Ln-30: $R^1 = R^2$ = OC_5H_{11}, Ln = Eu, Gd
Ln-31: $R^1 = R^2$ = OC_8H_{17}, Ln = La-Nd, Sm-Tm
Ln-32: $R^1 = R^2$ = $OC_{10}H_{21}$, Ln = Er, Lu
Ln-33: $R^1 = R^2$ = $OC_{12}H_{25}$, Ln = Ce-Nd, Eu-Lu
Ln-34: $R^1 = R^2$ = $OC_{14}H_{25}$, Ln = Lu
Ln-35: $R^1 = R^2$ = $OC_{16}H_{33}$, Ln = Lu
Ln-36: $R^1 = R^2$ = $OC_{18}H_{37}$, Ln = Er
Ln-37: $R^1 = R^2$ = OPh, Ln = Eu, Ho, Lu
Ln-38: $R^1 = R^2$ = SPh, Ln = Eu
Ln-39: $R^1 = R^2$ = Ph, Ln = Eu, Er, Lu

Scheme 23. Synthetic pathway for the preparation of $Ln(Pc)_2$.[88,97,128–147]

The *homoleptic double-decker* $LnPcs_2$ could also be prepared by metal insertion into (un)substituted H_2Pcs or by transmetalation of labile (un)substituted Li_2Pcs. Compared with direct template synthesis, this approach avoids unwanted side reactions such as the formation of linear oligomers of phthalonitrile. However, the formation of the desired $LnPcs_2$ is complicated by steric interactions between the initial H_2Pcs and $LnPc_2$. Using this method, a series of (un)substituted $LnPcs_2$ (**Ln-19, Ln-23, Ln-25, Ln-27, Ln-39**) has been obtained (Scheme 23. Table 12).[68,126,127,145–147]

Homoleptic triple-decker LnPcs can be prepared *via* three routes: (i) by direct template condensation of the corresponding phthalonitrile with a lanthanide salt; (ii) by metalation of the corresponding H_2Pc and (iii) by sublimation of the *single-decker* LnPc *in vacuo*. The direct template condensation of (un)substituted phthalonitriles and lanthanide(III) acetates led to the formation of a mixture of double-decker $LnPc_2$ and the target dinuclear triple-decker Ln_2Pc_3 (**Ln-40, Ln-41**) (Scheme 24, Table 13).[128,148]

On the other hand, *triple-decker* Ln_2Pc_3 (**Ln-41** – **Ln-44**) could be prepared *via* metalation of the corresponding H_2Pc (Scheme 24. Table 13).[68,126,127,147] Indeed, these complexes have been obtained in good yield by using high-boiling solvents (*e.g.* TCB, 1-ClN, *n*-octanol) (Table 13). In addition, triple-decker Ln_2Pc_3 (**Ln-40**)

Table 12. Preparation of *double-decker* $Ln(Pcs)_2$.

	Conditions	Yield (%)	Ref.
Ln-19	$Ln(OAc)_3$, melting, 0.5–4 h	5–60	97,129–132
Ln-19	$Lu(OAc)_3$, DBU, hexanol, refluxing, 5 h	10	137
Ln-19	H_2Pc, $Lu(OAc)_3$, quinoline, 240 °C, 22 h	9	145
Ln-19	Li_2Pc, $Ln(acac)_3$, TCB, 220 °C, 4–10 h	12–33	136,146
Ln-20	$Ln(OAc)_3$, melting, 280–290 °C, 2 h	60–80	129,130
Ln-21	$Lu(OAc)_3$, melting, 280–290 °C, 2 h	70	129
Ln-22	$Lu(OAc)_3$, melting, 280–290 °C, 2 h	70	129
Ln-23	$Lu(OAc)_3$, melting, 265 °C, 4 h	35	135
Ln-23	H_2PcR_8, $Ln(OAc)_3$, DBU, 1-ClN, 1.5–2 h	15–80	127,147
Ln-24	$Ln(OAc)_3$ or $Ln(acac)_3$, melting, 2–3.5 h	n.r.	133
Ln-25	$Ln(OAc)_3$ or $Ln(acac)_3$, melting, 2–8 h	22–49	134
Ln-25	$Ln(OAc)_3$ or $Ln(acac)_3$, DBU, *i*-pentanol, refluxing, 15–20 h	6–46	134
Ln-25	H_2PcR_8, $Ln(OAc)_3$, hexanol, 230 °C, 1 h	85–90	126
Ln-26	$Ln(OAc)_3$, melting, 0.5–4 h	n.r.	133
Ln-27	$Ln(OAc)_3$ or $Ln(acac)_3$, DBU, *i*-pentanol, refluxing, 15–20 h	11–53	134
Ln-27	H_2PcR_8, $Ln(OAc)_3$, hexanol, 230 °C, 1 h	89–90	126
Ln-28	$Ln(acac)_3$, DBU, pentanol, refluxing, 13 h	68–72	136
Ln-29	$Ln(OAc)_3$ or $Ln(HCO_2)_3$, DBU, pentanol, refluxing, 8 h	61–66	138
Ln-30	$Ln(acac)_3$, DBU, pentanol, refluxing, 8 h	75–79	136
Ln-31	$Ln(acac)_3$, DBU, pentanol, refluxing, 12 h	9–63	88
Ln-32	$Ln(OAc)_3$, DBU, hexanol, refluxing, 20–24 h	20–24	139,140
Ln-33	$Ln(OAc)_3$ or $Ln(acac)_3$, DBU, hexanol or pentanol, refluxing, 20 h	15–30	139,141,143
Ln-34	$Lu(OAc)_3$, DBU, hexanol, refluxing, 24 h	14	140
Ln-35	$Ln(OAc)_3$, DBU, hexanol, refluxing, 24 h	11	140
Ln-36	$Ln(OAc)_3$, DBU, hexanol, refluxing, 20 h	27	139
Ln-37	$Ln(acac)_3$, DBU, pentanol, refluxing, 8 h	36–43	144
Ln-38	$Ln(acac)_3$, DBU, pentanol, refluxing, 8 h	42	144
Ln-39	H_2PcR_8, $Ln(OAc)_3$, MeOLi, $C_{16}H_{33}OH$:TCB (1:1), 250 °C, 5–10 min	49–96	68

could be prepared *via* heating of unsubstituted Lu(OAc)Pc to 400°C *in vacuo* (*ca.* 1 mmHg) (Scheme 25, Table 13).[149]

Phthalocyanine complexes of *actinides*, in particular with *uranium* and *thorium*, have been known since 1964.[150] Indeed, Bloor *et al.* have published a

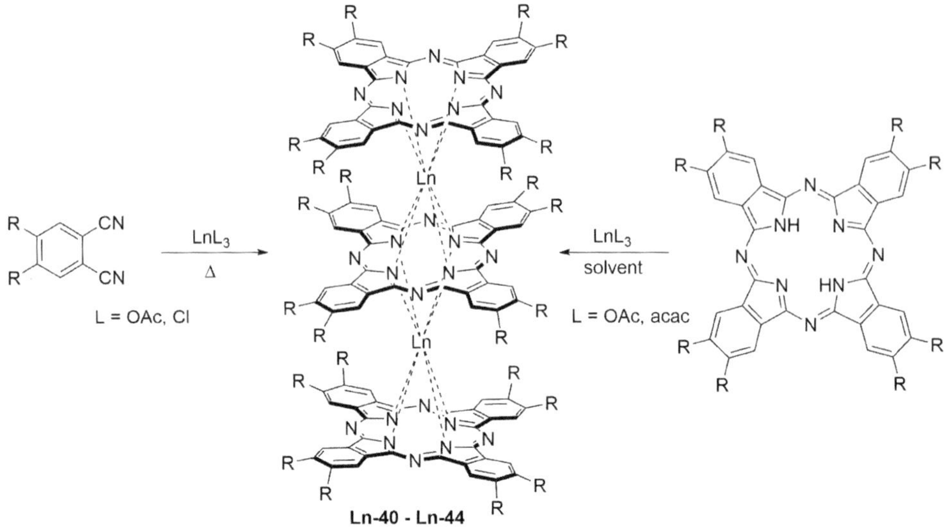

Ln-40 - Ln-44

Ln-40: R = H, Ln = La, Pr, Nd, Sm-Lu
Ln-41: R = 15-crown-5, Ln = Nd, Gd, Tb, Yb, Lu
Ln-42: R = Et, Ln = Eu, Er, Lu
Ln-43: R = Bu, Ln = Eu, Er, Lu
Ln-44: R = Ph, Ln = Eu, Er, Lu

Scheme 24. Synthetic approaches for preparation of the *homoleptic triple-decker* Ln_2Pcs_3 (**Ln-40** – **Ln-44**).[68,126–128,147,148]

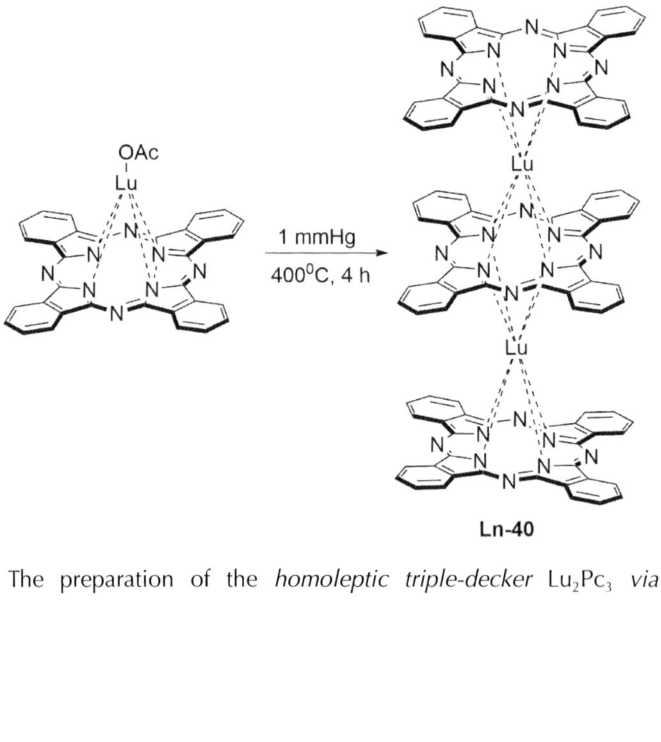

Ln-40

Scheme 25. The preparation of the *homoleptic triple-decker* Lu_2Pc_3 *via* sublimation *in vacuo*.[149]

Table 13. Preparation of triple-decker Ln_2Pcs_3.

	Conditions	Yield (%)	Ref.
Ln-40	Ln(OAc)3, melting, 280–290 °C, 1 h	n.r.	128, 148
Ln-40	Lu(OAc)Pc, 1 mm Hg, 400 °C, 4 h	40	149
Ln-41	H_2PcR_8, $Ln(OAc)_3$ or $Ln(acac)_3$, TCB or 1-ClN, 1.5–2 h	10–68	127, 147
Ln-42	H_2PcR_8, $Ln(OAc)_3$, *n*-octanol, 190 °C, 1–1.5 h or $C_{16}H_{33}OH$, 280 °C, 1 h	53–92	126
Ln-43	H_2PcR_8, $Ln(OAc)_3$, *n*-octanol, 190 °C, 1–1.5 h or $C_{16}H_{33}OH$, 280 °C, 1 h	49–94	126
Ln-44	H_2PcR_8, $Ln(OAc)_3$ or $Ln(acac)_3$, MeOLi, $C_{16}H_{33}OH$:TCB	71–80	68

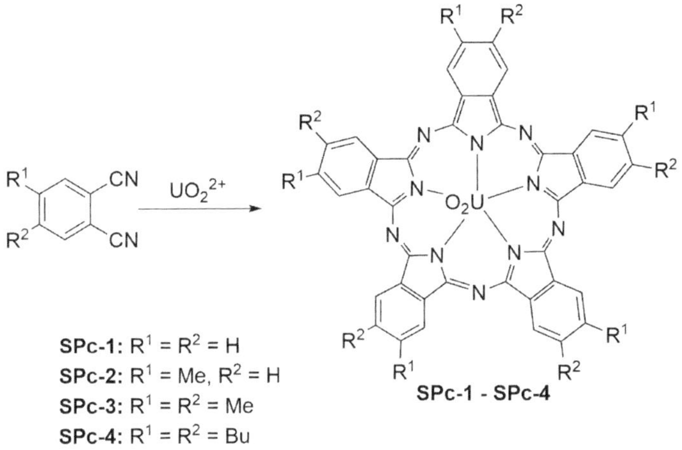

Scheme 26. Synthesis of (un)substituted SuperPcs.[152,153]

procedure for the preparation of uranyl phthalocyanine in 11.6% yield by template condensation of the phthalonitrile and anhydrous uranyl chloride.[150] However, later, Marks and coworkers re-examined Bloor's data by X-ray crystallography and found that the structure contained five isoindoline units instead of four.[151] Since then, the structures containing five isoindoline units have been known as "superphthalocyanines" (SuperPcs). Marks and coworkers published the relevant procedures for preparing (un)substituted SuperPcs (**SPc-1**, Scheme 26, Table 14).[152,153] The yield of unsubstituted SuperPc depends on the uranium salt used (Table 14).[152] In addition, a procedure for the preparation of **SPc-1** under high-pressure conditions in the presence of a strong organic base and reducing agent was reported (Table 14).[42]

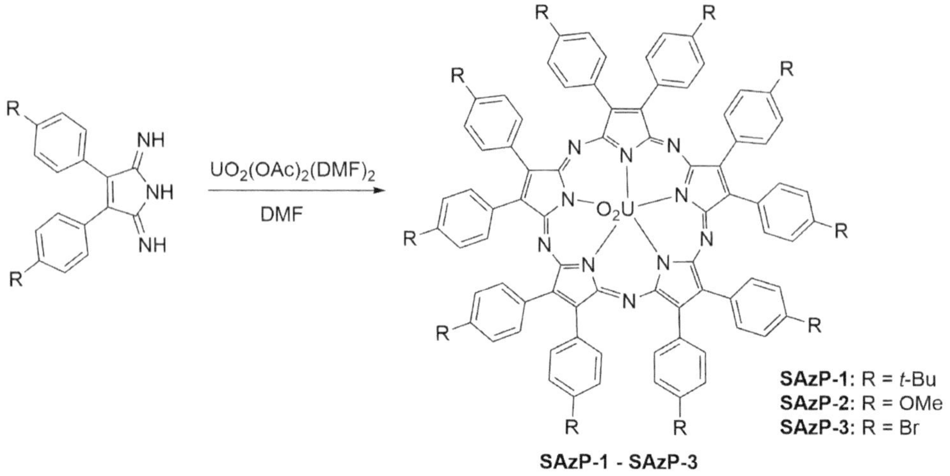

Scheme 27. Synthesis of substituted superazaporphyrins.[154]

Table 14. Preparation of SuperPcs and their analogs.

	Conditions	Yield (%)	Ref.
SPc-1	UO_2Cl_2, DMF, 170 °C, 80–90 min	24	152
SPc-1	$UO_2(OAc)_2(DMF)_2$, DMF, 170 °C, 80–90 min	7.1	152
SPc-1	$UO_2(NO_3)_2(DMF)_2$, DMF, 170 °C, 80–90 min	3.7	152
SPc-1	UO_2Cl_2, MeOH, DBU, hydroquinone, 9 kbar, 90 °C, 22 h	27	42
SPc-2	UO_2Cl_2, DMF, 170 °C, 40 min	7.6	152
SPc-3	UO_2Cl_2, DMF, 175 °C, 45 min (glovebox)	0.36	153
SPc-4	UO_2Cl_2, quinoline, 170 °C, 40 min	2.5	153
SAzP-1	$UO_2(OAc)_2(DMF)_2$, DMF, 190 °C, 1 h	6	154
SAzP-2	$UO_2(OAc)_2(DMF)_2$, DMF, 190 °C, 1 h	2	154
SAzP-3	$UO_2(OAc)_2(DMF)_2$, DMF, 190 °C, 1 h	6	154

Kobayashi and coworkers described a procedure for preparing expanded analogs of porphyrazines – superazaporphyrins (SAzP, **SAzP-1 – SAzP-3**) using Marks' approach (Scheme 27, Table 14).[154]

Single-decker *uranium* (**Ac-1**) and *thorium* (**Ac-2**) Pcs can be prepared *via* metal exchange of Li_2Pc with acetylacetonates of the corresponding metals (Scheme 28, Table 15).[155]

The *homoleptic double-decker* actinide Pcs can be prepared by direct template condensation of phthalonitrile with a metal salt[156,157] or *via* metal exchange of Na_2Pc with uranium and thorium(IV) chlorides (Scheme 29, Table 15).[158]

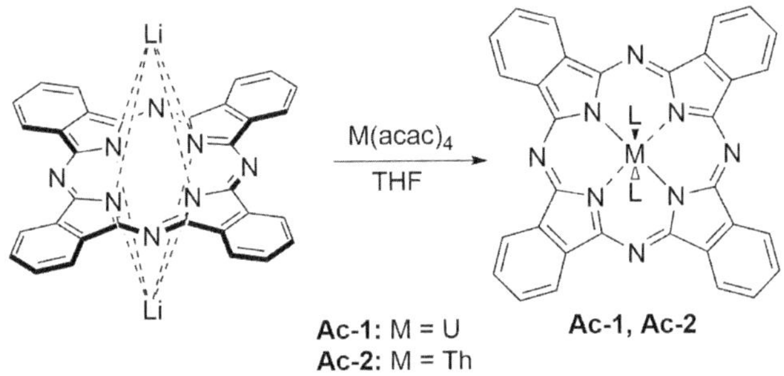

Scheme 28. Preparation of the *single-decker* actinide phthalocyanines.[155]

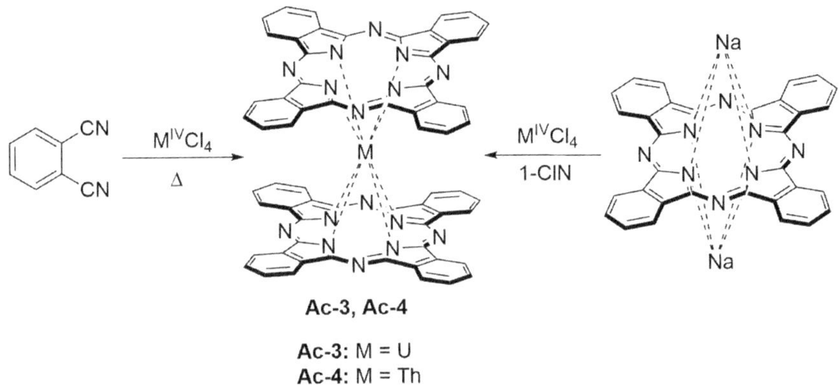

Scheme 29. The synthetic approaches for preparation of the double-decker actinide Pcs.[156–158]

Table 15. Preparation of SuperPcs and their analogs.

	Conditions	Yield (%)	Ref.
Ac-1	U(acac)$_4$, THF	45	155
Ac-2	Th(acac)$_4$, THF	31	155
Ac-3	PN, UCl$_4$, melting, 240 °C, 30 min	38	156
Ac-3	PN, UCl$_4$(Et$_3$N), melting, 300–350 °C, 30–40 min	45	157
Ac-3	Na$_2$Pc, UCl$_4$, 1-ClN, refluxing, 12 h	7.4	158
Ac-4	PN, ThCl$_4$, melting, 240 °C, 30 min	43	156

Synthesis of Lu(III)(OAc)Pc (Ref. 115)

A mixture of phthalonitrile (1.55 g), lutetium acetate pre-dried at 100 °C under vacuum (5 Torr) (0.7 g) and DBU (0.9 mL) in *n*-hexanol (15 mL) was heated under reflux for 75 min. After cooling to room temperature, the blue solution was filtered.

The precipitate was washed with acetic anhydride (10 mL), cold acetone (10 mL) and finally with pentane, and the solid extracted with dichloromethane. The soluble fraction (1.3 g) was purified by chromatography over deactivated silica gel (deactivation was achieved by addition of 20 ml of CH_3OH in 100 g of silica and stirring for 30 min). The first fraction, containing $LuPc_2$, was collected with a CH_2Cl_2:CH_3OH mixture (99:1, v:v). The desired Lu(OAc)Pc was then obtained by elution with CH_2Cl_2 containing from 3 to 5% of CH_3OH. Yield: 1.2 g (75%).

Synthesis of Eu(III)(OAc)PcCl$_{16}$ (Ref. 75)

Tetrachlorophthalonitrile (213.0 mg), $Eu(OAc)_3 \times 3H_2O$ (76.4 mg) and DBU (119 μL) were stirred in boiling isoamyl alcohol (3 mL) for 4 h (TLC-control:SiO_2, toluene). The reaction mixture was cooled to room temperature and a MeOH:H_2O (4:1 v/v) mixture added. The precipitate was filtered and washed with a MeOH:H_2O (4:1 v/v) mixture and dried at 70°C. Yield: 110.8 mg (44%).

Synthesis of Er(III)(OAc)PcCl$_{16}$ (Ref. 75)

1,2,3,4,8,9,10,11,15,16,17,18,22,23,24,25-Hexadecachlorophthalocyanine (78 mg), $Er(OAc)_3 \times 4H_2O$ (29 mg) and DBU (18 μL) were stirred in 3 mL boiling *o*-DCB for 4 h (TLC:SiO_2, toluene). The reaction mixture was cooled to room temperature and a MeOH:H_2O (4:1 v/v) mixture added. The precipitate was filtered, washed with a MeOH:H_2O (4:1 v/v) mixture, and dried at 70 °C. Yield: 61 mg (68%).

Synthesis of Nd(III)[Pc(t-Bu)$_4$]$_2$ (Ref. 130)

A mixture of $Nd(OAc)_3$ (0.31 g) and 4-*tert*-butylphthalonitrile (1.47 g) was fused in an evacuated glass ampule with gradual raising of the temperature to 290 °C, and this temperature was then maintained for 2 h. The cooled melt was ground, and purified by column chromatography on Al_2O_3, using benzene as eluent, to obtain a blue powder. Yield: 1.03 g (60%).

Synthesis of Lu[Pc(OPr)$_8$]$_2$ (Ref. 138)

A mixture of 4,5-dipropyloxyphthalodinitrile (340 mg), $Lu(OAc)_3$ dihydrates (60 mg), and DBU (106 mg) was refluxed in amyl alcohol (5 mL) for 8 h under an Ar atmosphere. After cooling, the solvent was removed *in vacuo*, the residue was reprecipitated using a water-DMF mixture, washed with aqueous MeOH, and dried in air. After separation with column chromatography (Al_2O_3, using $CHCl_3$:MeOH (10:1) as the eluent), the desired $Lu[Pc(OPr)_8]_2$ was obtained as a dark green, fine crystalline powder. Yield: 182 mg (61%).

Synthesis of $Eu[Pc(OC_5H_{11})_8]_2$ (Ref. 136)

Under a slow stream of nitrogen, a mixture of 4,5-bis(pentyloxy)phthalonitrile (300 mg), DBU (78 mg) and $Eu(acac)_3 \times nH_2O$ (30 mg) was heated in amyl alcohol (4 mL) under reflux for 8 h, to give a dark greenish-blue solution. The volatiles were removed *in vacuo*, then the residue was chromatographed with $CHCl_3$:hexane (2: 3, v/v) as eluent. A green band was first developed which contained a small amount of the metal-free $H_2Pc(OC_5H_{11})_8$ as determined by UV-vis spectroscopy. Then, a dark greenish-blue band containing the sandwich compound was developed, which was collected and evaporated under reduced pressure. The obtained crude product was further purified by the same chromatographic procedure. Yield: 121 mg (79%).

Synthesis of $Ce[Pc(OC_{12}H_{25})_8]_2$ (Ref. 142)

A mixture of $Ce(acac)_3 \times nH_2O$ (22 mg), 4,5-bis(dodecyloxy)phthalonitrile (0.30 g), and DBU (40 mg) in *n*-pentanol (2 mL) was refluxed under a slow stream of nitrogen for 12 h. The mixture was cooled briefly, and the volatiles removed under reduced pressure. The residue was chromatographed on a silica gel column using $CHCl_3$ as eluent. A green band of the target double-decker was developed, followed by a fraction of the metal-free $H_2Pc(OC_{12}H_{25})_8$. The crude product was further purified by repeated chromatography, followed by re-precipitation from a mixture of $CHCl_3$ and MeOH. Yield: 45 mg (22%).

Synthesis of $Lu[PcPh_8]_2$ (Ref. 68)

A mixture of 2,3,9,10,16,17,23,24-octaphenylphthalocyanine (15 mg), $Lu(OAc)_3 \times 4H_2O$ (3 mg) and a catalytic amount of MeOLi was gradually heated from 200 to 255 °C in a $C_{16}H_{33}OH$:TCB (300 μL, 1:1, v/v) mixture, and then kept at 255 °C for 5–10 min. The reaction mixture was cooled to room temperature and a $MeOH:H_2O$ (10:1, v/v) mixture added. The precipitate was filtered and washed with water and MeOH. The desired compound was separated using gel permeation chromatography with THF as eluent. The solvent was evaporated, and the resulting precipitate repeatedly washed with a $MeOH:H_2O$ (10:1, v/v) mixture. Yield: 12 mg (96%).

Synthesis of $Lu_2(Pc)_3$ (Ref. 149)

Lutetium phthalocyanine acetate (0.5 g) was heated at 400 °C for 4 h *in vacuo* (*ca.* 1 mmHg) in a conventional sublimation tube equipped with a water-cooled inner trap. The solid remaining in the tube was extracted with dichloromethane. The solution was identified as a mixture of $Lu_2(Pc)_3$ and a small amount of $Lu(Pc)_2$

by UV-vis spectral measurement. The solution was treated with hydrazine monohydrate to reduce $Lu(Pc)_2$ to $[Lu(Pc)_2]^-$, and was passed through a silica gel column (*ca.* 5 cm length) to remove the anionic $[Lu(Pc)_2]^-$. A hydrazine monohydrate layer was kept on top of the solution in the column to avoid oxidation of $[Lu(Pc)_2]^-$. The solution was washed with water, dried by magnesium sulfate, concentrated by evaporation, and added to hexane to give a microcrystalline powder of $Lu_2(Pc)_3$. Yield: 0.20 g (40%).

Synthesis of $U(O_2)$SuperPc (Ref. 152)

A mixture of anhydrous powdered uranyl chloride (2.5 g), phthalonitrile (14.0 g), and dry dimethylformamide (1.1 mL) was heated under nitrogen with stirring at 170 °C. Within 80–90 min, the tan mixture abruptly turned dark green. On cooling to room temperature, the reaction mixture solidified. The dark green solid was crushed, washed with ethanol, water, and acetone, and then dried under vacuum to yield a mixture of $U(O_2)$SuperPc and H_2Pc in a molar ratio of 1.9:1 (by UV-vis spectrum). Soxhlet extraction of this material with benzene yielded 1.37 g (24%) of a dark blue, microcrystalline product which was ≥ 97% $U(O_2)$SuperPc by spectrophotometry (≤3% H_2Pc). Larger crystals can be grown by slow evaporation of 1,2,4-trichlorobenzene solutions. A slightly purer product can also be obtained by sublimation at 400 °C (10^{-3} mm); however, the yield is greatly reduced.

Synthesis of Th(IV)$(acac)_2$Pc (Ref. 155)

A THF solution (40 mL) containing $Th(acac)_4$ (1.28 g) was added to Li_2Pc (1.16 g) in THF (50 mL). The reaction mixture was evaporated to dryness and Soxhlet extracted with toluene (50 mL). After concentrating to 20 mL, the solution was cooled to –20°C. This workup yielded 0.65 g of dark blue crystals (31%), and elemental analysis indicated one solvent molecule per molecule of compound.

Synthesis of $ThPc_2$ (Ref. 156)

A mixture of thorium iodide (1.83 g) and phthalodinitrile (6.3 g) was stirred in an atmosphere of inert gas at 240 °C for 30 min, iodine being evolved and the melt becoming dark green. After cooling to room temperature, the solidified melt was powdered, and extracted with benzene. The residue was dried in a high vacuum, enclosed in a platinum net, and heated in a quartz sublimation apparatus at 10^{-5} torr; at *ca.* 200 °C, a rise in pressure and evolution of iodine were observed; the temperature was then raised to 550 °C. The material separating in the lower zone of the formed sublimate was similarly sublimed twice more for purification. Yield: (43%).

Synthesis of UPc_2 (Ref. 158)

Na_2Pc (3.0 g) and UCl_4 (0.9 g) were allowed to react in boiling 1-chloronaphthalene (30 mL) for 12 h. The solution was filtered while hot and, after cooling, the solid was precipitated from solution with heptane. The crude product was purified by column chromatography on silica gel, using a mixture of toluene: heptane (1:1, v:v) as eluent. Yield: 0.25 g (7.4%).

5. *Group 4: TiPc, $TiPc_2$, ZrPc, $ZrPc_2$, HfPc, and $HfPc_2$*

Titanium phthalocyanines ($Ti(Cl_2)PcR_4$, **Ti-1** – **Ti-3**) can be prepared by melting phthalonitrile with titanium(IV) chloride (Scheme 30)[159] or *via* refluxing phthalonitrile with titanium(III)- or titanium(IV) chlorides in 1-chloronaphthalene[160,161] as well as 1-chloronaphthalene: benzene,[162] 2-bromonaphthalene: 2-methylnapthalene[163] and 1-chloronaphthalene:2-methylnapthalene mixtures (Scheme 30, Table 16).[164]

More soluble peripherally-substituted titanyl phthalocyanines (Ti(=O)Pc) can be prepared by two approaches. Indeed, the complexes **Ti-4**, **Ti-5** could be prepared by hydrolysis of initial complexes **Ti-2** and **Ti-3** (Scheme 31, Table 16).[161,164]

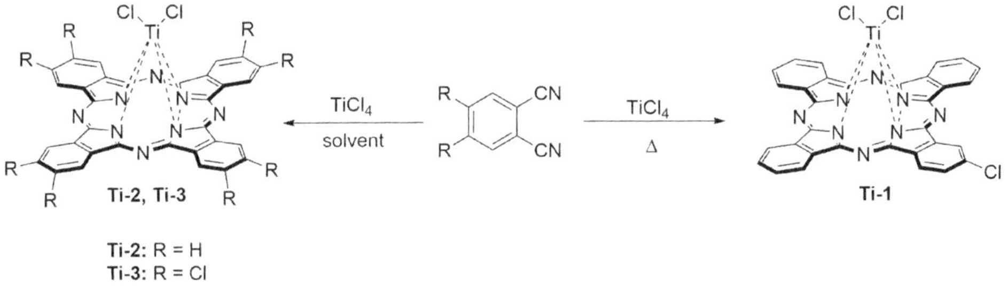

Scheme 30. Syntheses of $Ti(Cl_2)Pcs$.[159–164]

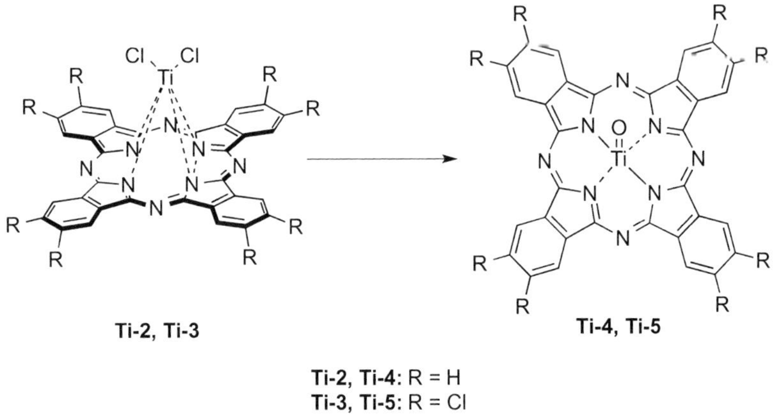

Scheme 31. Hydrolysis of $Ti(Cl_2)Pcs$ to Ti(O)Pcs.[161,164]

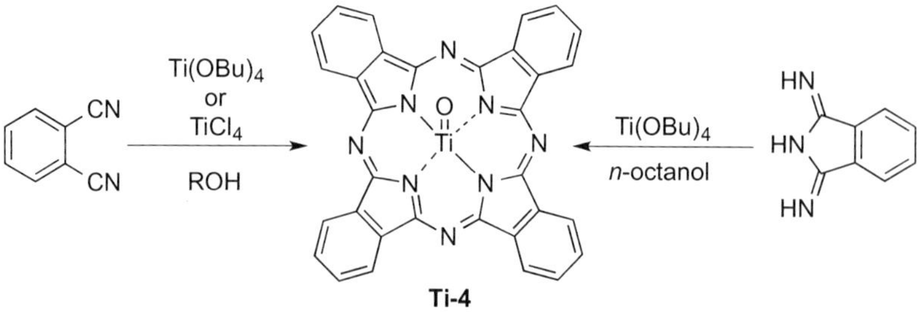

Scheme 32. Synthetic approaches of preparation of the unsubstituted titanyl-Pc**Ti-4**.[163,165]

Table 16. Preparation of Ti(L)Pcs.

	Conditions	Yield (%)	Ref.
Ti-1	$TiCl_4$, melting, 170–190 °C, 1 h	35	159
Ti-2	$TiCl_3$, 1-ClN, 252 °C, 0.5 h	95	160
Ti-2	$TiCl_4$, 1-ClN, 220 °C, 10 h	60	161
Ti-2	$TiCl_4$, 1-ClN:benzene, 250 °C	n.r.	162
Ti-2	$TiCl_4$, 2-bromonaphthalene:2-methylnaphthalene, 220 °C, 1 h	77	163
Ti-3	$TiCl_4$, 1-ClN:2-methylnaphthalene, 250 °C, 5 h	n.r.	164
Ti-4	Et_3M, EtOH:H_2O (95:5, v/v) or Py:NH_4OH (80:20, v/v)	n.r.	161
Ti-4	$TiCl_4$, urea, *n*-heptanol, 170 °C, 6 h	60	163
Ti-4	$Ti(OBu)_4$, urea, *n*-butanol, 150 °C, 6 h	87	165
Ti-4	DI, $Ti(OBu)_4$, urea, *n*-butanol, 140 °C, 6 h	86.2	165
Ti-5	Py: H_2O (95:5, v/v), refluxing, 2 h	68	164

On the other hand, Ti(=O)Pcs could be obtained by template condensation of phthalonitriles or 1,3-diiminoisoindolines, as well as phthalimide with commercially-available titanium(IV) alkoxides (butoxide or isopropoxide)[165–177] or titanium(IV) chloride[171] as metal sources (Schemes 32–36, Table 16).

Unsubstituted titanyl Pc **Ti-4** could be prepared by template condensation of phthalonitrile or 1,3-diiminoisoindoline with titanium(IV) butoxide in *n*-octanol[165] (Scheme 32) or by treatment of phthalonitrile with titanium(IV) chloride in *n*-heptanol (Table 16).[163]

Titanyl 1,8(11),15(18),22(25)-tetra-substituted Pcs (**Ti-6** – **Ti-10**) were obtained by template condensation of the corresponding 3-substituted phthalonitriles[166–168] or 3-substituted phthalimide[11] with titanium(IV) butoxide in *n*-pentanol (Scheme 33, Table 17). In the case of tetraamino-substituted **Ti-10**, 3-nitrophthalimide was used as an initial compound, and by its template condensation, and the

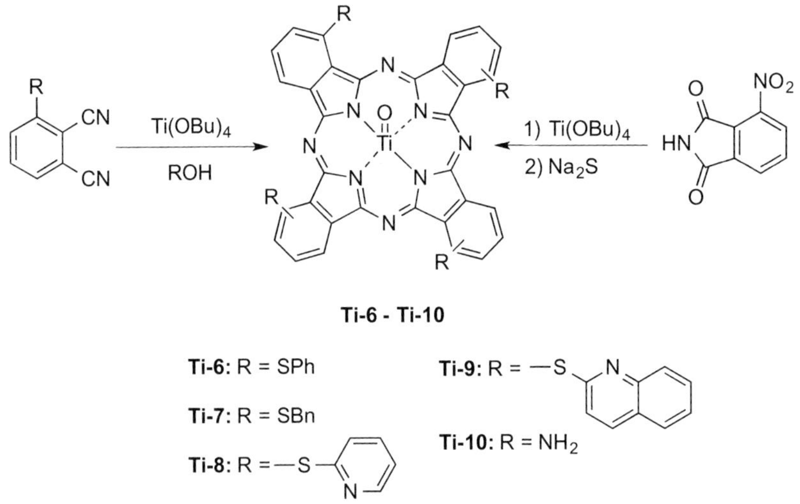

Scheme 33. Preparation of titanyl 1,8(11),15(18),22(25)-tetra-substituted Pcs.[166–169]

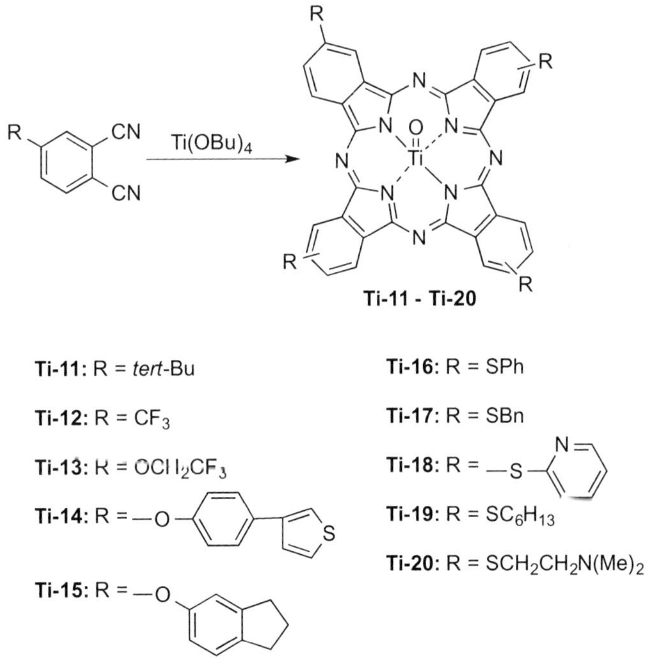

Scheme 34. Titanyl 2,9(10),16(17),23(24)-tetra-substituted Pcs.

subsequent reduction of the generated intermediate tetranitroPc by sodium sulfide, the desired **Ti-10** was obtained.[169]

Titanyl 2,9(10),16(17),23(24)-tetra-substituted Pcs (**Ti-11** – **Ti-20**) were prepared by template condensation of the corresponding 4-substituted phthalonitriles

Table 17. Preparation of TiPcs.

	Conditions	Yield (%)	Ref.
Ti-6	$Ti(OBu)_4$, urea, *n*-octanol, 150 °C, 14 h	40	166
Ti-7	$Ti(OBu)_4$, urea, *n*-octanol, 150 °C, 14 h	43	166
Ti-8	$Ti(OBu)_4$, DBU, *n*-pentanol, 140 °C, 5 h	38	167
Ti-9	$Ti(OBu)_4$, DBU, *n*-pentanol, refluxing, 12 h	21	168
Ti-10	1) 3-Nitrophthalimide, $Ti(OBu)_4$, MOA, $PhNO_2$, 185 °C, 2 h 2) Na_2S, 50 °C, 5 h	75	169
Ti-11	$Ti(OBu)_4$, DBU, urea, *n*-pentanol, 155 °C, 7 h	35	170
Ti-12	$Ti(OBu)_4$, DBU, urea, *n*-pentanol, 110 °C, 6 h	32	171
Ti-13	$Ti(OBu)_4$, sealed tube, 115 °C, 7 h	55	171
Ti-14	$Ti(OBu)_4$, DBU, 1-ClN, refluxing, 12 h	16	172
Ti-15	$Ti(OBu)_4$, urea, *n*-octanol, 150 °C, 6 h	94.1	173
Ti-16	$Ti(OBu)_4$, urea, *n*-octanol, 150 °C, 14 h	52	166
Ti-17	$Ti(OBu)_4$, urea, *n*-octanol, 150 °C, 14 h	48	166
Ti-18	$Ti(OBu)_4$, DBU, *n*-pentanol, 120 °C, 5 h	31	167
Ti-19	$Ti(OPr^i)_4$, urea, *n*-pentanol, refluxing, 24 h	24	174
Ti-20	$Ti(OPr^i)_4$, urea, *n*-pentanol, refluxing, 24 h	18	175

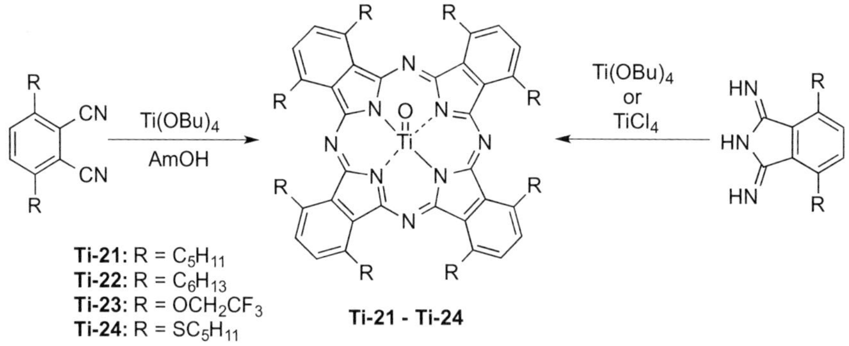

Scheme 35. Syntheses of 1,4,8,11,15,18,22,25-octa-substituted Ti(O)Pcs.[171,176]

with titanium(IV) butoxide[166,167,170–173] or -isopropoxide[174,175] in high-boiling solvents (alcohols or 1-ClN) (Scheme 34, Table 17).

Similarly, titanyl 1,4,8,11,15,18,22,25-octa-substituted Pcs (**Ti-21 – Ti-24**) were obtained by condensation of 3,6-disubstituted 1,3-diiminoisoindoline[171] or 3,6-disubstituted phthalonitrile in the presence of the titanium(IV) butoxide as the template (Scheme 35, Table 18).[176]

Titanyl 2,3,9,10,16,17,23,24-octa-substituted Pcs (**Ti-25** – **Ti-33**) were synthesized by template condensation of the corresponding 4,5-disubstituted phthalonitriles[174,175,177] or 4,5-disubstituted 1,3-diiminoisoindoline[171] with titanium(IV) butoxide[171,177] or -isopropoxide[174,175] in *n*-pentanol (Scheme 36, Table 18).

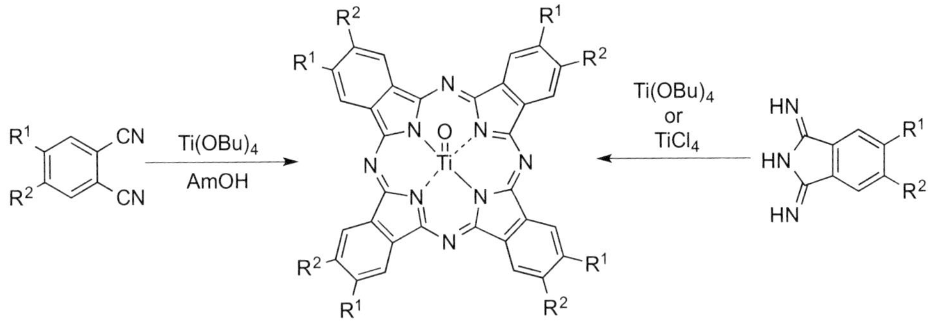

Ti-25 - Ti-32

Ti-25: $R^1 = R^2 = Pr$ **Ti-29:** $R^1 = R^2 = OC_5H_{11}$
Ti-26: $R^1 = R^2 = Bu$ **Ti-30:** $R^1 = R^2 = OCH_2CF_3$
Ti-27: $R^1 = R^2 = C_5H_{11}$ **Ti-31:** $R^1 = OC_8H_{17}$, $R^2 = OEt$
Ti-28: $R^1 = R^2 = C_7H_{15}$ **Ti-32:** $R^1 = R^2 = SC_6H_{13}$
Ti-33: $R^1 = R^2 = SCH_2CH_2N(Me)_2$

Scheme 36. Preparation of the titanyl 2,3,9,10,16,17,23,24-octa-substituted Pcs.[171,174,175,177]

Table 18. Preparation of TiPcs.

	Conditions	Yield (%)	Ref.
Ti-21	3,6-disubstituted-DI, $Ti(OBu)_4$, DBU, 180 °C, 6 h, sealed ampule	28	171
Ti-22	$Ti(OBu)_4$, DBU, 180 °C, 6 h, sealed ampule	44	171
Ti-23	$TiCl_4$, naphthalene, 1-ClN,190 °C, 24 h, sealed ampule	21	171
Ti-24	$Ti(OBu)_4$, urea, *n*-pentanol, refluxing, 24 h	22	177
Ti-25	$Ti(OBu)_4$, *n*-pentanol, refluxing, 6 h	32	177
Ti-26	$Ti(OBu)_4$, *n*-pentanol, refluxing, 6 h	25	177
Ti-27	$Ti(OBu)_4$, *n*-pentanol, refluxing, 6 h	40	177
Ti-28	$Ti(OBu)_4$, *n*-pentanol, refluxing, 6 h	63	177
Ti-29	$Ti(OBu)_4$, *n*-pentanol, refluxing, 6 h	26	177
Ti-30	$Ti(OBu)_4$, DBU, *n*-pentanol, refluxing, 6 h	15	171
Ti-31	4,5-disubstituted-DI, $Ti(OBu)_4$, *n*-pentanol, refluxing, 6 h	54	177
T-32	$Ti(OPr^i)_4$, urea, *n*-pentanol, refluxing, 24 h	38	174
T-33	$Ti(OPr^i)_4$, urea, *n*-pentanol, refluxing, 24 h	19	175

The synthesis and X-ray structure of the first, and so far the only, so-called "stapled" double-decker $TiPc_2$ (**Ti-34**) was published by Ercolani *et al.* (Figure 6) in 1990.[178] The reaction was carried out by treating **Ti-2** and Na_2Pc in 1-chloronaphthalene at 190 °C. The carbon–carbon σ-bond-stapled double-decker complex **Ti-34** was obtained in 16% yield.[178]

The syntheses of *zirconium* as well as *hafnium* phthalocyanines were published in 1964.[179] The reaction was carried out by melting phthalonitrile with zirconium(IV)- or hafnium(IV) chlorides (Table 19), to give the corresponding metal complex **Zr-1** or **Hf-1** containing one chlorine atom in the Pc core (Scheme 37).[179,180] In addition, monochlorinated **Zr-1** and **Hf-1** were yielded by treatment of phthalonitrile with chlorides of the corresponding metals in 1,2,4-trichlorobenzene (Scheme 37).[181] Complexes **Zr-2** and **Hf-2** (without a chlorine atom in the Pc core) could be prepared by reacting phthalonitrile with zirconium(IV)- and hafnium(IV) chlorides in high-boiling solvents such as 1-chloronaphthalene,[182,183] and mixtures of 2-methylnaphthalene:2-bromonaphthalene[163] or 2-methylnaphthalene:1,2,4-trichlorobenzene (Table 19).[181]

Complexes **Zr-3** and **Hf-3** containing two hydroxyl groups as axial ligands have been obtained by treating phthalonitrile with the corresponding metal(IV)

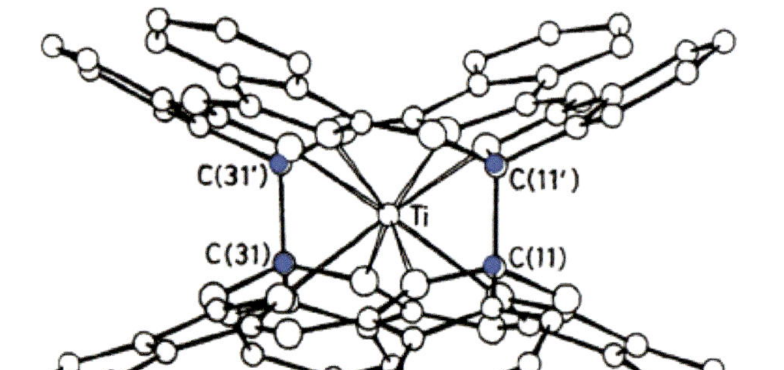

Figure 6. The X-ray structure of carbon–carbon σ-bonds stapled **Ti-34** (hydrogen atoms are omitted for clarity).[178] Adapted from Ref. 178.

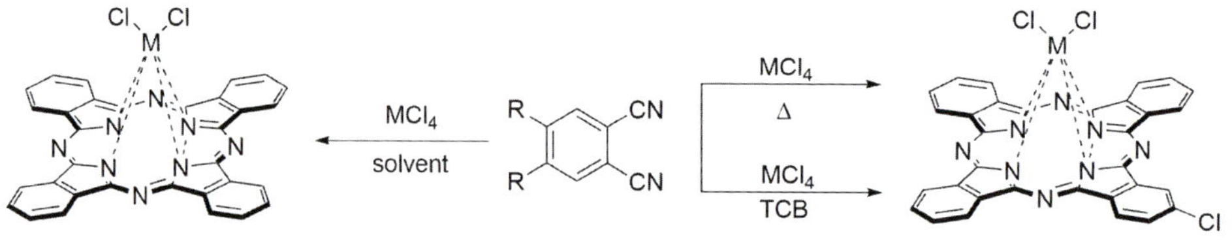

Zr-2, Hf-2

Zr-1: M = Zr, L = Cl
Zr-2: M = Zr, L = Cl

Hf-1: M = Hf, L = Cl
Hf-2: M = Hf, L = Cl

Zr-1, Hf-1

Scheme 37. Preparation of ZrPcs and HfPcs.[179–183]

chloride using *n*-heptanol as a solvent (Scheme 38)[163] or by using zirconium(IV) butoxide as the metal source (Table 19).[165] Unsubstituted zirconyl(IV) phthalocyanine was obtained under microwave irradiation of a mixture of phthalonitrile and zirconium(IV) oxychloride in 89% yield.[184]

Tomilova *et al.* published double-decker sandwich-like phthalocyanine complexes of zirconium ($ZrPc_2$) and hafnium ($HfPc_2$) in 1987.[185] Indeed, complexes **Zr-4** and **Hf-5** were obtained by melting phthalonitrile with $ZrCl_4$ and $HfCl_4$.[185] **Zr-4** can also be obtained by refluxing 1,3-diiminoisoindolines with $ZrCl_4$ in 1-chloronapthalene (Scheme 39, Table 20).[186] The more organic solvent-soluble *tert*-butyl substituted **Zr-5** and **Hf-5** were obtained later, by melting 4-*tert*-butylphthalonitrile and zirconium(IV) or hafnium(IV) chlorides (Scheme 39, Table 20).[187]

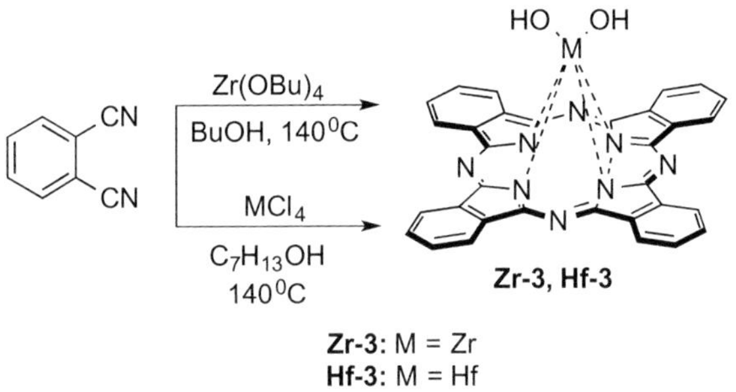

Scheme 38. Syntheses of **Zr-3** and **Hf-3** complexes.[163,165]

Table 19. Preparation of ZrPcs and HfPcs.

	Conditions	Yield (%)	Ref.
Zr-1	$ZrCl_4$, melting, 170–190 °C, 1 h	n.r.	179
Zr-1	$ZrCl_4$, TCB, 220 °C, 1 h	71	181
Zr-2	$ZrCl_4$, 1-ClN, 220 °C, 10 min	n.r.	182
Zr-2	$ZrCl_4$, 1-ClN, 220 °C, 15 min	48	183
Zr-2	$ZrCl_4$, 2-methylnaphthalene: 2-bromonaphthalene, 220 °C, 1 h	73	163
Zr-2	$ZrCl_4$, 2-methylnaphthalene: TCB, 170–230 °C, 1 h	73	181
Zr-3	$ZrCl_4$, urea, *n*-heptanol, 170 °C, 6 h	40	163
Zr-3	$Zr(OBu)_4$, urea, *n*-butanol, 140 °C, 6 h	51	165
Hf-1	$HfCl_4$, melting, 170–190 °C, 1 h	n.r.	179
Hf-1	$HfCl_4$, TCB, 220 °C, 1 h	77	181
Hf-2	$HfCl_4$, 2-methylnaphthalene: 2-bromonaphthalene, 220 °C, 1 h	76	163
Hf-2	$HfCl_4$, 2-methylnaphthalene: TCB, 170–230 °C, 1 h	76	181
Hf-3	$HfCl_4$, urea, *n*-heptanol, 170 °C, 6 h	28	163

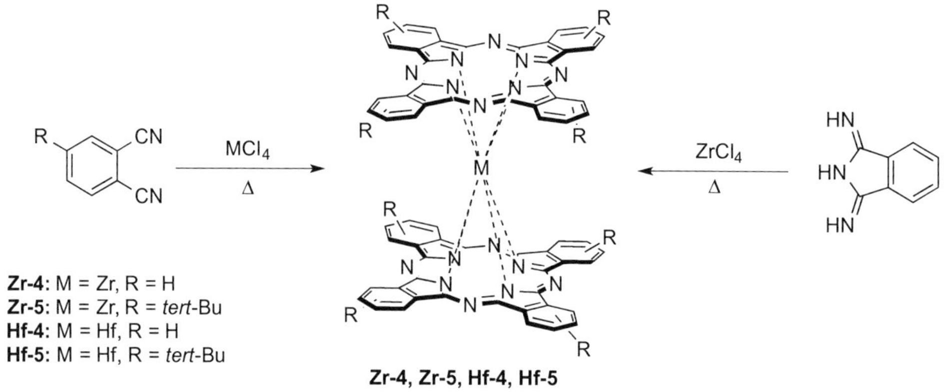

Scheme 39. Preparation of double-decker phthalocyanine complexes **Zr-4**, **Zr-5** and **Hf-4**, **Hf-5**.[185–187]

Table 20. Preparation of ZrPcs and HfPcs.

	Conditions	Yield (%)	Ref.
Zr-4	$ZrCl_4$, melting, 315 °C, 4.5 h	n.r.	185
Zr-4	DI, $ZrCl_4$, quinoline, refluxing, 2 h	n.r.	186
Zr-5	$ZrCl_4$, melting, 220 °C, 2–3 h	42	187
Hf-4	$HfCl_4$, melting, 315 °C, 4.5 h	64	185
Hf-5	$HfCl_4$, melting, 220 °C, 2–3 h	n.r.	187

Synthesis of Ti(IV)(Cl_2)Pc (Ref. 161)

Phthalodinitrile (10.0 g) was dissolved in a hot 1-chloronaphthalene (10 mL), and then $TiC1_4$ (0.5 g) was introduced *via* a syringe while maintaining an N_2 atmosphere. The reaction mixture was heated at 220 °C for 10 h with stirring. After the solution was cooled and filtered, the obtained solid was washed with methanol until the filtrate was colorless. Yield: (60%).

Synthesis of Ti(IV)(=O)Pc (Ref. 165)

A mixture of phthalonitrile (5.00 g), $Ti(OBu)_4$ (3.65 g), urea (1.17 g), and *n*-octanol (6.00 g) was heated at 150 °C under N_2 for 6 h. After addition of methanol (30 mL) to the reaction mixture, followed by refluxing for 0.5 h, fine blue crystals were collected by filtration, washed with toluene, methanol, and water, and then dried at 100 °C under 1 Torr for 3 h, to give 4.92 g (87.6%) of pure Ti(O)Pc.

Synthesis of Ti(IV)(=O)Pc(tert-Bu)$_4$ (Ref. 170)

4-*tert*-Butylphthalonitrile (1.0 g), urea (0.18 mg) and three drops of DBU were mixed in 1-pentanol (15 mL), and heated to 120 °C, then, Ti(OBu)$_4$ (0.51 mL) added by syringe, and the reaction mixture refluxed for 7 h at 155 °C. After cooling to room temperature, the mixture was poured into 400 mL methanol, and precipitation was induced by adding dropwise approximately 50 mL water. The crude product was collected by centrifugation and dried *in vacuo*. The obtained solid was purified by column chromatography on silica gel. Elution with a mixture of toluene:chloroform (1:1) gave the H_2Pc(*tert*-Bu)$_4$. The main product Ti(O)Pc(*tert*-Bu)$_4$ was obtained by changing the eluent to pure chloroform. After recrystallization from dichloromethane by addition of methanol, the pure Ti(O)Pc(*tert*-Bu)$_4$ was obtained. Yield: 380 mg (35%).

Synthesis of Zr(Cl$_2$)Pc (Ref. 183)

1-Chloronaphthalene (10 mL) was added to phthalonitrile (4.08 g) and zirconium(IV) chloride (0.932 g), and the reaction mixture heated to 230 °C. During heating, the color of the solution changed from beige to yellow, then red, purple and finally deep blue. The reaction mixture was maintained at 230 °C for 15 min, and immediately filtered through a glass frit. After the dark blue powder had cooled, it was rinsed with dichloromethane (5 × 20 mL or until the filtrate ran clear). After sublimation at 160 °C for 4 h, the pure Zr(Cl$_2$) Pc was obtained as a dark blue powder. Yield: 1.29 g (48%).

6. *Group 5: VPc, NbPc and TaPc*

The first example of a phthalocyanine complex with *vanadium* was described in 1936 by Linstead and coworkers.[21] Subsequently, numerous articles on the preparation of vanadyl phthalocyanines have been published.

Nowadays, V(=O)Pcs can be obtained by template condensation of phthalic acid derivatives (nitriles or anhydrides) in the presence of vanadium ions as templates[188–199] or by metal insertion into the corresponding H_2Pcs.[203,204] The template condensation reaction was carried out using high-boiling solvents (nitrophenol, quinoline, 1,2,4-trichlorobenzene, *etc.*) or without solvent (by fusion). The procedures for preparation of V(=O)Pcs under microwave irradiation have also been published (Table 21).[184,202] Vanadium(V) oxide,[21,191,192,199,200,204] vanadium(II)-,[197] vanadium(III)-[188,189,190,192,194,196,198,201] and vanadium(IV) chlorides,[167,193] vanadyl sulfate,[203] vanadium oxytrichloride[184,202] as well as vanadium(III)[195] acetate were used as the metal source. Unsubstituted **V-1** could

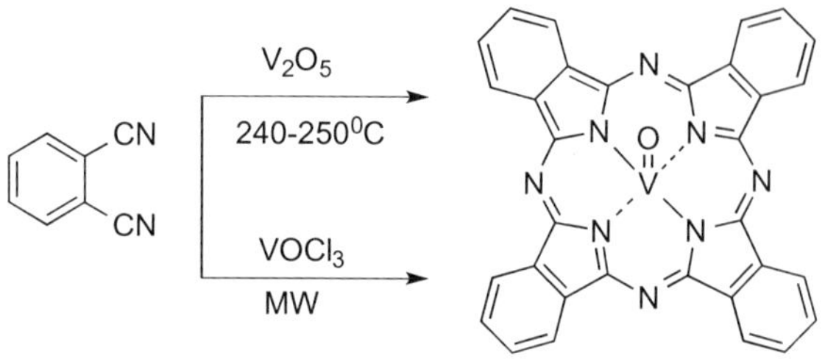

V-1

Scheme 40. Syntheses of unsubstituted V(=O)Pc.[21,184]

Table 21. Preparation of unsubstituted vanadylPc.

	Conditions	Yield (%)	Ref.
V-1	V_2O_5, 240–250 °C, melting, 0.5 h	27	21
V-1	$VOCl_3$, MW (560 W), 10 min	81	184

be obtained by condensing phthalonitrile with vanadium pentoxide at 240–250 °C (Table 21).[21] **V-1** was also prepared in good yield under microwave irradiation (Scheme 40. Table 21).[184]

Several vanadyl 1,8(11),15(18),22(25)-tetra-substituted Pcs (**V-2** – **V-7**) were prepared by template condensation of the corresponding 3-substituted phthalonitriles (**V-2** – **V-5**)[167,188,189,191] or phthalic anhydride[190] (**V-5**) with vanadium(III)-[188,189] and vanadium(IV) chlorides[167,190] in high-boiling solvents, as well as *via* melting of 3-substituted phthalonitrile with vanadium(V) oxide (**V-6**, **V-7**) (Scheme 41, Table 22).[191] Vanadyl 2,9(10),16(17),23(24)-tetra-substituted Pcs (**V-8** – **V-12**) were synthesized by template condensation of the corresponding 4-substituted phthalonitriles (**V-8** – **V-11**) or 4-clorophthalic anhydride (**V-12**) and urea with vanadium chlorides or vanadium pentoxide (Scheme 42, Table 23).[167,189,190,192, 195]

Vanadyl 2,9(10),16(17),23(24)-tetra-substituted Pcs (**V-8** – **V-12**) has been obtained by template condensation of the corresponding 4-substituted phthalonitriles (**V-8** – **V-11**) or 4-clorophthalic anhydride (**V-12**) with vanadium chlorides or vanadium pentoxide (Scheme 42, Table 23).[167,189,190,192,195]

The syntheses of a few examples of octa-substituted V(=O)Pcs (**V-13** – **V-16**) have also been described (Scheme 43).[189,190,193] Vanadyl 1,2,8,9(10,11),15,16 (17,18),22,23(24,25)-octachloro-Pc **V-16** was obtained *via* template condensation of 3,4-dichlorophthalic anhydride with vanadium(III) chloride in the presence of urea (Table 24).[190] Template condensation of 5-*tert*-butyl-3-bromophthalonitrile

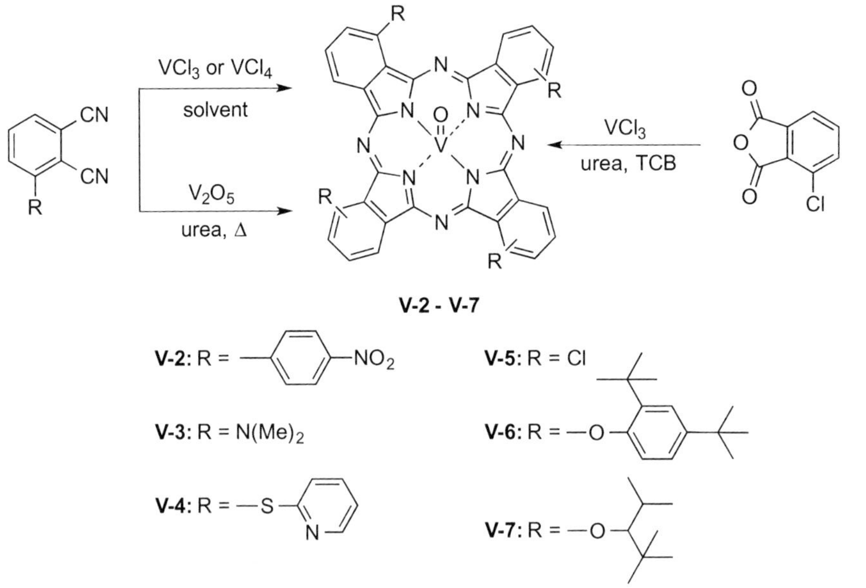

Scheme 41. Preparation of vanadyl 1,8(11),15(18),22(25)-tetra-substituted Pcs.[167,189–191,195]

Table 22. Preparation of vanadyl 1,8(11),15(18),22(25)-tetra-substituted Pcs (**V-2** – **V-7**).

	Conditions	Yield (%)	Ref.
V-2	VCl_3, urea, $PhNO_2$, 180–190 °C, 4 h	40	188
V-3	VCl_3, urea, MOA, quinoline, 200–205 °C, 2 h	55	189
V-4	VCl_4, DBU, 140–145 °C, 5 h	63.6	167
V-5	VCl_3, urea, MOA, TCB, 195–200 °C, 4 h	29	190
V-6	V_2O_5, NH_4Cl, urea, MOA, 150 °C for 0.5 h, then 190 °C for 4 h	48	191
V-7	V_2O_5, NH_4Cl, urea, MOA, 150 °C for 0.5 h, then 190 °C for 4 h	17	191

with vanadium(III) chloride in the presence of urea provided vanadyl 3,9(10),16(17),23(24)-tetra-*tert*-butyl-1,8(11),15(18),22(25)- tetrabromo-Pc **V-14** (Scheme 43, Table 24).[4] Vanadyl 1,4,8,11,15,18,22,25-octa-subsituted V(=O)Pcs could be prepared by cyclotetramerization of 3,6-dichlorophthalic anhydride in the presence of urea (**V-15**)[190] or by using 3,6-di(pentylsulfanyl)phthalonitrile as a starting material (**V-16**) (Scheme 43, Table 24).[193]

Methyl-,[194] phenyl-,[195] chloro-,[190] dimethylamino,[189] and alkoxy-[196,197] 2,3,9,10, 16,17,23,24-octa-substituted vanadyl Pcs (**V-17** – **V-24**) were prepared from the corresponding 4,5-disubstituted phthalonitriles (**V-17**, **V-18**, **V-20** – **V-25**)[189,194–197] or 4,5-dichlorophthalic anhydride[190] (**V-19**) (Scheme 44, Table 25). Vanadium(II)[197]-

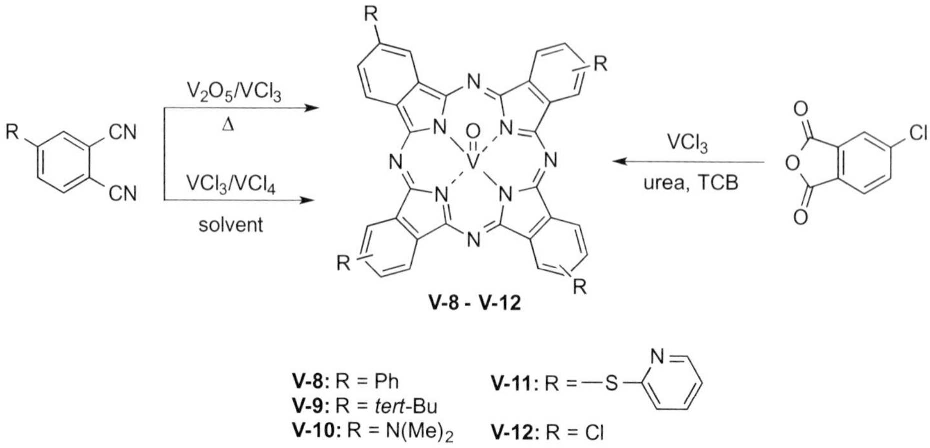

Scheme 42. Synthesis of vanadyl 2,9(10),16(17),23(24)-tetra-substituted Pcs.[167,189,190,192,195]

Table 23. Preparation of vanadyl 2,9(10),16(17),23(24)-tetra-substituted Pcs **V-8** – **V-12**.

	Conditions	Yield (%)	Ref.
V-8	V_2O_5, 260–280 °C, melting, 2 h	49	195
V-9	VCl_3, 240–260 °C, melting, 2 h	46	192
V-10	VCl_3, urea, MOA, quinoline, 200–205 °C, 2 h	62	189
V-11	VCl_4, DBU, *n*-pentanol, 120 °C, 5 h	47	167
V-12	VCl_3, urea, MOA, TCB, 195–200 °C, 4 h	48	190

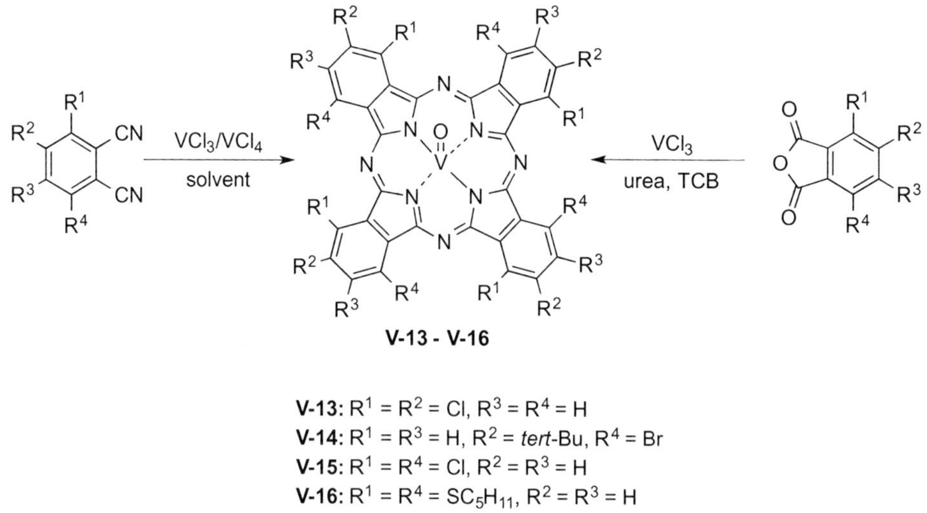

Scheme 43. Synthetic pathways for preparation of octa-substituted V(=O)Pcs.[189,190,193]

Table 24. Preparation of octa-substituted **V-13** – **V-16**.

	Conditions	Yield (%)	Ref.
V-13	VCl_3, urea, MOA, TCB, 195–200 °C, 4 h	53	190
V-14	VCl_3, urea, MOA, quinoline, 200–205 °C, 2 h	34	189
V-15	VCl_3, urea, MOA, TCB, 195–200 °C, 4 h	47	190
V-16	VCl_4, DBU, *n*-pentanol, refluxing, 6 h	42	193

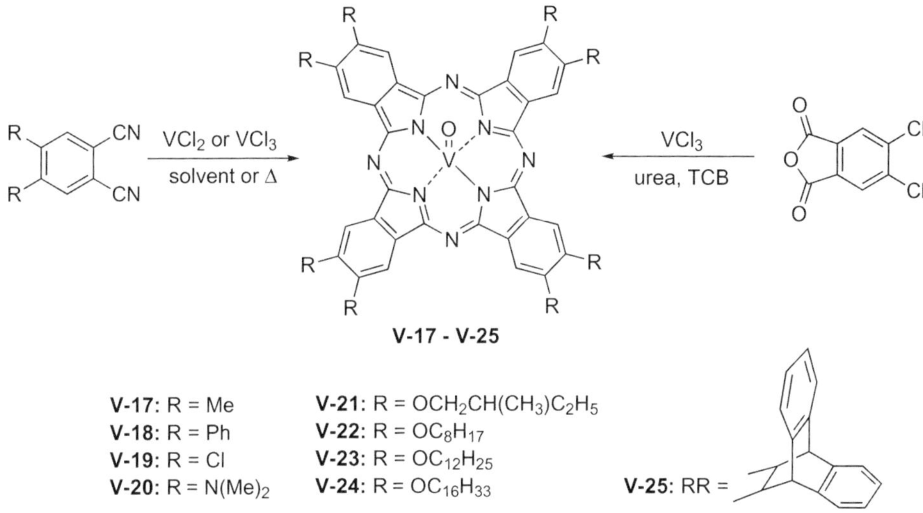

Scheme 44. Syntheses of 2,3,9,10,16,17,23,24-octa-substituted **V-17–V-26**.[189,190,194–197]

Table 25. Preparation of 2,3,9,10,16,17,23,24-octa-substituted V(O)Pcs **V-17–V-25**.

	Conditions	Yield (%)	Ref.
V-17	VCl_3, melting, 210–220 °C, 5 h	64	194
V-18	$V(OAc)_3$, 1-bromonaphthalene, 240–250 °C, 6 h	44	195
V-19	VCl_3, urea, MOA, TCB, 195–200 °C, 4 h	42	190
V-20	VCl_3, urea, MOA, quinoline, 200–205 °C, 2 h	80	189
V-21	VCl_3, urea, melting, 150 °C	n.r.	196
V-22	VCl_2, DBU, DMAE, 150 °C, 48 h	29	197
V-23	VCl_2, DBU, DMAE, 150 °C, 48 h	12	197
V-24	VCl_2, DBU, DMAE, 150 °C, 48 h	4	197
V-25	VCl_3, MOA, melting, 260–270 °C, 2 h	24	198

and vanadium(III) chlorides[62,189,190,194,196] as well as vanadium(III) acetate[195] were used as the metal source.

The vanadyl hexadeca-substituted Pcs **V-26** – **V-29** were prepared using 3,4,5,6-tetra-substituted phthalonitriles (Scheme 45).[190,199–201] The vanadyl 1,2,3, 4,8,9,10,11,15,16,17,18,22,23,24,25-hexadecacloro- (**V-26**),[190] -fluoro- (**V-27**),[199,200] -ethoxy- (**V-29**)[201] and -(2,2,2-trifluoroethoxy) (**V-30**)[201] Pcs were synthesized *via* template condensation of the corresponding phthalonitriles with vanadium(III) chloride[190,201] or vanadium pentaoxide as a metal source (Table 26).[199,200]

In 2009, the synthesis of a vanadyl Pc possessing two types of substituents in the peripheral (β, β) and nonperipheral (α, α) sites **V-30** was achieved (Scheme 45).[202] The preparation of **V-30** was carried out under microwave irradiation of 1,4-difluoro-

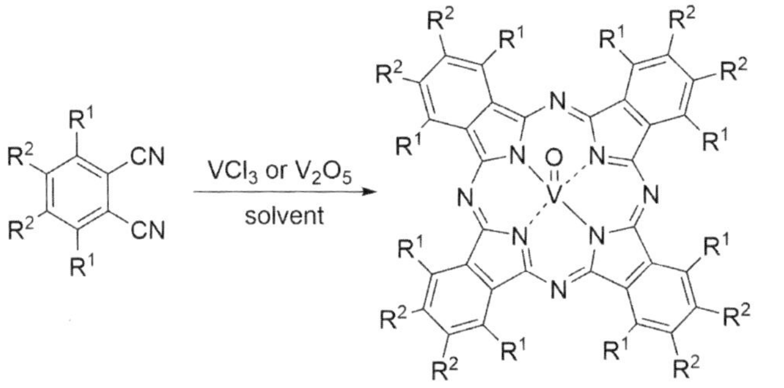

V-26 - V-30

V-26: $R^1 = R^2 = Cl$,
V-27: $R^1 = R^2 = F$
V-28: $R^1 = R^2 = OEt$
V-29: $R^1 = R^2 = OCH_2CF_3$
V-30: $R^1 = F$, $R^2 = C_3F_7$

Scheme 45. Syntheses of hexadeca-subsituted V(=O)Pcs.

Table 26. Preparation of vanadyl 1,2,3,4,8,9,10,11,15,16,17,18,22,23,24,25-hexadeca-substituted Pcs **V-26** – **V-30**.

	Conditions	Yield (%)	Ref.
V-26	VCl_3, urea, $PhNO_2$, 200–205 °C, 4–5 h	60.5	190
V-27	V_2O_5, *n*-pentanol, refluxing, 20 h	16	199
V-27	V_2O_5, vacuum-sealed glass tube, 10^{-3} Torr, 220 °C, 7 h	15	200
V-28	VCl3, urea, DBU, melting, 190–200 °C, 2 h	12.5	201
V-29	VCl3, urea, DBN, abs. ethanol, refluxing, 45 h	52	201
V-30	$VOCl_3$, DMF, 225 °C, MW, 10 min, sealed tube	56	202

4,5-di(*iso*propylperfluoro)phthalonitrile with vanadium oxytrichloride in a sealed tube (Table 26).[202]

The vanadyl tetracrown-substituted (**V-31**)[203] as well as 1,8(11),15(18),22(25)-tetra(hexyloxy) (**V-32**)[204] Pcs could be prepared by complexation of the corresponding H_2Pcs with vanadyl sulfate (**V-31**) or vanadium pentoxide (**V-32**) (Scheme 46, Table 27).

In contrast to *vanadium*, only a few examples of *niobium* Pcs have been described.[79,195,205] Indeed, unsubstituted niobium(V) Pc can be prepared by refluxing phthalonitrile with niobium(V) bromide[79] (**Nb-1**) or niobium(V) chloride[205] (**Nb-2**) in 1-chloronaphthalene (Scheme 47, Table 28). Unsubstituted niobium(IV) Pc (**Nb-3**) was prepared in the same synthetic route using quinoline as a solvent (Scheme 47).[205] Niobium 2,9(10),16(17),23(24)-tetraphenyl-Pc (**Nb-4**) can be prepared *via* melting of 4-phenylphthalonitrile with niobium(III) chloride (Table 28).[195]

Two types of double-decker sandwich-like phthalocyanine complexes of niobium ($NbPc_2$) were described by Gingl and Strähle[206] (**Nb-5**) and Ercolani and coworkers[207–209] (**Nb-6**). μ-Bromo-μ-dioxo-bis(phthalocyaninato- niobium(V)) tribromide (**Nb-5**) was obtained as the solvate form **Nb-5**•1-ClN by reaction of

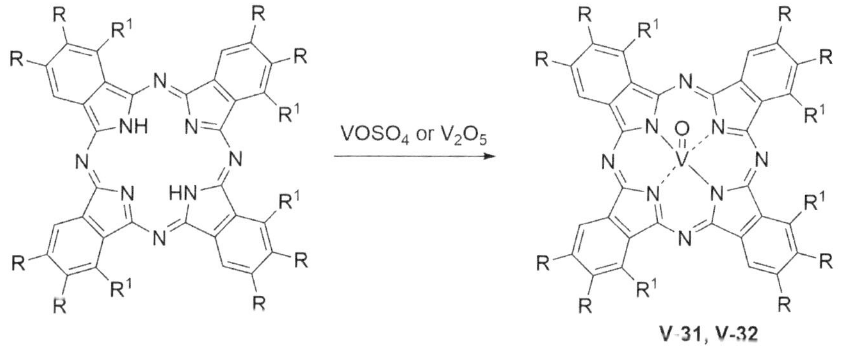

V-31: R^1 = H, RR = 15-crown-5
V-32: R^1 = $O(CH_2)_5CH_3$, R = H

Scheme 46. Synthesis of the vanadyl Pcs *via* metal insertion into metal-free Pc core.[203,204]

Table 27. Preparation of **V-31** – **V-32**.

	Conditions	Yield (%)	Ref.
V-31	$VOSO_4$, DMF, refluxing, 7 d	30	203
V-32	V_2O_5, urea, (1) melting, 140 °C, 1 h; (2) DMF, refluxing, 17.5 h	66	204

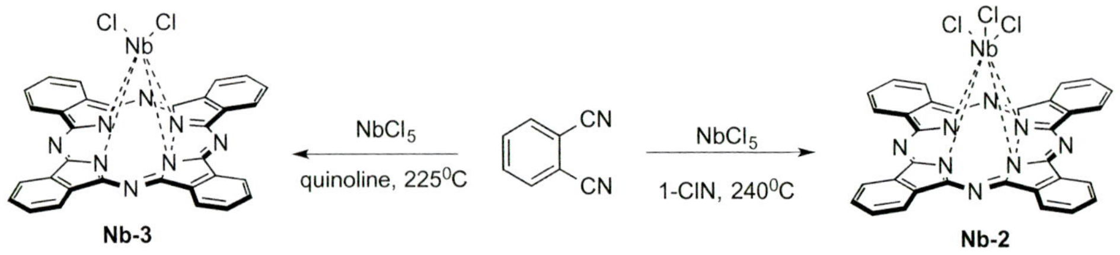

Scheme 47. Preparation of the Nb(V)Pc (**Nb-2**) and Nb(IV)Pc (**Nb-3**).[205]

Table 28. Preparation of **Nb-1 – Nb-6.**

	Conditions	Yield (%)	Ref.
Nb-1	$NbBr_5$, 1-ClN, refluxing, 5 h	n.r.	79
Nb-2	$NbCl_5$, 1-ClN, 240 °C, 7 h	71	205
Nb-3	$NbCl_5$, quinoline, 225 °C, 5 h	55	205
Nb-4	$NbCl_3$, MOA, melting, 265–280 °C, 5 h	48.5	195
Nb-5	$NbOBr_3$, 1-ClN, 230 °C, 3 h	52	206
Nb-6	**Nb-3**, Na_2Pc, 1-ClN, 220 °C, 4 h	14	209

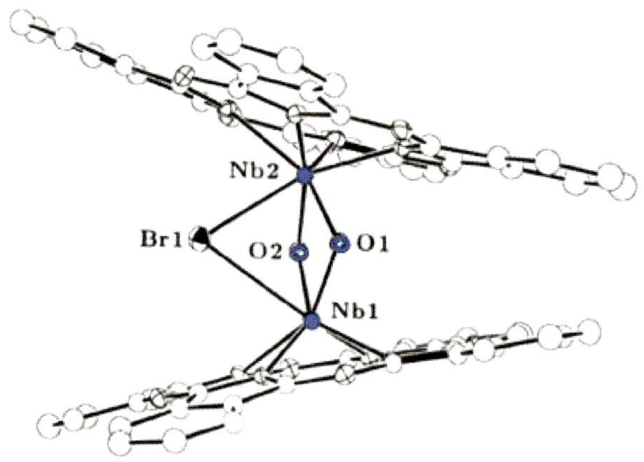

Figure 7. X-ray structure of **Nb-5**; the solvate molecule and hydrogen atoms are omitted for clarity.[206] Adapted from Ref. 206.

$NbOBr_3$ with phthalonitrile or as the solvate form **Nb-5•**[phthalonitrile] when melting conditions were adopted (Figure 7).[206]

The synthesis and X-ray structure of the so-called "stapled" double-decker $NbPc_2$ (**Nb-6**) (Figure 8) was published by Ercolani and coworkers.[207–209] They

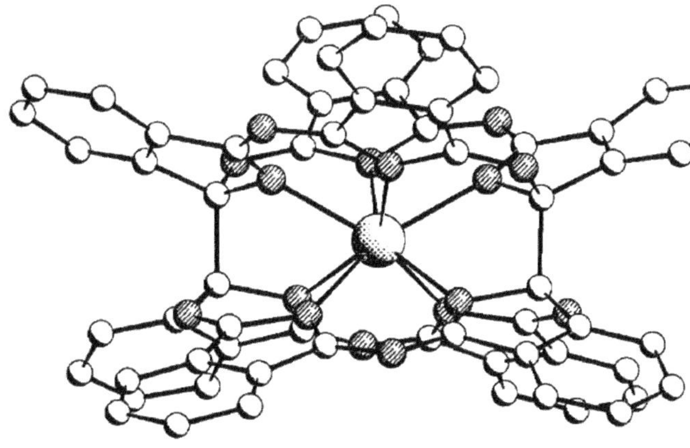

Figure 8. X-ray structure of **Nb-6**, hydrogen atoms are omitted for clarity.[209] Adapted from Ref. 209.

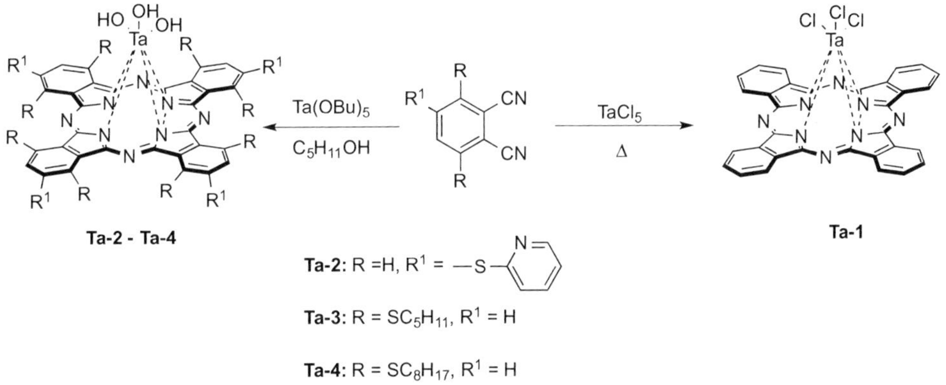

Scheme 48. Preparation of tantalum(V) phthalocyanines **Ta-1** – **Ta-4**.[211–214]

used the same approach as described earlier for the synthesis of titanium "stapled" double-decker **Ti-34**.[178] The reaction was carried out by treating **Nb-3** and Na_2Pc in 1-chloronaphthalene at 220 °C. The carbon–carbon σ-bond stapled double-decker complex **Nb-6** was obtained as a brown solid in 14% yield.[209] Another procedure for the preparation of **Nb-6** was also published.[210]

Unsubstituted *tantalum*(V) phthalocyanine (**Ta-1**) was synthesized by melting phthalonitrile with tantalum(V) chloride (Scheme 48. Table 29).[211,212] Tantalum(V) 2,9(10),16(17),23(24)-tetra- and 1,4,8,11,15,18,22,25-octasulfanyl-substituted Pcs (**Ta-2** – **Ta-4**) were prepared by Nyokong and coworkers (Scheme 49).[213,214] The reaction was carried out by refluxing sulfanylsubstituted phthalonitriles with tantalum(V) butoxide in *n*-pentanol (Table 29).

Table 29. Preparation of **Ta-1** – **Ta-5**.

	Conditions	Yield (%)	Ref.
Ta-1	$TaCl_5$, melting, 240 °C, 24 h	72	211
		68	212
Ta-2	$Ta(OBu)_5$, DBU, *n*-pentanol, refluxing, 5 h	42	213
Ta-3	$Ta(OBu)_5$, DBU, *n*-pentanol, refluxing, 6 h	42	214
Ta-4	$Ta(OBu)_5$, DBU, *n*-pentanol, refluxing, 6 h	48	214

Synthesis of V(IV)(=O)Pc(tert-Bu)$_4$ (Ref. 192)

4-*tert*-Butylphthalonitrile (0.72 g) and vanadium trichloride (0.74 g) were melted at 240–260 °C for 2 h. The product mixture was cooled to RT, and then boiled successively with 30 mL of 5% aqueous HCl solution and 30 mL of 5% aqueous NaOH solution. The obtained blue solid was washed thoroughly with water and methanol. The crude product was purified by column chromatography on Al_2O_3, (neutral, activity 1) using $CHCl_3$ as eluent. The blue fraction was collected and dried, yielding purple reflective blue crystals. Yield: 0.36 g (46%).

Synthesis of V(IV)(=O)PcF$_{16}$ (Ref. 199)

A suspension of tetrafluorophthalonitrile (1.22 g) and vanadium(V) oxide (0.125 g) in *n*-pentanol was refluxed for 20 h. The reaction mixture was chromatographed on a SiO_2 column with THF-acetone (4:1, v/v) as eluent. The eluate was reduced to a small portion, and chromatographed on a SiO_2 column with THF as eluent. The adsorbed band was eluted using acetone, then the eluate filtered and evaporated, to give a blue powder. Yield: 0.095 (16%).

Synthesis of V(IV)(=O)PcCl$_8^{\alpha\alpha}$ (Ref. 190)

A mixture of 3,6-dichlorophthalic anhydride (17 g), urea (6 g), vanadium(III) chloride (0.55 g) and a catalytic amount of ammonium molybdate in 1,2,4-trichlorobenzene (10 mL) was stirred at 195–200 °C for 4 h. After cooling, the solid residue was collected by filtration and washed with methanol. The crude product was washed with 3% HNO_3 at 80 °C, then boiling HCl(aq), and finally NaOH (aq). It was then dissolved in concentrated H_2SO_4 and precipitated by pouring the solution into ice-water, to give the target compound. Yield: 1g (47%).

Synthesis of V(IV)(=O)PcCl$_8^{\beta\beta}$ (Ref. 190)

A mixture of 4,5-dichlorophthalic anhydride (17 g), urea (6 g), vanadium(III) chloride (0.55 g) and a catalytic amount of ammonium molybdate in 1,2,4-trichlorobenzene (10 mL) was stirred at 195–200 °C for 4 h. After cooling, the solid residue was collected by filtration and washed with methanol. The crude product was washed with 3% HNO_3 at 80 °C, then boiling HCl(aq), and finally with NaOH(aq). It was then dissolved in concentated H_2SO_4 and precipitated by pouring the solution into ice-water, to give the target compound. Yield: 0.91 g (42%).

Synthesis of V(IV)(=O)Pc(15-crown-5)$_4$ (Ref. 203)

Metal-free tetra(15-crown-5)phthalocyanine (40 mg) and vanadyl sulfate (100 mg) were dissolved in dry DMF (100 mL) and the reaction mixture refluxed for 7 days. The solvent was removed under reduced pressure, and the obtained dark green powder washed with excess water to remove unreacted $VOSO_4$. The crude product was purified by column chromatography on basic alumina. The unreacted initial metal-free phthalocyanine moved as a light green band in a $CHCl_3$:CH_3OH (98:2, v/v) mixture. The target complex was eluted as a bluish-green band using a $CHCl_3$:CH_3OH (95:5, v/v) mixture. Yield: 13 mg (30%).

Synthesis of V(IV)(=O)Pc(n-OC$_6$H$_{13}$)$_4$ (Ref. 204)

A mixture of metal-free 1,8,15,22-tetrahexyloxyphthalocyanine (40.7 mg), vanadium(V) pentoxide (99 mg), and dried urea (547 mg) was stirred at 140 °C for approx.1 h, and after addition of DMF (1 mL), the reaction was continued for a further 17.5 h. After removal of the solvent, the residue was washed with water, dried, and purified by column chromatography on alumina using a CCl_4:methanol (10:1, v/v) mixture as eluent. Recrystallization from a chloroform:methanol mixture gave the desired V(O)Pc. Yield: 29.2 mg (66%).

Synthesis of Nb(V)(Cl$_3$)Pc (Ref. 205)

A mixture of $NbCl_5$ (1.01 g) and phthalonitrile (2.40 g) were suspended (partly dissolved) under nitrogen in freshly-distilled 1-ClN (20 mL), and the mixture heated at 240 °C, with stirring, for 7 h. The hot suspension was filtered, and the separated solid washed three or four times with benzene, and dried *in vacuo*, to give 1.88 g (71%) of Nb(Cl_3)Pc.

Synthesis of Nb(IV)(Cl_2)Pc (Ref. 205)

A mixture of $NbCl_5$ (3.20 g), phthalonitrile (6.07 g) and freshly-distilled quinoline (40 mL) was heated with stirring at 225 °C for 5 h in an inert atmosphere (N_2). After cooling, the reaction mixture was centrifuged, washed with methanol, then ethanol, and dried *in vacuo*, to give 4.37 g (55% yield) of Nb(Cl_2)Pc.

Synthesis of Ta(V)(Cl_3)Pc (Refs. 211, 212)

A mixture of phthalonitrile (13.0 g) and $TaCl_5$ (0.75 g) was melted at 240 °C for 24 h under an N_2 atmosphere. The resulting solid was Soxhlet extracted with acetone, then methanol, and finally benzene, and dried *in vacuo*. Yield: 1.2 g (72%, based on $TaCl_5$).

Synthesis of Ta(V)$(OH)_3$Pc (Ref. 214)

3,6-Dipentylthiophthalonitrile (0.8 g) in *n*-pentanol (7.0 mL) was refluxed under a nitrogen atmosphere, and tantalum(V) butoxide (0.138 g) added. After the addition of DBU (0.30 mL), the reaction was continued for 6 h. The reaction mixture was cooled and purified *via* column chromatography on SiO_2 using $CHCl_3$ as eluent. Yield: 0.25 g (42%).

7. *Group 6: CrPc, MoPc, and WPc*

In 1934, Linstead and coworkers reported the first *chromium* phthalocyanine, obtained by melting *o*-cyanobenzamide in the presence of chromium metal.[17] Subsequently, they attempted to prepare it by complexation of chromium ions with unsubstituted H_2Pc using 1-chloronaphthalene or quinoline as the solvent, but unfortunately, this ended in a failure.[21]

A Cr(II)Pc (**Cr-1**) was later obtained by template condensation of phthalonitrile and chromium(IV) carbonyl in 1-chloronapthalene[215,216] or tetraglyme (Scheme 49, Table 30).[217] Leznoff and coworkers proved that under the conditions described by Meloni *et al.*,[215] a mixture of Cr(II)Pc and Cr(III)(Cl)Pc was obtained.[216] They also described an adequate procedure for the synthesis of **Cr-1** (Table 30), as well as reporting its X-ray structure (Figure 9).[216]

Hanack and Datz described the preparation of the unstable β-form of **Cr-1** and its transformation to the stable α-form by refluxing in pyridine.[217] The preparation of **Cr-1** as well as polymeric MPcs (M = Cr, Mo (**Mo-1**), W (**W-1**)) by melting the phthalonitriles and metallocarbonyls in the presence of a mixture of SiO_2 and TiO_2 in a vacuum-sealed tube was reported by Wöhrle *et al.* (Table 30).[218] **Cr-1** could also be prepared under microwave irradiation of phthalonitrile and chromium(IV) oxide in the presence of ammonium formate as a catalyst (Table 30).[184]

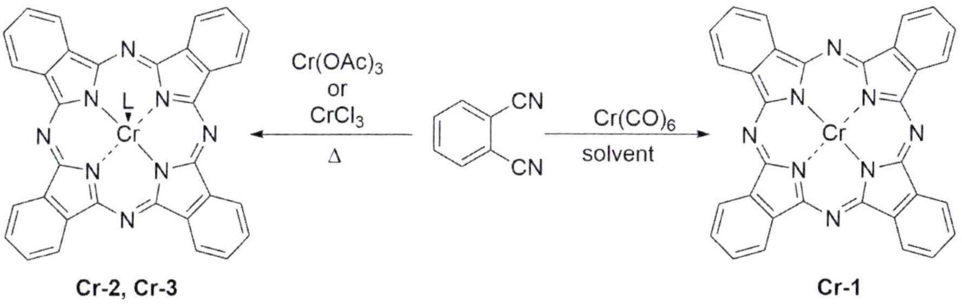

Scheme 49. Syntheses of **Cr-1–Cr-3**.[184,215–220]

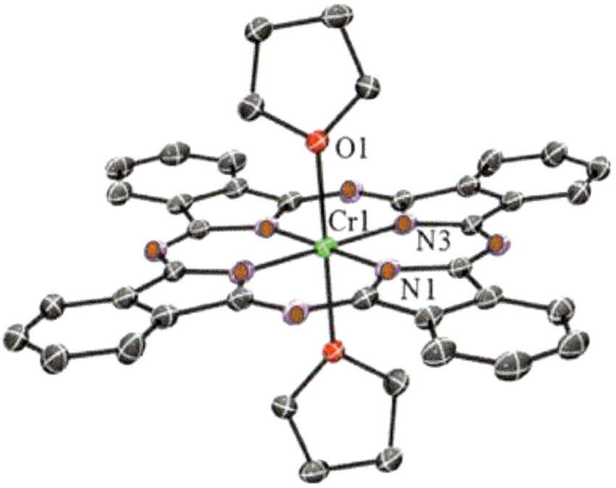

Figure 9. X-ray structure of **Cr-1**·THF_2.[216] The hydrogen atoms and free THF molecules are omitted for clarity. Adapted from Ref. 216.

Cr(III)Pcs with different axial ligands were obtained by melting phthalonitrile with chromium(III) acetate (**Cr-2**)[219] or chromium(III) chloride (**Cr-3**) (Scheme 49, Table 30).[220] Many solvent-soluble Cr(III)Pcs such as **Cr-4** can also be obtained by metal-insertion reaction to H_2Pcs (Scheme 50).

Leznoff and coworkers obtained the Cr(III) μ-oxodimer (**Cr-5**) of a Cr(III)Pc in good yield. Here, complex **Cr-5** was prepared *via* reaction of **Cr-1** (Figure 9) with iodozobenzene (Table 30)[216] and its structure confirmed by single-crystal X-ray diffraction analysis (Figure 10).[216]

Some publications have focused on the preparation of *molybdenum* and *tungsten* Pcs.[29,42,222–230] Indeed, oxomolybdenum(IV)Pc (**Mo-1**) was prepared in low yield by template condensation of phthalonitrile in the presence of molybdenum(VI) dichloride dioxide,[222] molybdenum(IV) chloride,[222] molybdenum powder,[223] molybdenum(VI) carbonyl,[224] and ammonium hexamolybdate [225] as the metal source (Table 31). **Mo-1** was also prepared by template condensation of phthalonitrile with

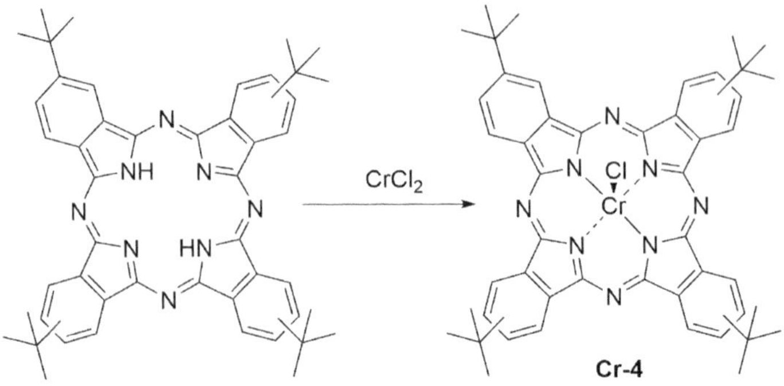

Scheme 50. The preparation of Cr(III)Pc *via* complexation of appropriate metal-freePc with metal salt.[221]

Table 30. Preparation of **Cr-1**–**Cr-5**.

	Conditions	Yield (%)	Ref.
Cr-1	$Cr(CO)_6$, 1-ClN, refluxing, 3–5 h	40	215
Cr-1	$Cr(CO)_6$, tetraglyme, 250 °C, 5 h	70.1	216
Cr-1	$Cr(CO)_6$, 1-ClN, refluxing, 6 h	42[a]	217
Cr-1	$Cr(CO)_6$, SiO_2 and TiO_2, vacuum sealed bomb, 250 °C	90	218
Cr-1	CrO_3, ammonium formate, MW,	92	185
Cr-2	$Cr(OAc)_3$, melting, 270 °C, 15 min	n.r.	219
Cr-3	$CrCl_3$x$4H_2O$, melting	n.r	220
Cr-4	$CrCl_2$, DMF, refluxing, 5 h	61	221
Cr-5	PhIO	49.6	216

[a]The yield to provide to β-form.

molybdenyl acetylacetonate in boiling *n*-butanol in the presence of a strong organic base, such as DBU.[29] The yield of **Mo-1** could be improved by using high-pressure conditions (Table 31).[42] To date, only one successful attempt at the syntheses of unsubstituted *molybdenum* (**Mo-1**) and *tungsten* (**W-1**) Pcs has been reported.[218] The complexes **Mo-1** and **W-1**, as well as their polymeric derivatives, were obtained by melting phthalonitrile with hexacarbonyls of appropriate metals in vacuum-sealed tube vessels at high temperature in the presence of a mixture of SiO_2 and TiO_2 (Scheme 51, Table 31).[218]

Dioxomolybdenum(VI)Pc (**Mo-2**) was prepared by complexation of unsubstituted H_2Pc with bis(acetylacetonato)dioxomolybdenum(VI) in chlorobenzene (Table 31).[226]

A series of unsubstituted and alkylsubstituted nitride-type M(≡N)Pcs (M = Mo, W) were prepared by Hanack and coworkers *via* template condensation

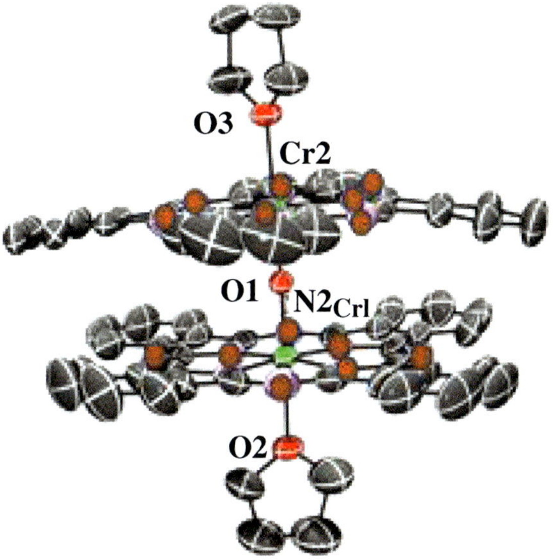

Figure 10. X-ray structure of **Cr-5**.[216] The hydrogen atoms and free THF molecules are omitted for clarity. Adapted from Ref. 216.

of the corresponding (un)substituted phthalonitriles with nitride trichlorides of molybdenum or tungsten, as well as ditosyliminomolybdenum dichloride as the metal source (Scheme 52, Table 31).[227,228]

The imido MPcs (M = Mo, W) were produced in good yield by template condensation of phthalonitrile with the corresponding bis(imido) complexes of molybdenum and tungsten.[229] Indeed, the complexes **Mo-8**, **Mo-9** as well as **W-10** – **W-12** could be prepared by melting a mixture of phthalonitrile and initial bis(imido) complexes at 210 °C (Scheme 53, Table 31).[229]

The so-called "expanded" Pc congeners with two *molybdenum* (**Mo-10** – **Mo-13**) or *tungsten* (**W-13**) central metal ions and four isoindole ring moieties have been synthesized using normal Pc formation conditions in the presence of urea (Scheme 54, Table 31).[230]

Synthesis of CrPc (Ref. 216)

Tetraglyme (15 mL) was added to phthalonitrile (2.89 g), the mixture heated to 180 °C, and the solid $Cr(CO)_6$ (1.09 g) was added to the beige solution. The reaction mixture was heated to 250 °C and maintained at this temperature for 5 h, during which time it turned deep purple and then dark blue. After allowing the reaction mixture to cool at ambient temperature, hexanes (10 mL) were added, and the mixture filtered through a glass frit under nitrogen. The obtained dark blue

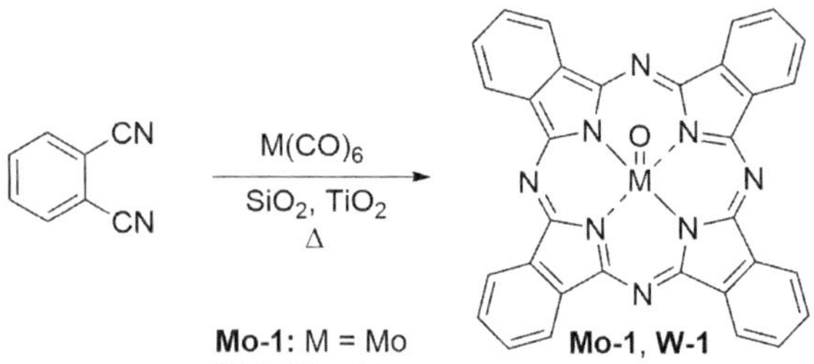

Mo-1: M = Mo
W-1: M = W

Scheme 51. Synthesis of **Mo-1** and **W-1**.[218]

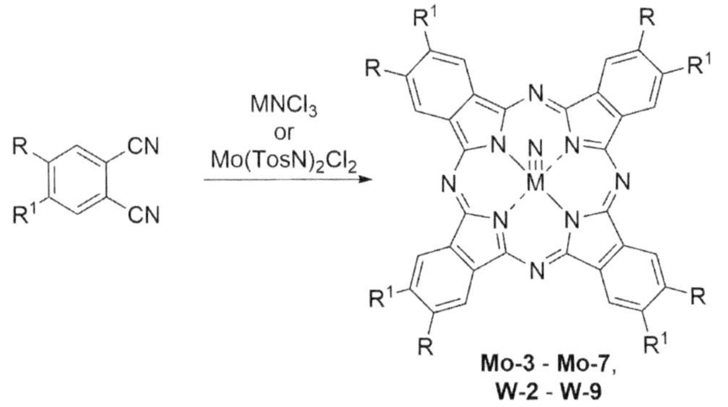

Mo-3: M = Mo, R = R^1 = H
Mo-4: M = Mo, R = R^1 = Me
Mo-5: M = Mo, R = R^1 = Pr
Mo-6: M = Mo, R = R^1 = Bu
Mo-7: M = Mo, R = R^1 = C_5H_{11}

W-2: M = W, R = R^1 = H
W-3: M = W, R = *tert*-Bu, R^1 = H
W-4: M = W, R = R^1 = Me
W-5: M = W, R = R^1 = Pr
W-6: M = W, R = R^1 = Bu
W-7: M = W, R = R_1 = C_5H_{11}
W-8: M = W, R = R_1 = C_6H_{13}
W-9: M = W, R = R_1 = C_7H_{15}

Scheme 52. Syntheses of M(≡N)Pcs (M = Mo, W).[227,228]

powder was washed with toluene (2 × 20 mL), and then with THF (2 × 10 mL). The solid was dried under vacuum, to give a blue powder of the target compound. Yield: 1.963 g (70.1%).

Synthesis of Cr(III)(Cl)Pc(tert-Bu)$_4$ (Ref. 221)

2,9(10),16(17),23(24)-Tetra(*tert*-butyl)phthalocyanine (2.22 g) was dissolved in refluxing DMF (150 mL). After 1–2 h, $CrCl_2$ (2 g) was added to the refluxing solution. The reaction mixture was refluxed for 5 h under Ar, and then allowed to

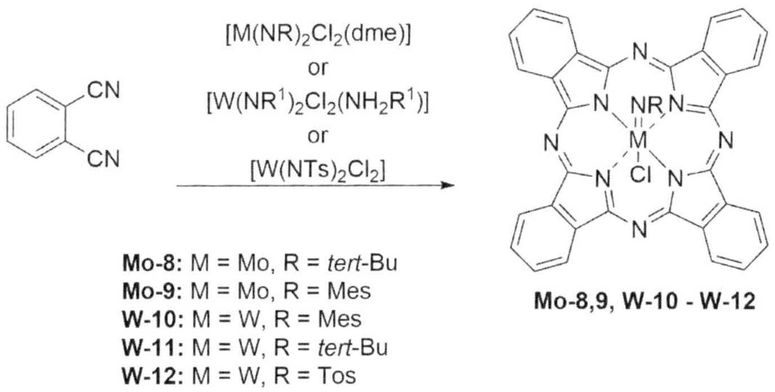

Scheme 53. Syntheses of organoimido MoPc and WPc.[229]

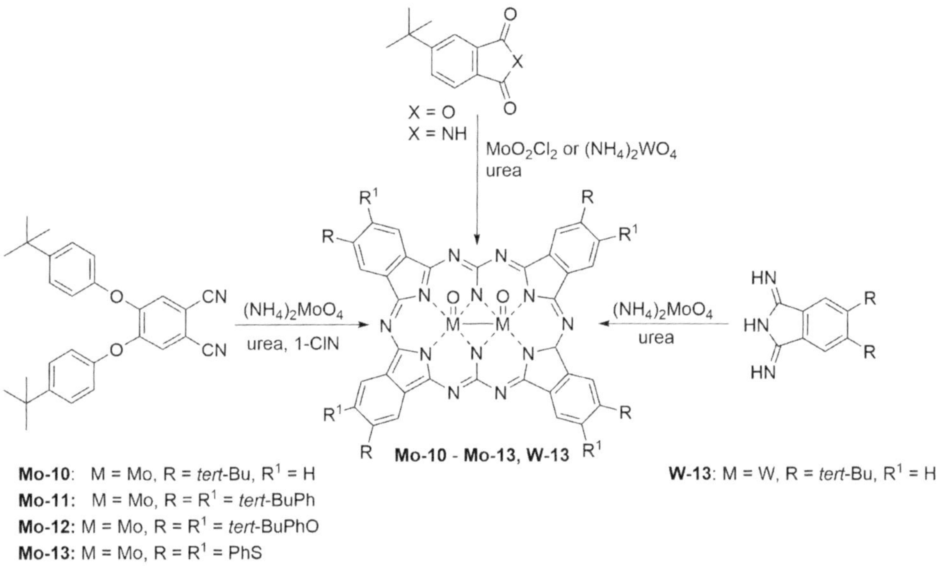

Scheme 54. Syntheses of expanded **Mo-10** – **Mo-13** and **W-13**.[230]

cool to RT. The volume of the solvent was reduced to *ca.* 50 mL under reduced pressure, poured into brine (1 L), and allowed to stand for 12 h at RT. The resulting precipitate was filtered and washed with water (20 mL × 3), to give a crude product. The crude product was dissolved in $CHCl_3$ (50 mL), and then filtered. The filtrate was concentrated under reduced pressure, and the residue purified by column chromatography on aluminum oxide (basic, activity stage I) using $CHCl_3$ as eluent. The desired compound was collected as a second green fraction. Then, this solution was concentrated to volume *ca.* 50 mL, washed with 1 M HCl, (100 mL × 2), dried over $MgSO_4$, and concentrated *in vacuo*. The resulting solid

Table 31. Preparation of **Mo-1** – **Mo-13** and **W-1** – **W-13**.

	Conditions	Yield (%)	Ref.
Mo-1	MoO_2Cl_2, urea, DMF, refluxing, 1 h	0.8	222
Mo-1	MoO_2Cl_2, DMF, refluxing, 1 h	0.9	222
Mo-1	$MoCl_4$, urea, DMF, refluxing, 19 h	1.6	222
Mo-1	Mo powder, melting, 6 h	n.r.	223
Mo-1	$Mo(CO)_6$, toluene, sealed tube, 180 °C, melting	n.r.	224
Mo-1	$(NH_4)_6Mo_7O_{24}{\times}4H_2O$, vacuum-sealed tube, 270 °C, melting	>70	225
Mo-1	$MoO_2(acac)_2$, BuOH, DBU, refluxing 24 h	2	29
Mo-1	$MoO_2(acac)_2$, DBU, MeOH, 9 kbar, 65 °C, 18 h	15	42
Mo-1	$Mo(CO)_6$, SiO_2 and TiO_2, vacuum-sealed bomb vessel, 250 °C	80	218
Mo-2	H_2Pc, $MoO_2(acac)_2$, chlorobenzene, 100 °C, 3 h	86	226
Mo-3	$MoNCl_3$, melting, 220 °C, 20 min	65	227
Mo-4	$Mo(TosN)_2Cl_2$, melting, 220 °C, 15 min	25	228
Mo-5	$MoNCl_3$, melting, 220 °C, 15 min	11	227
Mo-6	$MoNCl_3$, melting, 220 °C, 15 min	8	227
Mo-7	$MoNCl_3$, melting, 220 °C, 15 min	10	227
Mo-8	[Mo(N*tert*-Bu)$_2$Cl$_2$(dme)], melting, 210 °C, 20 min	67	229
Mo-9	[Mo(NMes)$_2$Cl$_2$(dme)], melting, 210 °C, 20 min	82	229
Mo-10	MoO_2Cl_2, urea, melting, 220 °C, 10 min	4.6	230
Mo-11	$(NH_4)_2MoO_4$, urea, 1-ClN, 220 °C, 30 min	7.6	230
Mo-12	$(NH_4)_2MoO_4$, urea, 1-ClN, 210 °C, 10 h	0.32	230
Mo-13	$(NH_4)_2MoO_4$, urea, melting, 250 °C, 10 min	8.3	230
W-1	$W(CO)_6$, SiO_2 and TiO_2, vacuum-sealed bomb vessel, 250 °C	80	218
W-2	$WNCl_3$, melting, 220 °C, 20 min	72	227
W-3	$WNCl_3$, melting, 220 °C, 15 min	14	227
W-4	$WNCl_3$, melting, 220 °C, 15 min	45	228
W-5	$WNCl_3$, melting, 220 °C, 15 min	7	228
W-6	$WNCl_3$, melting, 220 °C, 15 min	8	228
W-7	a) $WNCl_3$, melting, 220 °C, 15 min b) $W(TosN)_2Cl_2$, melting, 220 °C, 10 min	9 13	227
W-8	$WNCl_3$, melting, 220 °C, 15 min	22	228
W-9	$WNCl_3$, melting, 220 °C, 15 min	15	228
W-10	[W(NMes)$_2$Cl$_2$(dme)], melting, 210 °C, 20 min	71	229
W-11	[W(N*tert*-Bu)$_2$Cl$_2$(NH$_2$*tert*-Bu)], melting, 210 °C, 20 min	6	229
W-12	[W(NTos)$_2$Cl$_2$], melting, 210 °C, 20 min	45	229
W-13	$(NH_4)_2WO_4$, urea, melting, 240° C, 10 min	1.6	230

was recrystallized from CH_2Cl_2/hexane and dried at 100 °C under reduced pressure (< 1 mmHg) for 12 h to give target compounds. Yield: 1.50 g (61%).

Synthesis of [Cr(IV)(THF)Pc]$_2$(μ-O) (Ref. 216)

To a solution of iodozobenzene (42 mg) in THF (5 mL), a suspension of **Cr-1** (189 mg) in THF (15 mL) was added slowly, causing the solution to become dark green. After stirring for 40 h, the solvent was removed *in vacuo*, the residue was washed with ether (10 mL) and extracted sequentially with 4 mL THF and 2 mL hexanes, and the extracts filtered through Celite. Black crystals of [Cr(IV)(THF)Pc]$_2$(μ-O) were isolated by slow evaporation in three fractions. Yield: 107 mg (49.6%).

Synthesis of Mo(IV)(O)Pc (Ref. 222)

Molybdenum tetrachloride (2.76 g) was added to a solution of phthalonitrile (5.0 g) and urea (6.0 g) in DMF (30 mL), and the solution refluxed under argon for 19 h. The product was cooled, filtered off, and washed with DMF (50 mL). The blue precipitate was sublimed under nitrogen at 400 °C/0.01 mm. Yield: 0.1 g (1.6%).

Synthesis of Mo(V)(≡N)Pc (Ref. 227)

Phthalonitrile (393 mg) was heated with $MoNCl_3$ (148 mg) at 220 °C, and the temperature held constant for 20 min. After cooling the reaction mixture to room temperature, the solid was crushed in a mortar and Soxhlet extracted with acetone and toluene to remove impurities. Yield: 275 mg (65%), as a dark green powder.

Synthesis of Mo(V)(≡N)PcMe$_8$ (Ref. 228)

A mixture of 4,5-dimethylphthalonitrile (95.7 mg) and ditosyliminomolybdenum dichloride (65.0 mg) was melted for 15 min at 220 °C under Argon. The residue was purified by Soxhlet extraction with hexane, toluene and acetonitrile for 12 h. Yield: 25.0 mg (25%), as a dark green, amorphous solid.

Synthesis of W(V)(≡N)Pc (Ref. 227)

Phthalonitrile (260 mg) was mixed with $WNCl_3$ (138 mg) and melted at 220 °C for 20 min. The crude product was extracted with EtOH, acetone and toluene. Yield: 230 mg (72%), as a dark green solid.

Synthesis of W(V)(≡N)PcMe$_8$ (Ref. 228)

A mixture of 4,5-dimethylphthalonitrile (98.3 mg) and nitridotungsten trichloride (42.5 mg) was melted at 220 °C under argon for 15 min. The residue was extracted with hexane, toluene and acetonitrile in a Soxhlet for 12 h. Yield: 52.0 mg (45%), as a dark green, amorphous solid.

8. *Group 7: MnPc, TcPc, and RePc*

A *manganese* phthalocyanine (MnPc) obtained *via* melting of phthalonitrile or *o*-cyanobenzamide with manganese(IV) dioxide was first described by Linstead and coworkers in 1936.[21] In both cases, the yield of MnPc was very poor.[21] Subsequently, they attempted to obtain MnPc by transmetalation of Li_2Pc using manganous chloride. Although this approach appeared successful, unfortunately the yield was not reported.[52] The Mn(II)Pc (**Mn-1**) could be obtained in good yield by template condensation of phthalonitrile and manganous acetate in propylene glycol (Scheme 55, Table 32).[231,217] This complex was also prepared with a similar yield using high-pressure conditions (Scheme 55, Table 32).[42] **Mn-1** was obtained in high yield under a microwave irradiation (Scheme 55, Table 32).[184]

The MnPc (**Mn-2**) was prepared by template condensation of the phthalonitrile in amyl alcohol in the presence of DBU and manganous chloride or acetylacetonate (Scheme 56).[29] The yield of **Mn-2** depended on the metal source (Table 32).

Unsubstituted nitride Mn^{V}Pc (**Mn-3**) was obtained in good yield by oxidation of **Mn-2** with chlorine in the presence of excess ammonia (Table 32).[232]

A series of manganese 1,8(11),15(18),22(25)-tetra-substituted (**Mn-4** ~ **Mn-6**)-, 2,9(10),16(17),23(24)-tetra-substituted (**Mn-7** ~ **Mn-10**)- as well as 1,3,8,10(9,11), 15,17(16,18),22,24(23,25)-octasubstituted (**Mn-11** ~ **Mn-15**) Pcs has been prepared with good yields by melting the corresponding 3- or 4-substituted, as well as 3,5-disubstituted phthalonitriles and manganese acetate in the presence of a catalytic amount of ammonium molybdate (Scheme 57, Table 32).[233,234] The yield of **Mn-7** depended on the metal source (Table 32).[233,234]

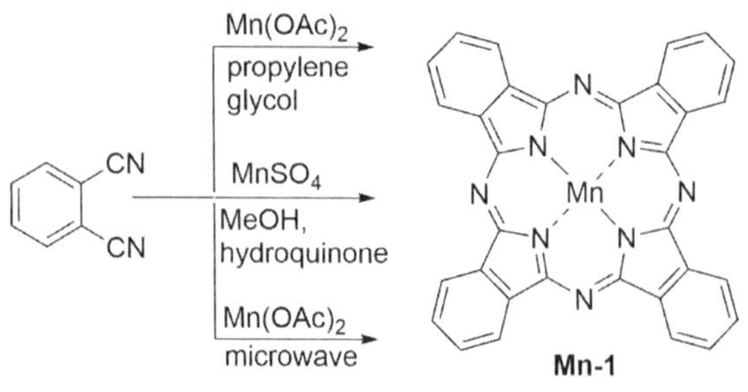

Scheme 55. Synthetic approaches for syntheses of the unsubstituted **Mn-1**.[42,184,217,231]

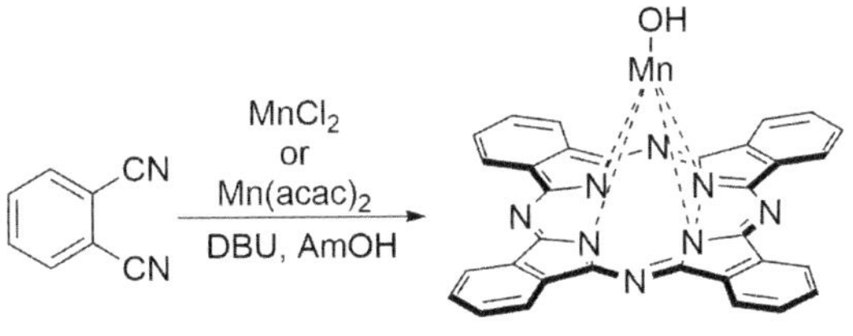

Mn-2

Scheme 56. Synthesis of **Mn-2**.[29]

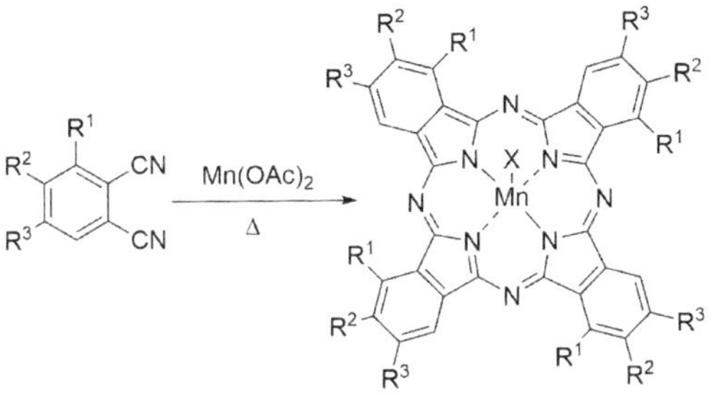

Mn-4 - Mn-15

Mn-4: R^1 = *o*-ClC$_6$H$_4$, R^2 = R^3 = H, X = OH
Mn-5: R^1 = SO$_2$Ph, R^2 = R^3 = H
Mn-6: R^1 = SPh, R^2 = R^3 = H, X = Cl
Mn-7: R^1 = R^3 = H, R^2 = *tert*-Bu, X = Cl
Mn-8: R^1 = R^3 = H, R^2 = O-*tert*-Bu, X = OH
Mn-9: R^1 = R^3 = H, R^2 = S-*tert*-Bu, X = OH
Mn-10: R^1 = R^3 = H, R^2 = C$_5$H$_{10}$N, X = Cl
Mn-11: R^1 = SO$_2$Ph, R^2 = H, R^3 = *tert*-Bu, X = Cl
Mn-12: R^1 = SPh, R^2 = H, R^3 = *tert*-Bu, X = Cl
Mn-13: R^1 = N(Me)$_2$, R^2 = H, R^3 = *tert*-Bu, X = Cl
Mn-14: R^1 = R^3 = NO$_2$, R^2 = H
Mn-15: R^1 = R^3 = SO$_2$Ph, R^2 = H

Scheme 57. Synthesis of manganese Pcs **Mn-4** – **Mn-15**.[233,234]

In addition, the manganese 1,8(11),15(18),22(25)-tetraamino-Pc **Mn-16** could be prepared in good yield by template condensation of 3-nitrophthalimide with manganese chloride in the presence of urea and ammonium molybdate, followed by treatment of sodium sulfide (Table 32).[169] 1,4,8,11,15,18,22,25-Octa(pentylsulfanyl)-substituted Mn(OAc)Pc (**Mn-17**) was prepared by template condensation of the 3,6-di(pentylsulfanyl)phthalonitrile with manganese acetate in the presence of DBU (Table 32).[176] *μ*-Oxodimers of substituted MnPcs were obtained in almost quantitative yields by treating pyridine solutions of **Mn-4** ~ **Mn-15** with hot water.[233]

Only a few articles have focused on the synthesis of *technetium* phthalocyanine (TcPc).[235,236] Indeed, ^{95}TcPc was obtained by recoil implantation of a CuPc and molybdenum metal.[235] Later, the nitrido-TcPc (**Tc-1**) was produced in good yield (94%) by melting phthalonitrile at 280 °C in the presence of ammonium pertechnetate as the metal source (Scheme 58).[236]

The first procedure for the preparation of *rhenium* phthalocyanine (RePc) was published in 1970.[237] The target RePc (**Re-1**) was obtained by melting phthalonitrile in the presence of rhenium(III) chloride, potassium hexachlororhenate(IV) or ammonium perrhenate as the metal source (Table 33). An oxo-RePc was prepared by refluxing phthalonitrile with rhenium(VII) oxide in 1-chloronaphthalene. Treatment of this oxo-RePc with triphenylphosphine gave rhenium(II) *trans*-di(triphenylphosphine)phthalocyanine (**Re-2**) (Table 33).[238]

Table 32. Preparation of **Mn-1** – **Mn-17**.

	Conditions	Yield (%)	Ref.
Mn-1	$Mn(OAc)_2 \times H_2O$, propylene glycol, refluxing[a]	46	231
Mn-1	$Mn(OAc)_2 \times 4H_2O$, propylene glycol, refluxing	40	217
Mn-1	$MnSO_4$, hydroquinone, MeOH, 11 kbar, 90 °C, 1 h	50	42
Mn-1	$Mn(OAc)_2$, MW, P = 560 W, 10 min	90	184
Mn-2	$MnCl_2$, DBU, *n*-pentanol, refluxing, 6 h	74	29
Mn-2	$Mn(acac)_2$, DBU, *n*-pentanol, refluxing, 6 h	49	29
Mn-3	1. $MnCl_2 \times 4H_2O$, ethylene glycol, 180 °C, 1 h; 2. Cl_2, DCM, r.t. 1 h; next NH_3, NH_4Cl, 15 min	85	232
Mn-4	$Mn(OAc)_2$, melting, 220 °C, 15 min; next 220 °C, 15 min	39	233
Mn-5	$Mn(OAc)_2$, melting, 240 °C, 40 min	44	233
Mn-6	$Mn(OAc)_2$, melting, 240 °C, 40 min	55	233
Mn-7	$Mn(OAc)_2$, melting, 270–280 °C, 1 h	60	233
Mn-7	$Mn_2(CO)_{10}$, MOA, melting, 270–280 °C, 15 min	40	234
Mn-7	$Mn(OAc)_2$, melting, 270–280 °C, 1 h	65	234
Mn-8	$Mn(OAc)_2$, melting, 180 °C, 1 h; next 200 °C, 1 h	30	233
Mn-9	$Mn(OAc)_2$, melting, 180 °C, 1 h; next 200 °C, 1 h	42	233
Mn-10	$Mn(OAc)_2$, melting, 240 °C, 40 min	30	233
Mn-11	$Mn(OAc)_2$, melting, 270 °C, 1.5 h	57	233
Mn-12	$Mn(OAc)_2$, melting, 280 °C, 45 min	50	233
Mn-13	$Mn(OAc)_2$, melting, 260 °C, 1.5 h	76	233
Mn-14	$Mn(OAc)_2$, melting, 200 °C, 1.5 h	50	233
Mn-15	$Mn(OAc)_2$, melting, 240 °C, 1 h	20	233
Mn-16	1. $MnCl_3$, urea, MOA, $PhNO_2$,185 °C, 4.5 h; 2. Na_2S, 50 °C, 5 h	84	169
Mn-17	$Mn(OAc)_3$, DBU, *n*-pentanol, 6 h	42	176

[a]Until the brown color was discharged and the solution became greenish black.

Tetra- and octa-substituted Re(V)Pcs have been prepared with low yields as nitrido-RePc **Re-3** – **Re-9** by melting the corresponding substituted phthalonitriles with ammonium perrhenate (Scheme 59, Table 33).[228,239,240]

Later, the oxo-rhenium(IV)Pc (**Re-10**) was prepared in good yield by Kaliya and coworkers using the same reaction conditions as described above (Scheme 59, Table 33).[241]

The preparation of μ-oxo RePc dimers ($(RePc)_2O$) was first described by Hanack and coworkers.[227,240,242] However, the μ-oxo dimer **Re-11** was obtained as a minor

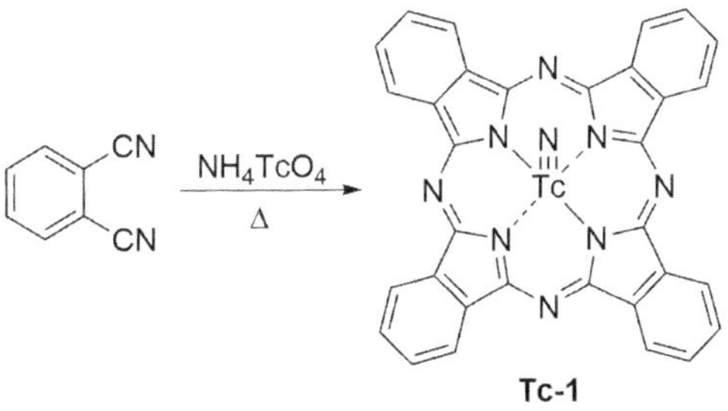

Scheme 58. Synthesis of **Tc-1**.[236]

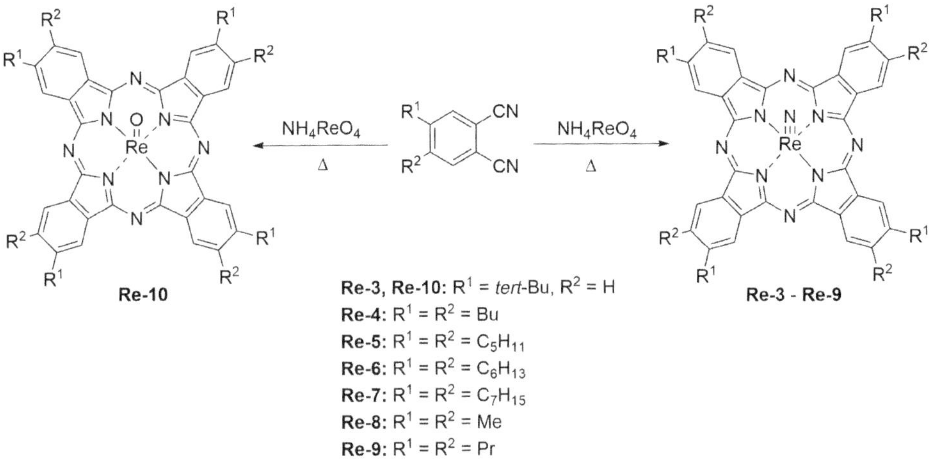

Scheme 59. Syntheses of Re(≡N)Pcs **Re-3** – **Re-9** and Re(=O)Pc **Re-10**.[228,239–241]

product in the direct template condensation reaction of 4-*tert*-butylphthalonitrile in the presence of rhenium(V) chloride, although the main product of this reaction was Re(V)Pc (**Re-12**) (Scheme 60, Table 33).[242] Complex **Re-12** was converted to μ-oxo dimer **Re-11** by treating it with pyridine (Scheme 60).[242]

A μ-oxo dimer **Re-13** was obtained with a low yield by treatment of **Re-3** with acetone and hydrochloric acid (Scheme 61).[240] The main product of this reaction was **Re-14** (Table 33).[240]

μ-Oxo dimers **Re-15**, **Re-16** can obtained in moderate yield by treating the corresponding nitrido-RePcs **Re-3** and **Re-5** with triphenylphosphine under pression (Scheme 62, Table33).[227]

The imido-RePcs were prepared with good yields by template condensation of phthalonitrile in the presence of rhenium bis(imido) complexes.[229] The complexes **Re-17, Re-18** could be obtained by simply melting phthalonitrile and initial bis(imido) complexes at 210 °C (Scheme 63, Table 33).[229]

The double-decker *rhenium bisphthalocyanine* **Re-19** containing an Re–Re direct bond (Re_2Pc_2) was obtained by thermal decomposition of **Re-2** in boiling

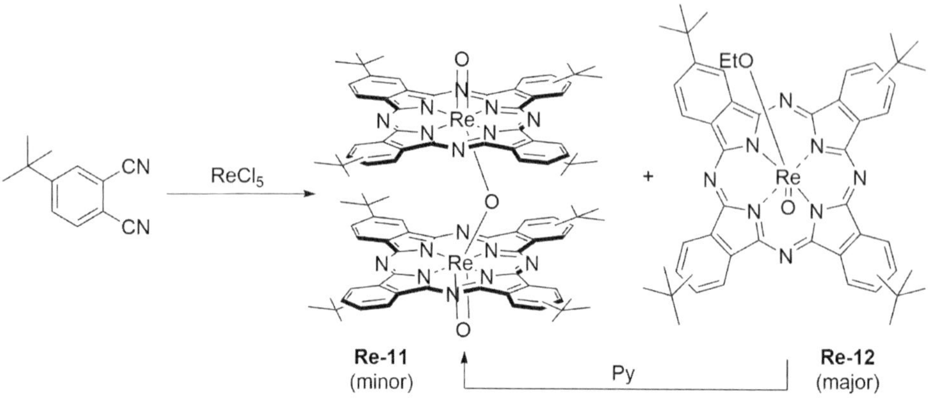

Scheme 60. Preparation of μ-oxo RePcs by direct template synthesis.[242]

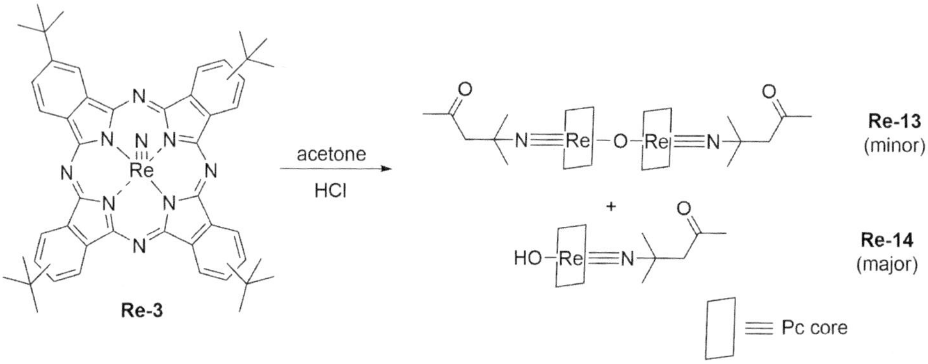

Scheme 61. Preparation of μ-oxo RePcs by treatment of **Re-3**.[240]

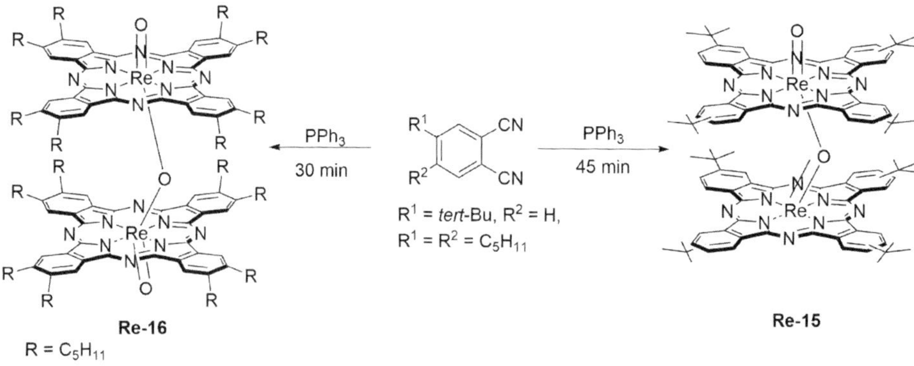

Scheme 62. Preparation of μ-oxo RePcs.[242]

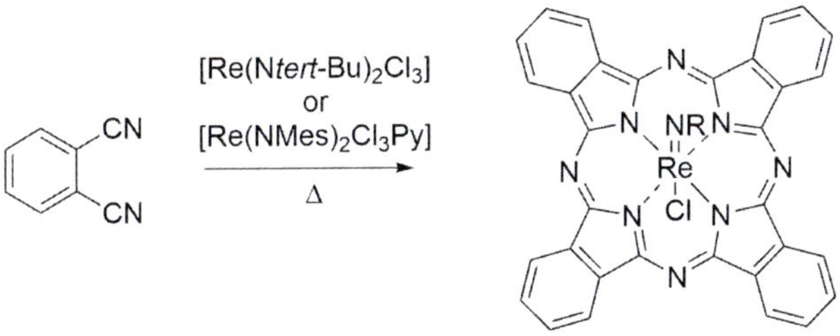

Re-17: R = *tert*-Bu
Re-18: R = Mes

Re-17, Re-18

Scheme 63. Synthesis of organoimido RePcs (**Re-17** and **Re-18**).[229]

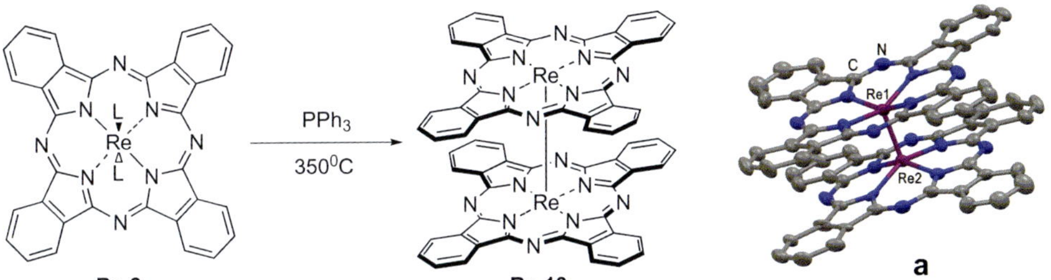

Re-2
L = PPh_3

Re-19

Scheme 64. Synthesis and X-ray structure (a) of **Re-19**.[243] The hydrogen atoms are omitted for clarity. Adapted from Ref. 243.

triphenylphosphine (Scheme 64).[243] The structure of **Re-19** was confirmed by single-crystal X-ray diffraction analysis (Scheme 64).[243]

Synthesis of Mn(III)(Cl)Pc(tert-Bu)$_4$ (Ref. 233)

A mixture of 4-*tert*-butylphthalonitrile (0.184 g), manganese acetate tetrahydrate (0.081 g) and a catalytic amount of ammonium molybdate was melted in a sealed tube for 1 h at 270–280 °C. The reaction mixture was purified by column chromatography on Al_2O_3, using benzene and chloroform as eluents. Yield: 0.124 g (60%).

Synthesis of Mn(III)(OH)Pc (Ref. 29)

A mixture of phthalonitrile (1.28 g) and manganese chloride (0.315 g) in the presence of DBU (1.5 mL) was heated to reflux in *n*-pentanol, and the reaction continued for 6 h. As the reaction proceeded, a crude product gradually precipitated. The precipitate was collected by filtration, washed with 3% hydrochloric acid solution, water and then ethanol, and purified by extraction with chloroform. Yield: 1.08 g (74%).

Table 33. Preparation of **Re-1** – **Re-18**.

	Conditions	Yield (%)	Ref.
Re-1	$ReCl_3$ or K_2ReCl_6 or NH_4ReO_4, melting, 290 °C, 30 min	n.r.	237
Re-2	1. Re_2O_7, 1-ClN, refluxing, 2 h; 2. PPh_3, 225 °C	n.r.	238
Re-3	NH_4ReO_4, melting, 290 °C, 45 min	6	239
Re-4	NH_4ReO_4, melting, 280 °C, 30 min	5	240
Re-5	NH_4ReO_4, melting, 280 °C, 30 min	5	240
Re-6	NH_4ReO_4, melting, 280 °C, 30 min	4	240
Re-7	NH_4ReO_4, melting, 280 °C, 30 min	4	240
Re-8	NH_4ReO_4, melting, 280 °C, 50 min	34	228
Re-9	NH_4ReO_4, melting, 280 °C, 50 min	4	228
Re-10	NH_4ReO_4, melting in sealed tube, 270 °C, 1 h	50	241
Re-11	$ReCl_5$	n.r.	242
Re-12	$ReCl_5$	n.r.	242
Re-13	Acetone, HCl, 4 d in the dark at RT	2	240
Re-14	Acetone, HCl, 4 d in the dark at RT	54	240
Re-15	PPh_3, melting, 250 °C, $P = 6 \times 10^{-2}$ mbar, 45 min	11	227
Re-16	PPh_3, melting, 220 °C, $P = 6 \times 10^{-2}$ mbar, 30 min	18	227
Re-17	$[Re(N\textit{tert}\text{-}Bu)_2Cl_3]$, melting, 210 °C, 20 min	42	229
Re-18	$[Re(NMes)_2Cl_3Py]$, melting, 210 °C, 20 min	83	229

Synthesis of Mn(III)(OAc)Pc(SC_5H_{11})$_8$ (Ref. 176)

3,6-Di(pentylsulfanyl)phthalonitrile (0.4 g) in *n*-pentanol (3.5 mL) was refluxed under a nitrogen atmosphere, and manganese(III) acetate (0.069 g) added. After the addition of DBU (0.13 ml), the reaction was continued for 6 h, then the mixture cooled, and column chromatography over silica carried out with $CHCl_3$ as eluent. Yield: 0.23 g (42%).

Synthesis of Re(V)(≡N)PcMe$_8$ (Ref. 228)

A mixture of 4,5-dimethylphthalonitrile (156.2 mg) and ammonium perrhenate (53.6 mg) was melted at 280 °C for 50 min. The solid residual melt was finely powdered, and extracted with acetonitrile, toluene and hexane, respectively, in a Soxhlet apparatus for 5 d each. Yield: 56.0 mg (34%).

Synthesis of Re(=O)Pc(tert-Bu)$_4$ (Ref. 241)

A finely powdered mixture of 4-*tert*-butylphthalonitrile (0.670 g) and ammonium perrhenate (0.482 g) was melted in a sealed tube at 270 °C for 1 h. The solid residual melt was washed with hot water and hexane, and purified by column chromatography on Al_2O_3 using benzene as eluent. Yield: 0.426 g (50%).

9. Group 8: FePc, RuPc, and OsPc

Iron phthalocyanine (FePc) was first obtained as a by-product of the industrial manufacturing of phthalimide from phthalic anhydride and ammonia.[1] Later, Fe(II)Pc (**Fe-1**) was prepared by Linstead and coworkers using *o*-cyanobenzamide as a starting material[4,17,21] and iron metal,[17,21] ferric oxide,[17] as well as ferrous sulfate[17] as the metal source (Scheme 65, Table 34).

Currently, Fe(II)Pcs are generally prepared *via* two routes: (i) by direct template condensation of the so-called phthalogenes (anhydrides, imides, dinitriles of phthalic acid) with a metal source[24,29,42,53,62,184,215,245–247,249–253] or (ii) by metal insertion into an H_2Pc ligand.[246] Substituent-free **Fe-1** can be prepared by template condensation of phthalonitrile and iron(0) pentacarbonyl using 1-chloronapthalene as a solvent (Scheme 66, Table 34).[215] **Fe-1** is also obtained by heating phthalonitrile with iron(II) chloride in the presence[29] or absence[53] of DBU (Scheme 66. Table 34). In addition, **Fe-1** can be prepared by using high-pressure[42] as well as microwave irradiation[184] conditions (Scheme 66, Table 34). **Fe-1** is also obtained by an exotic procedure *via* a ring-contraction reaction of the uranyl superphthalocyanine with ferric chloride (Scheme 67).[244]

2,9(10),16(17),23(24)-Tetra-substituted Fe(II)Pcs (**Fe-2** – **Fe-7**) were prepared by direct template condensation of the corresponding 4-substituted phthalonitriles (**Fe-2** – **Fe-6**)[24,62,245–247] as well as by metal insertion into H_2Pc (**Fe-7**)[246] (Scheme 68). Indeed, the complexes **Fe-2**, **Fe-3**, **Fe-5** and **Fe-6** were obtained by direct template condensation using different solvents (DMAE, *n*-pentanol, 1-ClN, TCB as well as

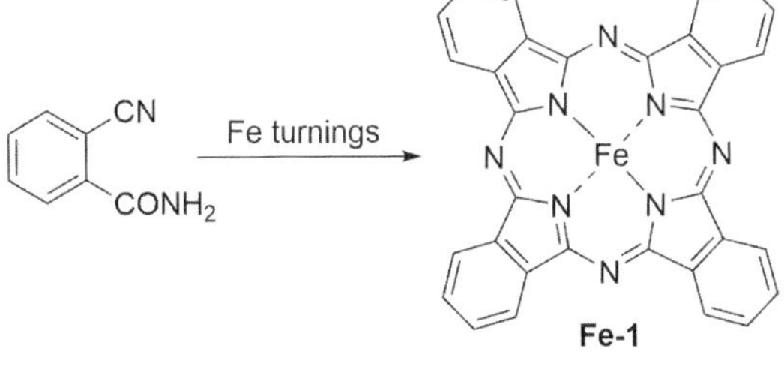

Scheme 65. Synthesis of **Fe-1**.[4,17,21]

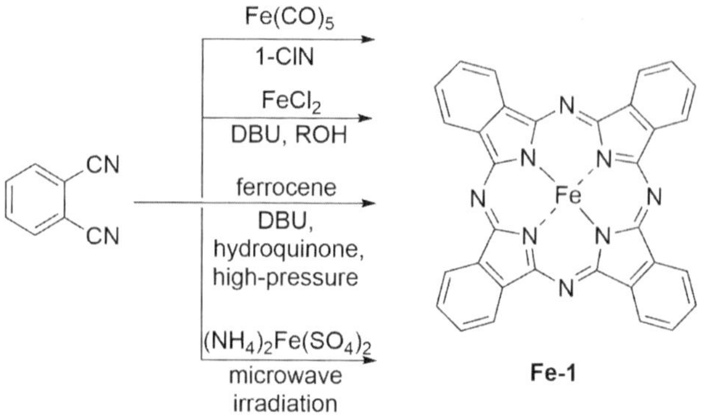

Scheme 66. Synthetic pathway for the preparation of **Fe-1** *via* direct template condensation.[29,42,53,184,215]

Table 34. Preparation of **Fe-1**.

	Conditions	Yield (%)	Ref.
Fe-1	CBA, Fe turnings, melting, 240 °C, 2 h	45	17
Fe-1	PN, $Fe(CO)_5$, 1-ClN, refluxing, 30 min	50	215
Fe-1	PN, $FeCl_2$, DBU, 2-ethoxyethanol, refluxing, 6 h	57	29
Fe-1	PN, ferrocene, hydroquinone, MeOH, 12 kbar, 90 °C, 3.5 h	16–25	42
Fe-1	PN, $(NH_4)_2Fe(SO_4)_2$, MW, 10 min	87	184

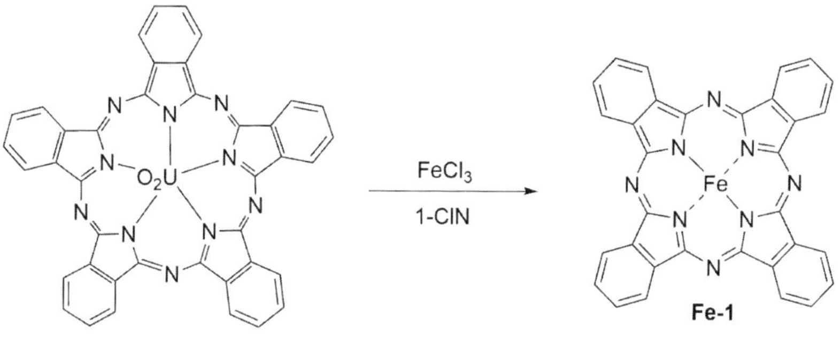

Scheme 67. The preparation of **Fe-1** from $U(O_2)$SuperPc.[244]

ethylene glycol) as reaction media, and ferrous sulfate,[245] iron(0) pentacarbonyl[24,246] or ferrous acetate[247] as the metal source (Table 34). By fusing 4-phenylphthalonitrile in the presence of iron powder, complex **Fe-4** was obtained (Scheme 68, Table 35).[62] Iron insertion into the corresponding H_2Pc, gave **Fe-7** (Scheme 68, Table 35).[246] In addition, an analytically pure sample of **Fe-3** was obtained by vacuum sublimation (5×10^{-5} mmHg) of the μ-oxo-bridged diiron tetra-*tert*-butylphthalocyanine.[248]

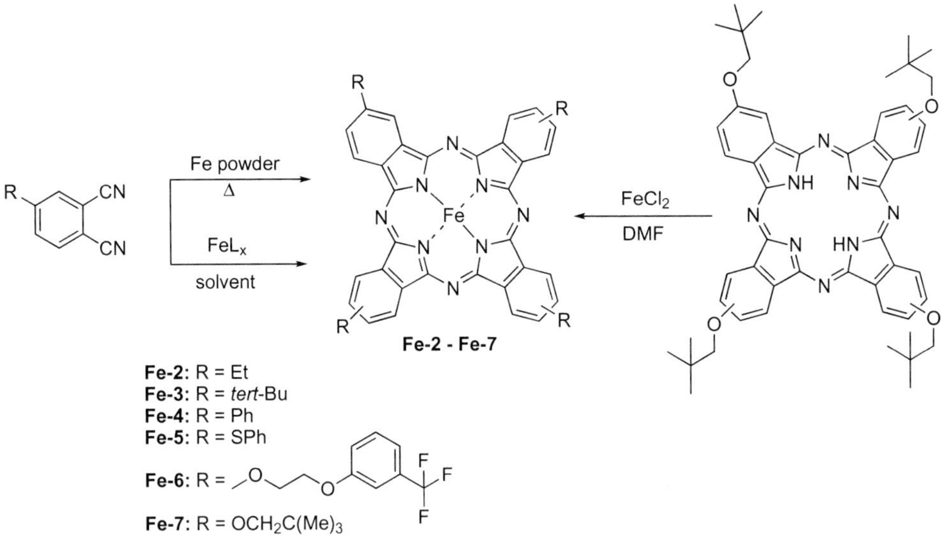

Scheme 68. Preparation of the 2,9(10),16(17),23(24)-tetra-substituted **Fe-2** – **Fe-7**.[24,62,245–247]

Table 35. Preparation of 2,9(10),16(17),23(24)-tetra-substituted Fe^{II}Pcs (**Fe-2** – **Fe-7**).

	Conditions	Yield (%)	Ref.
Fe-2	$FeSO_4 \times H_2O$, DMAE, refluxing	60	245
Fe-3	$Fe(CO)_5$, TCB, 215 °C, 2 h	33	24
Fe-3	$Fe(CO)_5$, ethylene glycol, 200 °C, 1.5 h	38	24
Fe-4	Fe powder, MOA, melting	27	62
Fe-5	$Fe(CO)_5$, ethylene glycol, refluxing, 2 h	18.2	246
Fe-6	$Fe(OAc)_2$, DBU, *n*-pentanol, sealed tube,160 °C, 24 h	70	247
Fe-7	$FeCl_2$, DMF, refluxing, 3 h	70.2	246

2,3,9,10,16,17,23,24-Octa-substituted FePcs (**Fe-8** – **Fe-10**), as well as tetra-*ββ*-fused FePcs (**Fe-11** – **Fe-13**) were obtained by direct template condensation of the corresponding 4,5-disubstituted[24,249] or *ββ*-fused phthalonitriles[246,250,251] with iron(0) pentacarbonyl,[24,246,251] ferrous chloride,[249] and ferrocene[250] as the metal source (Scheme 69, Table 36).

Hexadeca-substituted Fe(II)Pcs have also been known for many years.[24,252,253] Indeed, the complexes **Fe-14** – **Fe-16** were prepared *via* direct template condensation of the corresponding tetra-substituted phthalonitriles[252,253] or tetra-substituted phthalic anhydride[24] (Scheme 70, Table 37).

In addition, dinuclear dimeric phthalocyanine systems, PcFe–X–FePc, linked through the X (X = O, N, C) bridged atom have been reported based on FePcs.[254–263]

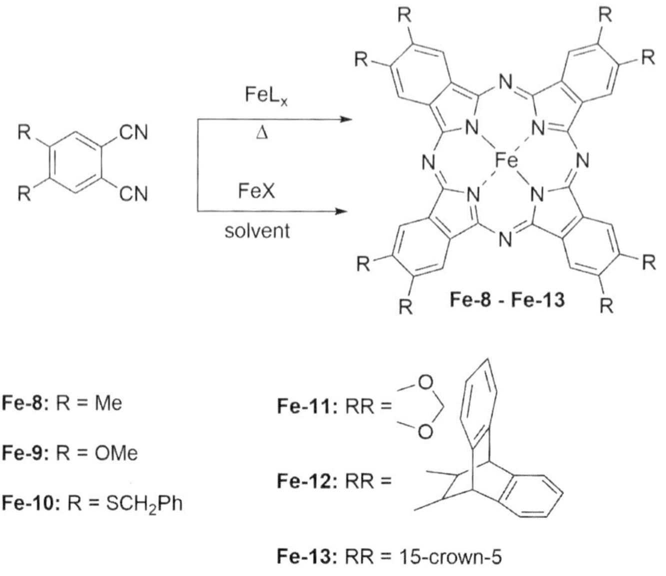

Scheme 69. Preparation of the **Fe-8** – **Fe-13** complexes.[24,246,249–251]

Table 36. Preparation of 2,3,9,10,16,17,23,24-octa-substituted and tetra-$\beta\beta$-fused FePcs (**Fe-8** – **Fe-13**).

	Conditions	Yield (%)	Ref.
Fe-8	$Fe(CO)_5$, 1-ClN, refluxing, 45 min; then 250 °C, 45 min	41	24
Fe-9	$Fe(CO)_5$, 1-ClN, refluxing, 45 min; then 250 °C, 45 min	37	24
Fe-10	$FeCl_2$x$4H_2O$, urea, melting, 150 °C, 1.5 h	55	249
Fe-11	$Fe(CO)_5$, 1-ClN, 270–280 °C, 1 h	43	246
Fe-12	Ferrocene, sealed tube, melting	n.r.	250
Fe-13	$Fe(CO)_5$, ethylene glycol, 190–210 °C, 3 h	16.2	251

Thus, μ-oxo (X = O) binuclear complexes of iron (**Fe-17** – **Fe-32**) were prepared by reaction of the corresponding substituted phthalonitriles with ferrous acetate or ferrous sulfate in various solvents (Scheme 71, Table 38).[256–259]

μ-Nitrido (X = N, **Fe-33**, **Fe-34**)[260,261] while μ-carbido (X = C, **Fe-35**)[262,263] binuclear complexes of iron could be prepared by reaction of (un)sunstituted FePcs with sodium azide (**Fe-33**, **Fe-34**) or tetraiodomethane (**Fe-35**) in the presence of reducting agents (sodium dithionite or sodium borohydride), respectively (Scheme 72, Table 38).

Ruthenium phthalocyanines (RuPc) can be prepared *via* two synthetic approaches: (i) by the direct template condensation of (un)substituted *o*-cyanobenzamides[264,265,271] or phthalonitriles,[184,265–268,270,272–274,278–283] as well as 1,3-diiminoisoindoline[275] with metal

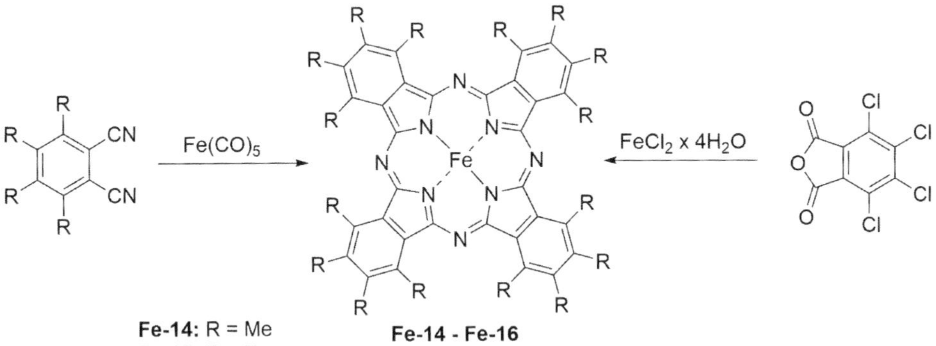

Fe-14: R = Me
Fe-15: R = F
Fe-16: R = Cl

Scheme 70. The preparation of the hexadeca-substituted **Fe-14** – **Fe-16**.[24,252,253]

Table 37. Preparation of hexadeca-substituted FePcs (**Fe-14** – **Fe-16**).

	Conditions	Yield (%)	Ref.
Fe-14	$Fe(CO)_5$, 1-ClN, 270 °C, 3 h	17	252
Fe-15	$Fe(CO)_5$, 1-methylnaphthalene, refluxing, 1 h	64	253
Fe-16	$FeCl_2$x$4H_2O$, urea, MOA, $PhNO_2$, 190 °C, 4.5 h	60	24

sources such as ruthenium(III) chloride[184,264–268,270–275,278,279,281,282] or ruthenium carbonyl,[265,276,277,280,282,283] and (ii) by metal insertion reaction into H_2Pc.[265,276,277,279–281] Indeed, the unsubstituted RuPc (**Ru-1**) was obtained by direct template condensation of unsubstituted *o*-cyanobenzamide[264,265] or phthalonitriles[266–268] with ruthenium(III) chloride[264–268] or ruthenium carbonyl[266] (Scheme 73, Table 39). In addition, **Ru-1** was prepared from phthalonitriles using microwave irradiation[184] and ionic liquids[269,270] as a reaction media, as well as by thermal decomposition of **Ru-1a**[265] (Table 39).

2,9(10),16(17),23(24)-Tetra-substituted RuPcs (**Ru-2**, **Ru-3**) were obtained by direct template condensation of the corresponding *o*-cyanobenzamide,[271] phthalonitriles[272–274] or 1,3-diiminoisoindoline,[275] as well as by metal insertion into H_2Pcs (Scheme 74, Table 40).[276,277] **Ru-3** without axially bridged groups was obtained by thermal decomposition of the appropriate complex (**Ru-3a**).[274]

1,4,8,11,15,18,22,25-Octa- (**Ru-4** – **Ru-7**), 2,3,9,10,16,17,23,24-octa-substituted (**Ru-8**, **Ru-10** – **Ru-12**), as well as tetra-*ββ*-fused RuPcs (**Ru-9**) can be prepared by direct template condensation of 3,6-, 4,5-disubstituted, as well as *ββ*-fused phthalonitriles (**Ru-4**, **Ru-8**, **Ru-9**),[278,281] or by metal insertion into an H_2Pc core (**Ru-5** – **Ru-7**; **Ru-10** – **Ru-12**)(Scheme 75, Table 41).[265,279–281] The yield of **Ru-9** depends on the metal source (Table 41).[282] It should be noted, that **Ru-10** – **Ru-12** can be obtained *via* metal exchange of Li_2PcR_8 pre-obtained *in situ* (Scheme 75).[265,281]

The *osmium* complex of phthalocyanine (**Os-1**) was prepared by direct template condensation of *o*-cyanobenzamide with osmium(III) chloride (Scheme 76,

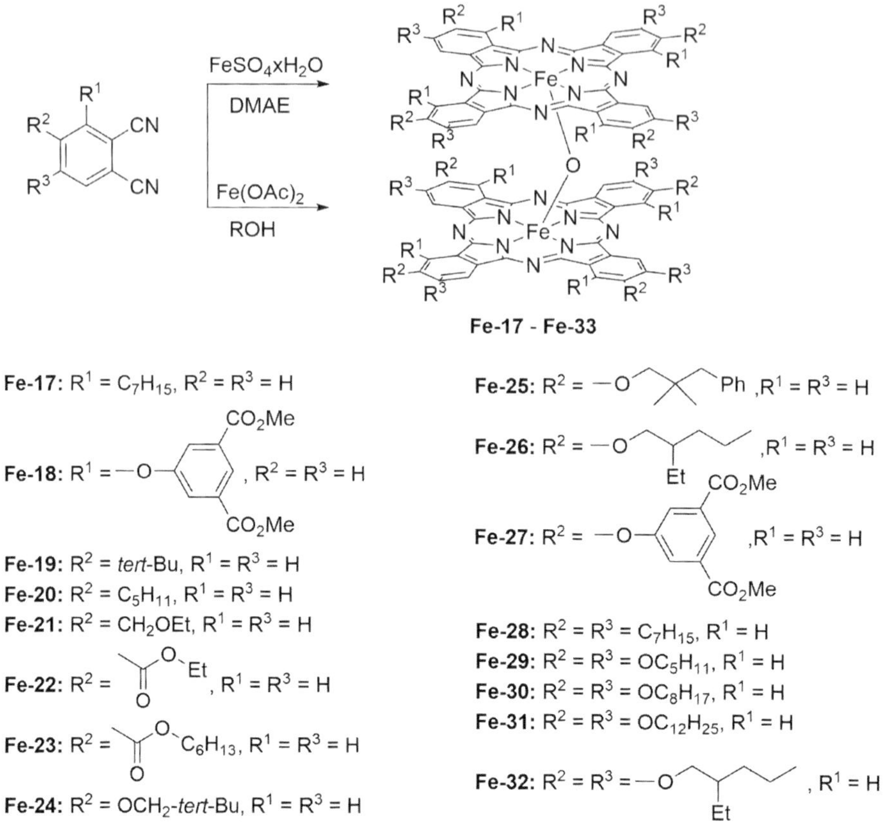

Scheme 71. The synthetic approaches of preparation of the μ-oxo-binuclear complexes of iron (**Fe-17** – **Fe-32**).[256–259]

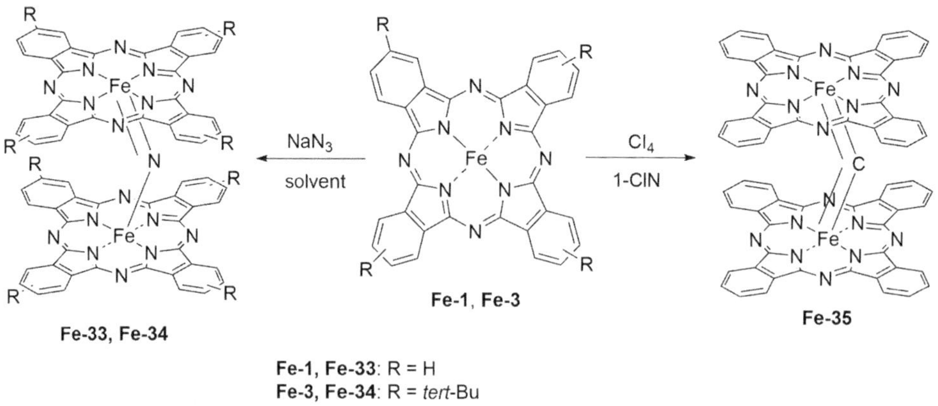

Scheme 72. Preparation of the μ-nitrido- (**Fe-33, Fe-34**)[260,261] and μ-carbido-(**Fe-35**)[262,263] binuclear complexes of iron.

Table 38. Preparation of μ-X(X = O, N, C)-binuclear complexes **Fe-17** – **Fe-35**.

	Conditions	Yield (%)	Ref.
Fe-17	$Fe(OAc)_2$, DBU, hexanol, 150 °C, 24 h	43	258
Fe-18	$Fe(OAc)_2$, DMAE, air, 170 °C, 5 h	16.7	259
Fe-19	$FeSO_4xH_2O$, DMAE, refluxed, 12 h	52	257
Fe-20	$Fe(OAc)_2$, DBU, hexanol, 150 °C, 12 h	43	258
Fe-21	$Fe(OAc)_2$, DBU, hexanol, 150 °C	32	257
Fe-22	$Fe(OAc)_2$, DBU, EtOH, refluxing, 8 h	32	257
Fe-23	$Fe(OAc)_2$, DBU, hexanol, 150 °C, 10 h	29	257
Fe-24	$FeSO_4xH_2O$, DMAE, refluxed, 24 h	58	257
Fe-25	$FeSO_4xH_2O$, DMAE, refluxed, 16 h	41	257
Fe-26	$FeSO_4xH_2O$, DMAE, refluxed, 24 h	57	257
Fe-27	$Fe(OAc)_2$, DMAE, air, 170 °C, 5 h	19.5	259
Fe-28	$Fe(OAc)_2$, DBU, hexanol, 150 °C, 24 h	41	258
Fe-29	Li, $Fe(OAc)_2$, TEBAC, pentanol, refluxing	27.3	256
Fe-30	Li, $Fe(OAc)_2$, TEBAC, pentanol, refluxing	32.6	256
Fe-30	$FeSO_4xH_2O$, DMAE, refluxed, 24 h	53	257
Fe-31	Li, $Fe(OAc)_2$, TEBAC, pentanol, refluxing	49.9	256
Fe-32	$FeSO_4xH_2O$, DMAE, refluxed, 24 h	38	257
Fe-33	NaN_3, 1-ClN, 265 °C, 1 h	90	260
Fe-34	NaN_3, xylene, 150 °C, 6 h	62	261
Fe-35	CI_4, 1-ClN, $Na_2S_2O_4$, 140–150 °C, 30 min	66	263

Table 42).[284,285] OsPc can also be prepared in high yield as bis-axially coordinated complexes by direct template condensation of the phthalonitrile with osmium(IV) oxide or osmium carbonyl.[265]

Synthesis of Fe(II)Pc (Ref. 4)

An intimate mixture of *o*-cyanobenzamide (20.0 g) and freshly-etched Fe (4.0 g) was heated at 250 °C for 6 h. On cooling, the phthalimide by-product was removed by extracting with concentrated aqueous NaOH. The product was collected by filtration and washed with H_2O and EtOH. Yield: 2.2 g (11%).

Synthesis of Fe(II)Pc (Ref. 215)

A solution of phthalonitrile (8.0 g) in 1-chloronaphthalene (100 mL) was heated to reflux under nitrogen with stirring, and then a solution of $Fe(CO)_5$ (3.0 g) in

Ru-1a-e

Ru-1a: R = H, $L^1 = L^2$ = Py
Ru-1b: R = H, L^1 = CO, L^2 = THF
Ru-1c: R = H, L^1 = CO, L^2 = Py
Ru-1d: R = Cl, $L^1 = L^2$ = Py
Ru-1e: R = H, $L^1 = L^2$ = isoquinoline
Ru-1: R = H, $L^1 = L^2$ = none

Scheme 73. The synthetic approaches of the preparation of **Ru-1**.[264–268]

Table 39. Preparation of unsubstituted **Ru-1**.

	Conditions	Yield (%)	Ref.
Ru-1a	CBA, $RuCl_3 \times 3H_2O$, naphthalene, 290 °C, 1 h	5	264
Ru-1a	CBA, $RuCl_3 \times 3H_2O$, naphthalene, 290 °C, 1 h	50	265
Ru-1b	PN, $RuCl_3 \times 3H_2O$, melting, 250 °C, 4 h	n.r.	266
Ru-1c	PN, $RuCl_3 \times 3H_2O$, melting, CO-atmosphere, 250 °C, 4 h	84	266
Ru-1c	PN, $Ru_3(CO)_{12}$, melting, air, 250 °C, 4 h	91	266
Ru-1d	PN, $RuCl_3 \times 3H_2O$, melting, 280 °C, 4 h	68	267
Ru-1e	PN, $RuCl_3 \times 3H_2O$, isoquinoline, refluxing, 2 h	5	268
Ru-1	PN, $RuCl_3 \times 3H_2O$, MW, 3 min	91	184
Ru-1	PN, $RuCl_3$, TMGT, sealed tube, 130 °C,10 h	35	270
Ru-1	**Ru-1a**, 0.01 mmHg, 250 °C, 6 h	97	265

l-chloronaphthalene (50 mL) was slowly added from a separatory funnel under nitrogen. Refluxing was continued for a further 30 min, then the cooled mixture filtered, and the insoluble product washed with benzene, chloroform, acetone, and diethyl ether, and dried. Yield: 4.3 g (50%).

Synthesis of Fe(II)Pc (Ref. 53)

Phthalonitrile (2.9 g) and iron powder (5.0 g) were intimately mixed and placed in a 25-mL conical flask. The mixture was heated in a Wood's metal or sand bath at

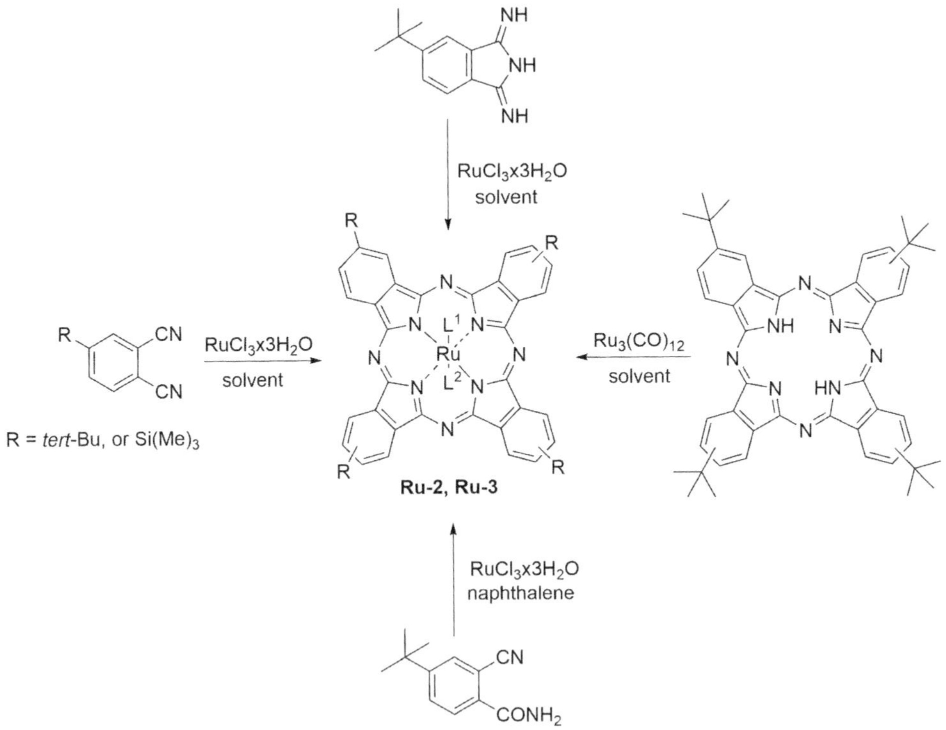

Ru-2: R = *tert*-Bu, $L^1 = L^2$ = Py (**a**), 2-picoline (**b**), 3-picoline (**c**), 4-picoline (**d**), 2,5-lutidine (**e**), 2,6-lutidine (**f**) 3-ClPy (**g**), pyz (**i**), dabco (**j**), bipyridine (**k**), NCPh (**l**)

Ru-2h: R = *tert*-Bu, L^1 = CO, L^2 = none

Ru-3a: R = $Si(Me)_3$, $L^1 = L^2$ = 3-ClPy

Ru-3: R = $Si(Me)_3$, $L^1 = L^2$ = none

Scheme 74. The 2,9(10),16(17),23(24)-tetra-substituted **Ru-2**, **Ru-3**.[271–277]

260 °C for 5 h. Phthalonitrile that condensed on the cooler upper portion of the flask was scraped down to the bottom periodically. The blue product was finely ground and extracted in a Soxhlet extractor with ethanol (200 mL) for 5 h. The ethanol was replaced by pyridine (200 mL), and the extraction was continued for a further 5 h. The pyridine extract was reduced in volume to 20 mL, and hexane added to precipitate the blue $Fe(II)(Py)_2Pc$, which was collected by filtration, washed with ethanol, and dried in vacuum, to give 0.8 g of $Fe(II)(Py)_2Pc$. A portion (0.1 g) of this material can be further purified by dissolving in pyridine (30 mL) and filtering the solution by suction through alumina (3 × 10 cm Woelm alumina grade I). Hexane (100 mL) was added to the filtrate and the blue precipitate was collected by filtration, washed with petroleum ether, and sublimed under vacuum at 300 °C, to give the dark-blue Fe(II)Pc. Yield: 0.05 g (50%).

Table 40. Preparation of 2,9(10),16(17),23(24)-tetra-substituted **Ru-2** and **Ru-3**.

	Conditions	Yield (%)	Ref.
Ru-2(a–f)	1. CBA, $RuCl_3x3H_2O$, naphthalene, 290 °C, 2. amine, refluxing, 5 h	20–40	271
Ru-2a	**Ru-2l**, Py, hexanes, refluxing, 1 h	79	276
Ru-2g	PN, $RuCl_3x3H_2O$, DBU, 3-ClPy, 2-ethoxyethanol, reflux. 24 h	45	272
Ru-2h	PN, $RuCl_3x3H_2O$, DBU, EtOH, 250 °C, 4560 kPa	75	273
Ru-2h	H_2Pc, $Ru_3(CO)_{12}$, PhOH, refluxing, 8 h	82	277
Ru-2i	DI, $RuCl_3x3H_2O$, 2-ethoxyethanol, pyrazine, refluxing, 2 d	36	275
Ru-2j	DI, $RuCl_3x3H_2O$, 2-ethoxyethanol, DABCO, reflux., 2 d	18	275
Ru-2k	DI,$RuCl_3x3H_2O$,2-ethoxyethanol, 4,4′-bipyridine, 160 °C, 36 h	34	275
Ru-2l	H_2Pc, $Ru_3(CO)_{12}$, benzonitrile, 198 °C, 24 h	43	276
Ru-3a	PN, $RuCl_3x3H_2O$, 2-ethoxyethanol, DBU, 3-ClPy, 140 °C, 2 d	30.6	274
Ru-3	**Ru-3a**, 250 °C, 10^{-2} mmHg	98	274

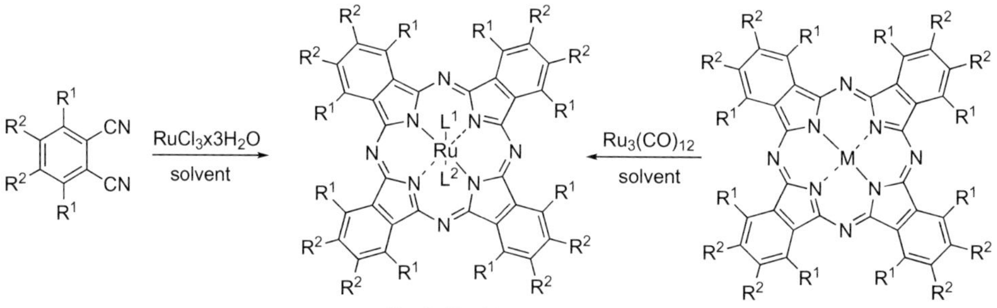

Ru-4 - Ru-12

M = 2H, 2Li

Ru-4: $R^1 = OC_5H_{11}$, R^2 = H. $L^1 = L^2$ = 3-ClPy
Ru-5: R^1 = OBu, R^2 = H. $L^1 = L^2$ = Py
Ru-6: $R^1 = C_8H_{17}$, R^2 = H. L^1 = CO, L^2 = none
Ru-7: $R^1 = C_{10}H_{21}$, R^2 = H. L^1 = CO, L^2 = none
Ru-8: R^1 = H, $R^2 = OC_5H_{11}$, $L^1 = L^2$ = 3-ClPy
Ru-9: R^1 = H, R^2 = 15-crown-5, L^1 = CO, L^2 = MeOH
Ru-10: R^1 = H, $R^2 = OC_5H_{11}$, $L^1 = L^2$ = none
Ru-11: R^1 = H, $R^2 = OC_5H_{11}$, L^1 = CO, L^2 = none
Ru-12: R^1 = H, $R^2 = OCH_2C(Et)(CH_2)_3CH_3$, L^1 = CO, L^2 = none

Scheme 75. The synthetic approaches of preparation of the octa-substituted as well as tetra-*ββ*-fused **Ru-4** – **Ru-12**.[265,278–283]

Synthesis of Fe(II)Pc(tert-Bu)$_4$ (Ref. 24)

4-*tert*-Butylphthalonitrile (18.4 g) was dissolved in 1,2,4-trichlorobenzene (15 mL) and heated to 215 °C , then $Fe(CO)_5$ (4.9 g) dissolved in 1,2,4-trichlorobenzene (80 mL) was added dropwise over a period of 50–60 min; the solution held at 215 °C

Table 41. Preparation of octa-substituted **Ru-4** – **Ru-12**.

	Conditions	Yield (%)	Ref.
Ru-4	$RuCl_3 \times 3H_2O$,DBU, 3-ClPy, 2-ethoxyethanol, refluxing, 7 h	12.7	278
Ru-5	H_2Pc, $RuCl_3 \times 3H_2O$, Py, refluxing, 3 h	20	279
Ru-6	$Ru_3(CO)_{12}$, benzonitrile, refluxing, 1 h	96	280
Ru-7	H_2Pc, $Ru_3(CO)_{12}$, benzonitrile, refluxing, 1 h	81	280
Ru-8	$RuCl_3 \times 3H_2O$, DBU, 3-ClPy, 2-ethoxyethanol	3.7	281
Ru-9	$Ru_3(CO)_{12}$, sealed tube, melting, 250 °C, 4 h	80–82	282, 283
Ru-9	$RuCl_3 \times 3H_2O$, sealed tube, melting, 250 °C, 4 h	10	282
Ru-10	(1) Li, *n*-pentanol, refluxing, 2 h; (2) $RuCl_3 \times 3H_2O$, 2-ethoxyethanol, refluxed, 6 h	36	265
Ru-10	H_2Pc, $RuCl_3 \times 3H_2O$, 2-ethoxyethanol, refluxing, 24 h	35	265
Ru-11	(1) Li, *n*-pentanol, (2)$RuCl_3 \times 3H_2O$, 2-ethoxyethanol, refluxing, 6 h	30	281
Ru-12	(1) Li, *n*-pentanol, (2) $RuCl_3 \times 3H_2O$, 2-ethoxyethanol, refluxing, 6 h	25	281

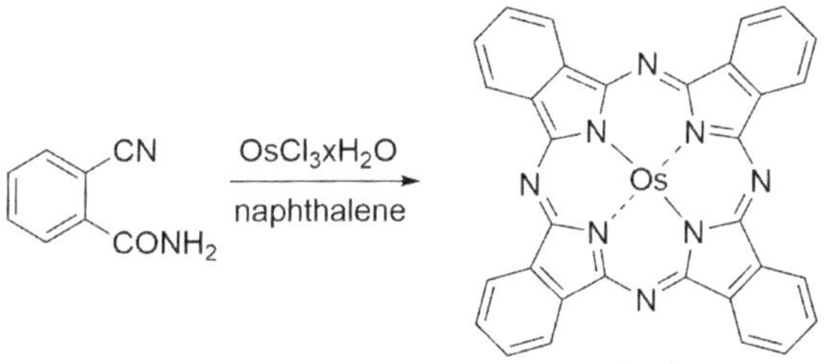

Scheme 76. The preparation of the **Os-1**.[284,285]

Table 42. Preparation of complex **Os-1**.

	Conditions	Yield (%)	Ref.
Os-1	$OsCl_3xH_2O$, naphthalene, 290 °C, 1 h	11	284

for an additional 1 h. After cooling, the mixture was diluted with an equal volume of methanol, and the ensuing precipitate vacuum-filtered and washed with a little MeOH. After drying at 60 °C *in vacuo*, a dark blue powder was obtained. Yield: 6.5 g (33%).

Synthesis of Fe(II)Pc(SPh)$_4$ (Ref. 246)

To refluxing 4-(phenylthio)phthalonitrile (2.37 g) in ethylene glycol (25 mL) under nitrogen, $Fe(CO)_5$ (0.33 mL) was added dropwise with a syringe. After about 2 min, the reaction mixture turned green, and was further refluxed for 2 h, then cooled and diluted with ethanol (50 mL). The precipitate was collected by filtration and washed with ethanol, water, dilute HCI, water, and ethanol. The residue was dissolved in CH_2Cl_2, which was washed with 5% HCl and then water, and the CH_2Cl_2 layer was dried by $MgSO_4$. After removal of $MgSO_4$, the CH_2Cl_2 solution was condensed to *ca.* 25 mL and methanol (280 mL) added. After 2 days, the precipitate was collected by filtration, and dried *in vacuo*. Yield: 429 mg (18.2%).

Synthesis of Fe(II)Pc(OCH$_2$C(CH$_3$)$_3$)$_4$ (Ref. 246)

Metal-free 2,9(10),16(17),23(24)-tetraneopentoxyphthalocyanine (342 mg) and $FeCl_2$ (340 mg) were refluxed in DMF (20 mL) for 3 h, the solvent removed using an evaporator, and then dried *in vacuo* at 100 °C. Silica gel chromatography of the residue with CH_2Cl_2 as eluent gave, upon evaporation of the solvent, a dark blue, shining solid. Yield: 255 mg (70.2%).

Synthesis of Fe(II)Pc(Me)$_8$ (Ref. 24)

To a solution of 4,5-dimethylphthalodinitrile (19.5 g) in 1-chloronaphthalene (200 mL) was added under reflux and dropwise a solution of $Fe(CO)_5$ (6.0 g) in 1-chloronaphthalene (50 mL) over 45 min. The reaction was allowed to reflux for an additional 45 min, then the mixture cooled, and the precipitate collected on a filter (suction) and washed with $CHCl_3$, ethanol, acetone, and finally ether. The crude product was boiled for 10 min each in 1% HCl and 1% NaOH solutions, washed with water until the washings were neutral, and then treated on a sintered glass disk with acetone and ether, and filtered. After the product was dried at 80 °C *in vacuo*, there remained a violet-black powder. Yield: 8.8 g (41%).

Synthesis of Fe(II)Pc(Me)$_{16}$ (Ref. 252)

A solution of $Fe(CO)_5$ (0.2 g) dissolved in 1-chloronaphthalene (2 mL) was added dropwise over 5 min to a solution of tetramethylphthalonitrile (0.73 g) in 1-chloronaphthalene (6 mL), under reflux conditions. The obtained black suspension was held at about 270 °C for a further 3 h, and then cooled, after which the precipitate was suction-filtered, followed by washing with chloroform, toluene, and acetone. After boiling in dilute HCl, followed by water, the product was placed in the thimble of a Soxhlet extractor, washed overnight with acetone, and then over another night with chloroform. After drying *in vacuo*, a blue-black powder of the desired compound was obtained. Yield: 0.13 g (17%).

Synthesis of [FePc(tert-Bu)$_4$]$_2$O (Ref. 257)

A mixture of 4-*tert*-butylphthalonitrile (5.0 g) and $FeSO_4 \times H_2O$ (1.42 g) in freshly-distilled *N,N*-dimethylaminoethanol (50 mL) was heated under reflux for 12 h. The mixture was cooled, and the product precipitated and washed with MeOH:H_2O (1:1, v/v). The obtained solid was dried and purified by column chromatography (neutral Al_2O_3, deactivated by 10% H_2O;eluent: toluene). To remove traces of toluene, the product was dissolved in CH_2Cl_2 and re-precipitated by adding MeOH. Finally, it was dried at 80 °C *in vacuo* for 6 h. Yield: 2.8 g (52%).

Synthesis of [FePc(tert-Bu)$_4$]$_2$N (Ref. 261)

A mixture of FePc(*tert*-Bu)$_4$ (0.4 g) and NaN_3 (2.0 g) were suspended in oxygen-free xylene (70 mL) under argon. The mixture was heated for 6 h at 150 °C under intensive stirring, then cooled and the resulting dark blue solution separated from undissolved material by filtration. The solution was chromatographed on Al_2O_3 (neutral) with CH_2Cl_2 to remove impurities. Then, the [FePc(*tert*-Bu)$_4$]$_2$N complex was collected using a CH_2Cl_2:EtOH (100:1, v/v) mixture as eluent. Evaporation of the solvent afforded a dark blue powder. Yield: 0.25 g (62%).

Synthesis of [FePc]$_2$C (Ref. 263)

A mixture of FePc (300 mg), CI_4 (500 mg), and sodium dithionite (500 mg) in 1-chloronaphthalene (10 mL) was heated in air, with stirring, at 140–150 °C for 30 min. After cooling and filtration, the solid residue was washed with water, then with acetone, and dried under vacuum (10^{-2} mmHg) at room temperature for 2 h. Yield: 200 mg (66%).

Synthesis of Ru(II)Pc (Ref. 265)

A mixture of $RuCl_3 \times H_2O$ (2 g), *o*-cyanobenzamide (20 g) and naphthalene (8 g) was heated for 45 min at 290 °C. The cooled residue was crushed in a mortar, and extracted with pyridine in a Soxhlet apparatus until the solvent was colorless (40 h). Excess pyridine was distilled off at reduced pressure, and the impure Ru(Py)$_2$Pc purified by column chromatography (neutral alumina, $CHCl_3$) and dried (80 °C, 0.01 mmHg). Yield of Ru(Py)$_2$Pc: 3 g (50%). Next, Ru(Py)$_2$Pc (0.5 g) was slowly heated (5 °C/min) under vacuum (0.01 mmHg) to a final temperature of 250 °C, which was maintained for 6 h. The desired Ru(II)Pc was obtained in nearly quantitative yield.

Synthesis of Ru(3-ClPy)$_2$Pc(tert-Bu)$_4$ (Ref. 272)

A mixture of 4-*tert*-butylphthalonitrile (2.75 g), $RuCl_3 \times 3H_20$ (1.0 g), 3-chloropyridine (3 mL), DBU (1.5 mL) and 2-ethoxyethanol (40 mL) was refluxed for 24 h. The cooled solution was poured into a methanol:water (1:l, v/v) mixture, and the precipitate centrifuged and dried. After purification by column chromatography (SiO_2; chloroform) the crude compound was dried at 60°C *in vacuo*. Yield: 1.82 g (45%) as a purple powder.

Synthesis of Ru(pyz)Pc(tert-Bu)$_4^{\beta}$ (Ref. 275)

A mixture of 5-*tert*-butyl-1,3-diiminoisoindoline (1.5 g), $RuC1_3 \times 3H_2O$ (0.5 g) and pyrazine (1.0 g) was heated under reflux in 2-ethoxyethanol (40 mL) for 2 days. After cooling, the mixture was poured into 200 mL methanol:H_2O (3:1, v/v) mixture, and the crude product extracted with $CHCl_3$ and dried with $MgSO_4$. The solvent was evaporated, and the product purified by column chromatography (SiO_2, $CHCl_3$) and dried at 50 °C *in vacuo*. Yield: 682 mg (36%) as a violet powder.

Synthesis of Ru(Py)$_2$Pc(OBu)$_8^{\alpha\alpha}$ (Ref. 279)

A mixture of $H_2Pc(OBu)_8$ (261 mg), $Ru_3(CO)_{12}$ (181 mg), and dry pyridine (130 mL) was refluxed under Ar for 3 h and cooled. The formed suspension was filtered (Florisil bed), the filter bed was washed with pyridine (100 mL), and the filtrate and washings were combined. While being purged with Ar, the solution was irradiated with a Hg lamp (medium pressure, 450 W) for 20 h. The resulting product was evaporated to a solid with a rotary evaporator (75 °C), which was then chromatographed (Al_2O_3 I, 1.5×8 cm, EtOAc followed by warm pyridine), recrystallized (toluene), and vacuum dried (60 °C). Yield: 60 mg (20%).

Synthesis of Ru(II)(CO)Pc(C$_8$H$_{17}$)$_8^{\alpha\alpha}$ (Ref. 280)

To a flame-dried flask under a nitrogen atmosphere were added $H_2Pc(C_8H_{17})_8^{\alpha\alpha}$ (0.49 g) and triruthenium dodecacarbonyl (0.45 g). Benzonitrile (20 mL) was added, and the reaction mixture heated under reflux for 100 min. The cooled reaction mixture was poured into cold methanol (400 mL), causing the formation of some crystals. After cooling at 5 °C for a further 12 h, the excess solvent was decanted. The crude product was purified by flash column chromatography (eluting with methanol to remove benzonitrile and then with petroleum ether (40–60 °C)). Re-precipitation (petroleum ether–methanol) gave the title compound as a dark blue, amorphous solid. Yield: 0.54 g (96%).

Synthesis of Ru(II)(CO)(MeOH)Pc(15-crown-5)$_4$ (Ref. 283)

A mixture of dicyano-benzo-15-crown-5 (100 mg), and $Ru_3(CO)_{12}$ (8.4 mg) was heated in an evacuated, sealed ampoule at 250 °C for 4 h. After cooling, the dark-blue reaction mixture was dissolved in chloroform and filtered, in order to remove unreacted salt. The desired complex was isolated in 80% yield by means of column liquid chromatography on neutral aluminum oxide, using a $CHCl_3$:MeOH (95:5, v/v) mixture as an eluent.

Synthesis of OsPc (Ref. 284)

A mixture of $OsCl_3$ (1 g) *o*-cyanobenzamide (15 g), and naphthalene (10 g) was heated under nitrogen at 290 °C for 1 h and then allowed to cool. The melt was crushed with a mortar and pestle, washed with ethanol, and extracted with glacial acetic acid for 3–4 days. The residue was dried *in vacuo* at 120 °C and chromatographed on Al_2O_3 (neutral, I) by eluting with $CHC1_3$, to yield 400 mg of crude $PcOsL_x$ as a purple powder. Next, the product was heated slowly to 400 °C under an inert-gas stream (2 K/min), and this temperature was maintained for 1 h. Yield: 270 mg (11%) as blue-black powder.

10. *Group 9: CoPc, RhPc, and IrPc*

Cobalt phthalocyanine (CoPc) was first obtained in 1936 by Linsted and coworkers.[21,52] Here, the unsubstituted CoPc (**Co-1**) was synthesized by fusing of phthalonitriles with cobalt metal[21] or *via* metal exchange reaction of Li_2Pc with cobalt(II) chloride (Scheme 77, Table 43).[52] Later, a simple method for preparing **Co-1** *via* direct template condensation of phthalonitrile with a cobalt salt (chloride, acetate or acetylacetonate) in the presence of a strong organic base, such as DBU, was published (Scheme 77, Table 43).[45,286] The yield of **Co-1** is almost independent of the nature of the cobalt salt (Table 43). **Co-1** can also be prepared by direct template condensation of a phthalimide with a cobalt(II) salt in the presence of *p*-toluenesulfonic acid and hexamethyldisilazane (Scheme 77, Table 43).[287] It should be noted that the yield of **Co-1** is almost independent of the reaction time (Table 43).

In addition, **Co-1** can be prepared by using high-pressure,[42] as well as microwave irradiation[184] conditions (Scheme 77, Table 43). Also, **Co-1** could be obtained by an exotic procedure *via* a ring-contraction reaction of the uranyl superphthalocyanine with cobalt(II) chloride (Scheme 77).[152]

1,8(11),15(18),22(25)-Tetra-substituted CoPcs (**Co-2** – **Co-4**) can be prepared by two general approaches: (i) *via* direct template condensation of mono-substituted

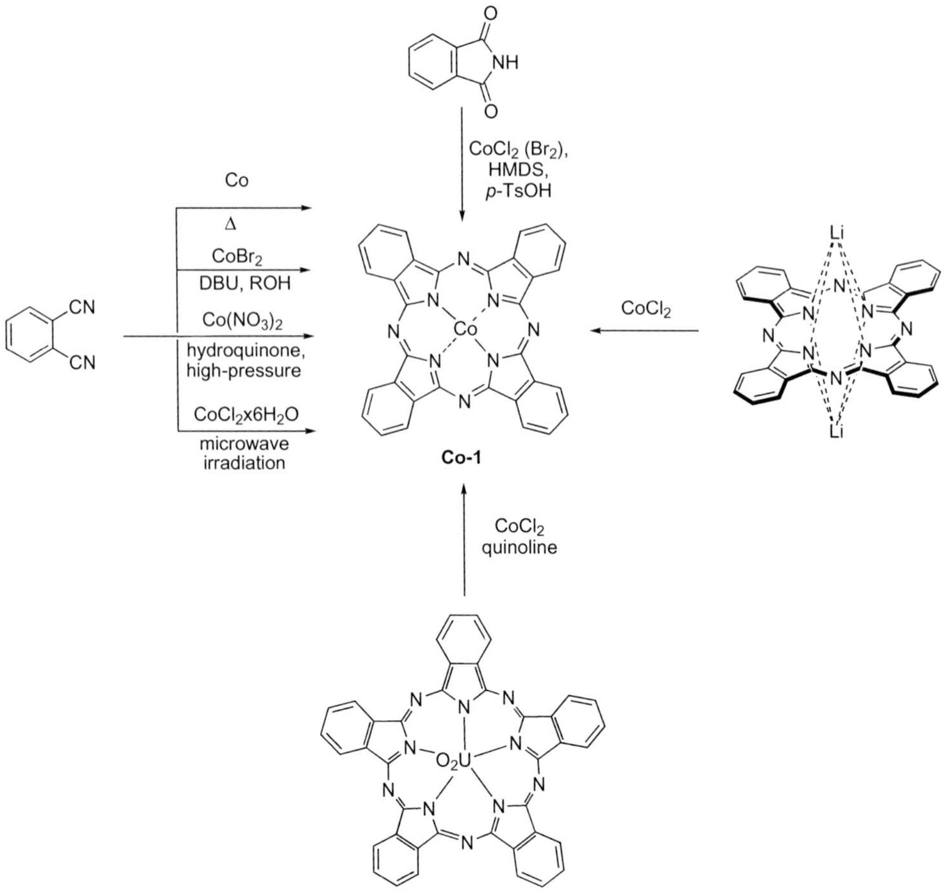

Scheme 77. The preparation of the **Co-1**.[21,42,45,52,152,184,286,287]

Table 43. Preparation of **Co-1**.

	Conditions	Yield (%)	Ref.
Co-1	PN, Co metal, melting	n.r.	21
Co-1	PN, $CoBr_2$, DBU, *n*-pentanol, refluxing, 6 h	60	45
Co-1	PN, $Co(acac)_2$, DBU, *n*-pentanol, refluxing, 6 h	53	45
Co-1	PN, $Co(OAc)_2$, DBU, MeOH, refluxing, 24 h	30	286
Co-1	PI, $CoCl_2$, *p*-TsOH, HMDS, DMF, 10 h	51	287
Co-1	PI, $CoBr_2$, *p*-TsOH, HMDS, DMF, 4 h	44	287
Co-1	PN, $Co(NO_3)_2$, *n*-octanol, hydroquinone, P = 0.001 kbar, 200 °C, 3 h	36	42
Co-1	PN, $CoCl_2$x$6H_2O$, MW (560 W), 10 min	91	184
Co-1	Li_2Pc, $CoCl_2$	n.r.	52
Co-1	$U(O_2)$SuperPc, $CoCl_2$, quinoline, 120 °C, 1.5 h	59	152

phthalogenes (phthalic acid and phthalic anhydride)[288,289] and (ii) *via* metal insertion into H_2Pc (**Co-5**) (Scheme 78, Table 44).[76]

2,9(10),16(17),23(24)-Tetra-substituted CoPcs (**Co-6** – **Co-20**) were prepared by direct template reaction of 4-substituted phthalonitriles or phthalic anhydride (**Co-7**) with cobalt(II) salts (chloride or acetate) in the presence of high-boiling solvents as reaction media, or in the absence of solvent (**Co-7**, **Co-8**, **Co-12**, **Co-14**, **Co-19**) (Scheme 79, Table 45).[24,62,64,66,189, 245,289,247,290–294] In addition, **Co-7**[290] and **Co-21**[295] were obtained *via* metal insertion into H_2Pc (Scheme 79, Table 45).

1,3,8,10(9,11),15,17(16,18),22,24(23,25)-Octa-, as well as 1,4,8,11,15,18,22, 25-octa-substituted CoPcs (**Co-22** – **Co-25**) were prepared by direct template condensation of 3,5-disubstituted phthalic acid (**Co-22**)[296] or 3,6-disubstituted phthalonitriles (**Co-23** – **Co-25**) (Schemes 80 and 81, Table 46).[297–299]

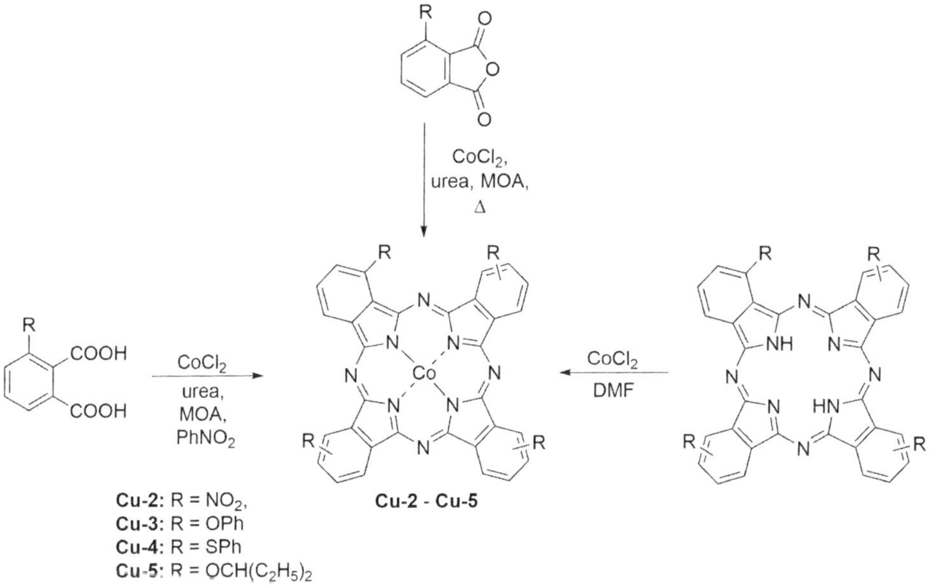

Scheme 78. Synthetic approaches for synthesis of 1,8(11),15(18),22(25)-tetra substituted **Co-2** – **Co-5**.[76,288,289]

Table 44. Preparation of 1,8(11),15(18),22(25)-tetrasubstituted **Co-2** – **Co-5**.

	Conditions	Yield (%)	Ref.
Co-2	$CoCl_2$, urea, NH_4Cl, MOA, $PhNO_2$, 190 °C, 1 h	70	288
Co-3	PA, $CoCl_2$, urea, Na_2SO_4, MOA, 160–170 °C, 1 h; 200–210 °C, 3 h	54	289
Co-4	PA, $CoCl_2$, urea, Na_2SO_4, MOA, 160–170 °C, 1 h; 200–210 °C, 3 h	38	289
Co-5	H_2PcR_4, $CoCl_2$, DMF, refluxing, 4 h	73	76

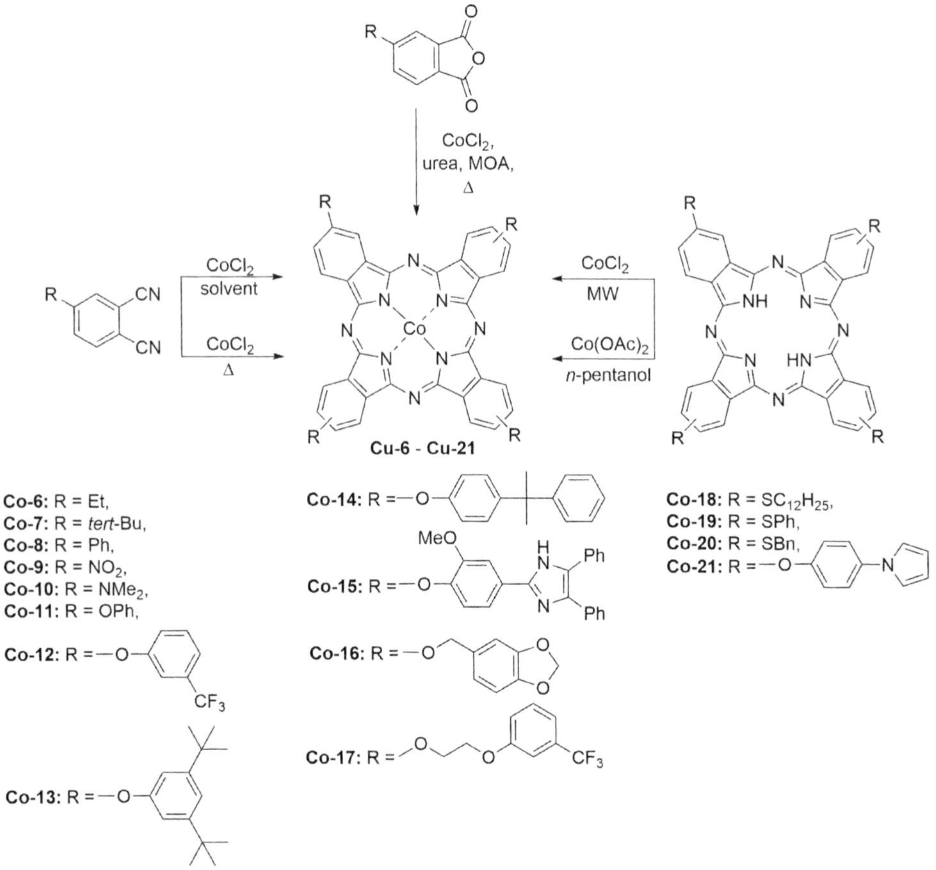

Scheme 79. Synthetic preparation of 2,9(10),16(17),23(24)-tetra-substituted **Co-6** – **Co-21**.[24,62,64,66,189,286,289,245,247,290–295]

2,3,9,10,16,17,23,24-Octa-substituted CoPcs (**Co-26** – **Co-33**) were obtained by direct template condensation of 4,5-disubstituted phthalonitriles and 5,6-disubstituted 1,3-diiminoisoindoline (**Co-30**) in the presence of cobalt(II) chloride or acetate (Scheme 82. Table 46).[24,249,251,286,297,300–302]

1,2,3,4,8,9,10,11,15,16,17,18,22,23,24,25-Hexadeca-substituted CoPcs (**Co-34**, **Co-35**) were obtained by direct template condensation of 3,4,5,6-tetra-substituted phthalonitriles in the presence of cobalt(II) chloride or acetate (Scheme 83, Table 47).[297,303]

The unsubstituted *rhodium* Pc **Rh-1** was prepared by melting phthalonitrile with rhodium(III) chloride (Scheme 84).[304] **Rh-1** could also be prepared in high yield using *o*-cyanobenzamide and rhodium(III) chloride as starting materials (Scheme 84, Table 48).[305,306] **Rh-1** was prepared using microwave irradiation[184] or an ionic liquid (*e.g.* tetrabutylammonium bromide) as the reaction medium.[269]

Table 45. Preparation of 2,9(10),16(17),23(24)-tetra-substituted **Co-6** – **Co-21**.

	Conditions	Yield (%)	Ref.
Co-6	$CoCl_2$, DMAE	56	245
Co-7	PN, $CoCl_2$, ethylene glycol, 190 °C, 5 h	73	24
Co-7	PA, $CoCl_2$, urea, MOA, TCB, 190 °C, 4 h	61	24
Co-7	H_2Pc, $CoCl_2$, MW (440 W), 10 min	56	290
Co-8	$CoCl_2$, MOA, melting, 265--280 °C, 5 h	46	62
Co-9	$Co(OAc)_2$, $PhNO_2$, 150 °C, 24 h	20	286
Co-10	$Co(OAc)_2$, urea, MOA, quinoline, 200–205 °C, 2 h	45	189
Co-11	PA, $CoCl_2$, urea, Na_2SO_4, MOA, 160--170 °C, 1 h; 200–210 °C, 3 h	56	289
Co-12	$CoCl_2$, DBU, DMF, refluxing, 9 h, sealed tube	42	292
Co-13	$CoCl_2$, DBU, DMF, refluxing, 9 h	13.8	291
Co-14	Co powder, melting, 280 °C, 12 h	35	64
Co-15	$CoCl_2$, DBU, DMAE, 160 °C, 24 h	33	293
Co-16	$CoCl_2$, hexanol, DMAE, 160 °C, 8 h	40	66
Co-17	$CoCl_2$, DBU, *n*-pentanol, 160 °C, 24 h, sealed tube	43	247
Co-18	$CoCl_2$, ethylene glycol, refluxing, 200 °C, 4 h	33	294
Co-19	PA, $CoCl_2$, urea, Na_2SO_4, MOA, 160--170 °C, 1 h; 200--210 °C, 3 h	48	289
Co-20	$CoCl_2$, ethylene glycol, refluxing, 200 °C, 4 h	47	294
Co-21	H_2Pc, $Co(OAc)_2$, *n*-pentanol, refluxing, 1.5 h	21	295

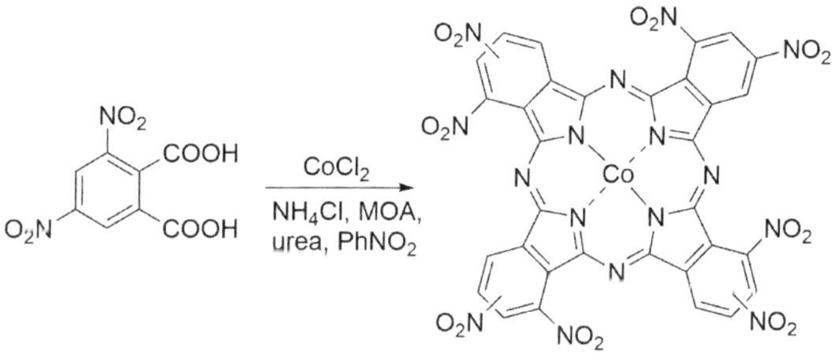

Co-22

Scheme 80. Synthetic preparation of 1,3,8,10(9,11),15,17(16,18),22,24(23,25)-octanitro-**Co-22**.[296]

Unsubstituted RhPcs also can be prepared as hexa-coordinated complexes (**Rh-2a–2d**) with nitrogenous bases (*e.g.* *n*-butylamine (**a**), Py (**b**), 4,4′-bipyridine (**c**), pyrazine (**d**)) *via* metal exchange reaction of Li_2Pc with di-μ-chloro-bis(1,5-cyclooctadiene)rhodium (Scheme 85. Table 48).[309]

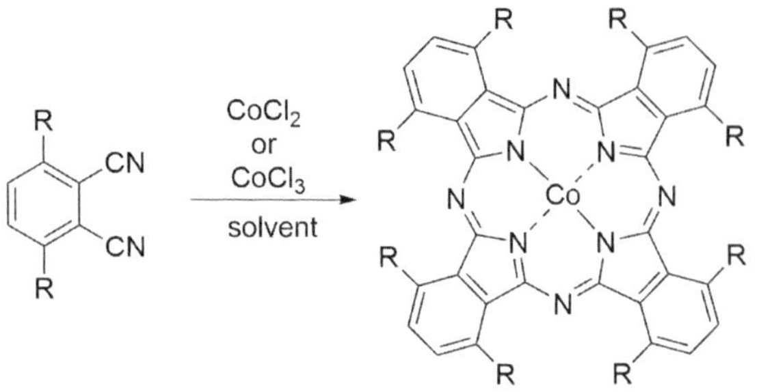

Co-23: R = C_7H_{15},
Co-24: R = $O(CH_2)_2CH(Me_2)$
Co-25: R = OBu

Co-23 - Co-25

Scheme 81. Synthetic preparation of 1,4,8,11,15,18,22,25-octa-substituted **Co-23** – **Co-25**.[297–299]

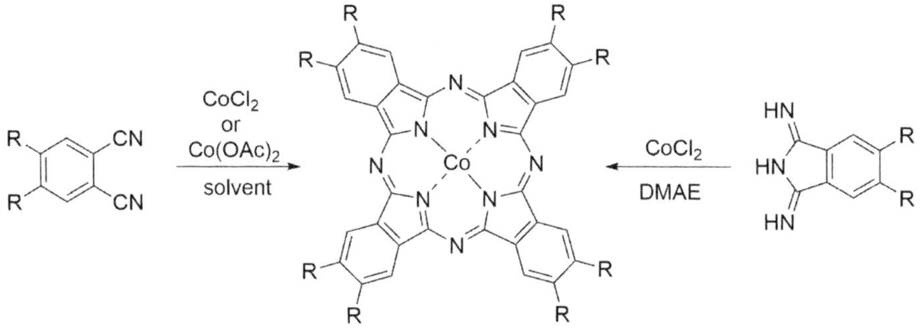

Co-26 - Co-33

Co-26: R = Me,
Co-27: R = C_7H_{15},
Co-28: R = CN,
Co-29: R = OMe,
Co-30: R = $OCH_2C(O)NH(C_8H_{17})_2$
Co-31: R = SC_6H_{13},
Co-32: R = SBn
Co-33: R = 15-crown-5

Scheme 82. Synthetic preparation of 2,3,9,10,16,17,23,24-octa-substituted **Co-26** – **Co-33**.[24,249,251,286,297,300–302]

2,9(10),16(17),23(24)-Tetra- (**Rh-3**, **Rh-4**) and 2,3,9,10,16,17,23,24-octa-substituted RhPcs (**Rh-5**) were obtained by template condensation of the corresponding mono-[304,307] or disubstituted[308] phthalonitriles using rhodium(III) chloride[307,308] or rhodium carbonyl chloride[304] as a metal source (Scheme 86, Table 48).

Unsubstituted *iridium(III)* Pc (**Ir-1**) can be obtained by template condensation of phthalonitrile with iridium(III) chloride (Scheme 87, Table 49).[310]

1,8(11),15(18),22(25)-Tetra-*tert*-butyl IrPc (**Ir-2**) was obtained *via* direct metalation of metal-free tetra-(*tert*-butyl)phthalocyanine with di-μ-chloro-bis-(1,5-cyclooctadiene)iridium (Scheme 88, Table 49).[311]

Table 46. Preparation of octa-substituted **Co-22 – Co-33**.

	Conditions	Yield (%)	Ref.
Co-22	$CoCl_2$, NH_4Cl, MOA, urea, $PhNO_2$, 180 °C, 30–45 min	50–60	296
Co-23	$CoCl_3$, ethylene glycol, refluxing, 24 h	58	297
Co-24	$CoCl_2$, DBU, *n*-pentanol, 160 °C, 10 h	20	298
Co-25	$Co(OAc)_2$, DBU, *n*-pentanol, 160 °C, 10 min	9	299
Co-26	$Co(OAc)_2$, ethylene glycol, refluxing, 24 h	30	286
Co-27	$CoCl_3$, ethylene glycol, refluxing, 24 h	63	297
Co-28	$Co(OAc)_2$, DBU, sulfolane, 135 °C, 2 h	69	300
Co-28	$Co(OAc)_2$, DMF, 150 °C, 2 h	50	286
Co-29	$CoCl_2$, ethylene glycol, 200 °C, 5 h	41	24
Co-30	DI,$CoCl_2 \times 6H_2O$, DMAE, refluxing, 7.5 h	30	301
Co-31	$CoCl_2$, ethylene glycol, 200 °C, 4 h	26	302
Co-32	$CoCl_2$, ethylene glycol, 200 °C, 4 h	55	249
Co-33	$CoCl_2 \times 6H_2O$, ethylene glycol, 185–190 °C, 5 h	45.5	251

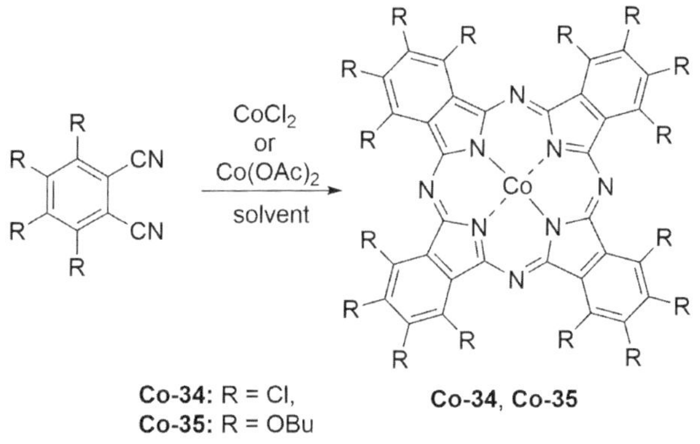

Scheme 83. Synthetic preparation of hexadeca-substituted **Co-34**, **Co-35**.[299,303]

Table 47. Preparation of hexadeca-substituted **Co-34**, **Co-35**.

	Conditions	Yield (%)	Ref.
Co-34	$CoCl_2 \times 4H_2O$, MOA, $PhNO_2$, 200 °C, 4–8 h	n.r.	303
Co-35	$Co(OAc)_2$, DBU, *n*-pentanol, 180 °C, 2 h	28	299

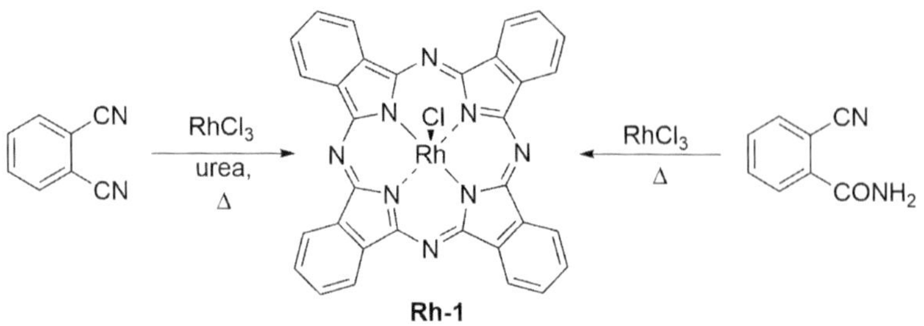

Scheme 84. The Preparation of unsubstituted rhodium phthalocyanine.[304–306]

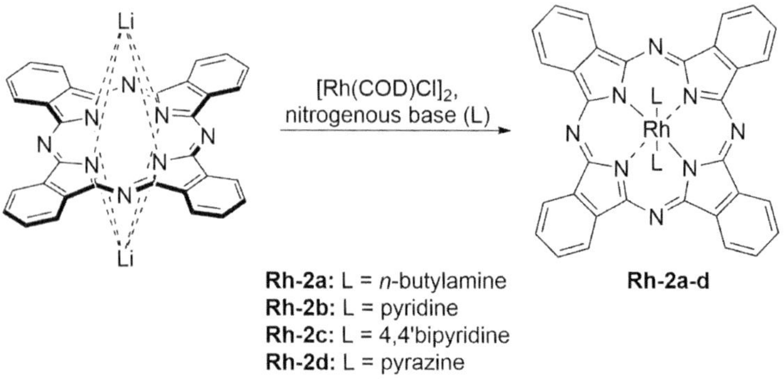

Scheme 85. Preparation of hexa-coordinated **Rh-2a–2d.**[309]

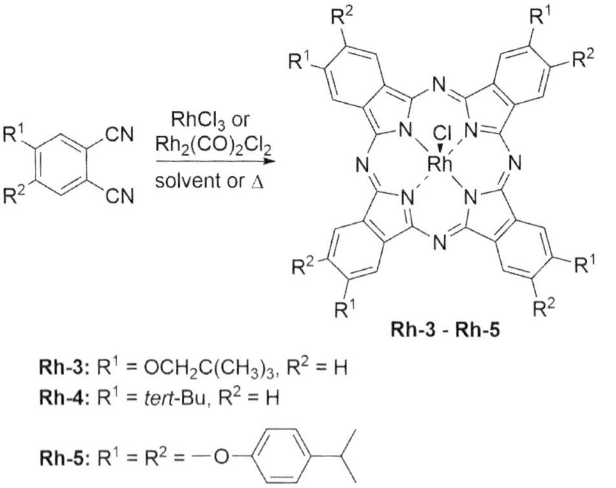

Scheme 86. Synthesis of tetra- and octa-substituted **Rh-3** – **Rh-5**.[304,307,308]

Table 48. Preparation of **Rh-1** – **Rh-5**.

	Conditions	Yield (%)	Ref.
Rh-1	PN, $RhCl_3 \times 2H_2O$, urea, melting, 15 min	n.r.	304
Rh-1	CBA, $RhCl_3 \times 3H_2O$, melting, 280 °C, 4 h	90	305
Rh-1	CBA, $RhCl_3$, naphthalene, melting, 1 h	60	306
Rh-1	PN, urea, MOA, TBAB, 10 min	98	269
Rh-2	Li_2Pc, $[Rh(COD)Cl]_2$, nitrogenous base, 100 °C, 24 h (**a**); 48 h (**b**); 72 h (**c**); 21 h (**d**)	45 (**a**), 71(**b**), 94 (**c**), 78 (**d**)	309
Rh-3	$Rh_2(CO)_2Cl_2$, DMAE, 165 °C, 48 h	7	304
Rh-4	$RhCl_3$, MOA, sealed tube, 270 °C, 2 h	n.r.	307
Rh-5	$RhCl_3 \times 3H_2O$, DBU, 2-ethoxyethanol, refluxing, 22 h	59	308

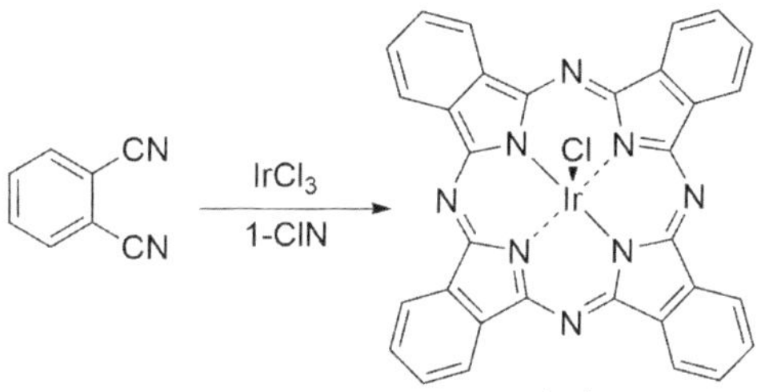

Scheme 87. Synthesis of unsubstituted iridium Pc (**Ir-1**).[310]

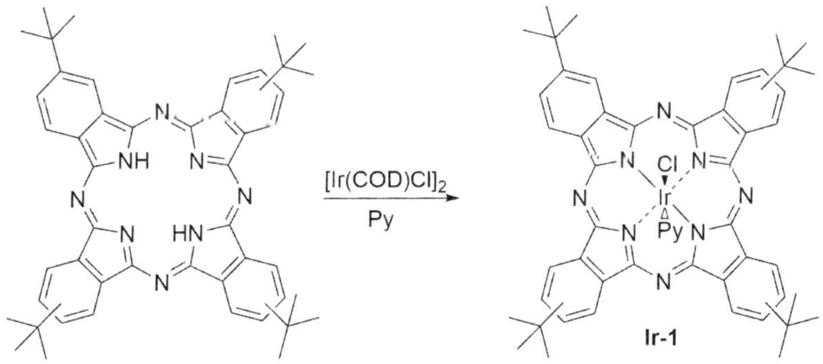

Scheme 88. Preparation of tetra-substituted IrPc (**Ir-2**).[311]

Table 49. Preparation of **Ir-1**, **Ir-2**.

	Conditions	Yield (%)	Ref.
Ir-1	$IrCl_3$, 1-ClN, refluxing, 2 h	<16	310
Ir-2	$[Ir(COD)Cl]_2$, Py, refluxing, 3 d	13	311

Synthesis of Co(II)Pc (Ref. 286)

Phthalonitrile (29.0 g) and cobalt(II) acetate (12.0 g) were dissolved in dehydrated MeOH (1000 mL). DBU (28 mL) was added dropwise into the methanol solution under stirring at r.t., then the solution was refluxed for 24 h. After cooling, the red-purple colored precipitate was filtered, and the crude product boiled in 10% HCl then in 10% NaOH aqueous solution and washed with water until the aqueous solution became neutral. The washed precipitate was dried *in vacuo*. Yield: 10 g (30%).

Synthesis of Co(II)Pc(tert-Bu)$_4^{\beta}$ (Ref. 24)

4-*tert*-Butylphthalodinitrile (13.5 g) and cobalt(II) chloride (1.5 g) were suspended in ethylene glycol (180 mL) and heated for 5 h at 190 °C. Finally, the mixture was cooled, then an equal volume of water added, and the solid material collected on a filter (vacuum) and washed with methanol. The crude product was boiled for 2 h, first with 1 N HCl (500 mL), and then with 1 N NaOH (500 mL), filtered, and washed with water each time until the filtrate was neutral. The solid material was then stirred in methanol, suction-filtered, and dried *in vacuo* for 6 h at 60 °C. Yield: 7.0 g (73.4).

Synthesis of Co(II)Pc(C_7H_{15})$_8^{\alpha\alpha}$ (Ref. 297)

3,6-Diheptylphthalonitrile (1.5 g) and cobalt trichloride (0.5 g) were refluxed for 24 h in ethylene glycol (25 mL). The cooled dark blue solution was poured into methanol (100 mL), and the formed precipitate centrifuged, washed with acetone, and dried. Further purification was performed by column chromatography (SiO_2, toluene). The blue fraction was collected, and the solvent evaporated. Yield: 0.910 g (58%) as waxy blue powder.

Synthesis of Co(II)Pc(CN)$_8^{\beta\beta}$ (Ref. 300)

A mixture of 1,2,4,5-tetracyanobenzene (3.0 g) and cobalt(II) acetate (0.89 g) was reacted in sulfolane (50 mL) at 135 °C for 2 h in the presence of a few drops of

DBU. The reaction mixture was precipitated from distilled methanol, and the resulting dark green solid collected, washed with ether, and dried *in vacuo*. The purified product was obtained by Soxhlet extraction with methanol, followed by gel permeation chromatography (Bio-beads SX-10) using DMF as eluent. Yield: (69%).

Synthesis of Co(II)Pc(OMe)$_4^{\beta\beta}$ (Ref. 24)

4,5-Dimethoxyphthalonitrile (13.77 g) and cobalt(II) chloride (1.50 g) were heated in ethylene glycol (180 mL) at 200 °C for 5 h, after which the mixture was cooled somewhat, and an equal volume of water added. The mixture was suction filtered while still hot, and the filtrate washed with hot water. Purification was carried out as described above with 1 N HCI and 1 N NaOH. Yield: 4 g (41.1%).

Synthesis of Co(II)Pc[(OCH$_2$C(O)NH(C$_8$H$_{17}$)$_2$]$_8^{\beta\beta}$ (Ref. 301)

1,3-Diimino-5,6-bis[((dioctylamino)carbonyl)methoxy)]isoindoline (170 mg) and $CoCl_2 \times 6H_2O$ (14.3 mg) in DMAE (0.17 mL) were heated at reflux temperature under argon for 7.5 h. After cooling, the crude product was extracted with a hexane:MeOH:H_2O (8:5:3, v/v/v) two-phase system, then purified by column chromatography on SiO_2 (AcOEt:MeOH, 10:1, v/v), and washed with CH_3CN:H_2O (5:1, v/v). Finally, the green-bluish product was dried at 140 °C/10^{-3} Torr. Yield: 52 mg (30%).

Synthesis of Co(II)Pc(SC$_6$H$_{13}$)$_8^{\beta\beta}$ (Ref. 302)

A mixture of 4,5-bis(hexsylsulfanyl)phthalonitrile (1.0 g), anhydrous $CoCl_2$ (0.091 g) and ethylene glycol (5.5 mL) was heated and stirred at 200 °C for 4 h under argon. After cooling to room temperature, the reaction mixture was treated with ethanol to precipitate the dark green product, and then filtered. The product was purified by column chromatography (SiO_2, chloroform). Yield: 0.270 g (26%).

Synthesis of Co(II)Pc(OBu)$_{16}$ (Ref. 299)

A mixture of 3,4,5,6-tetrabutoxyphthalonitrile (0.147 g) and cobalt chloride (0.065 g) in *n*-pentanol (20 mL) was refluxed under a nitrogen atmosphere. DBU (0.15 mL) was added, and reaction mixture heated to first distill off pentanol. After continuing the reaction for 2 h at 180 °C, water (50 mL) was added, and the product extracted with hexane. The solution was dried with anhydrous sodium sulfate, filtered, and the product purified by column chromatography (SiO_2, hexane:ether 1:1, v/v). Yield: 0.12 g (28%).

Synthesis of Rh(III)(Cl)Pc (Ref. 305)

A mixture $RhC1_3 \times 3H_2O$ (3.7 g) and *o*-cyanobenzamide (17.5 g) was heated at 280 °C for 4 h. Water and then hydrogen chloride were evolved from the molten reaction mixture. The purplish product was successively Soxhlet extracted with ethanol (1 h), benzene (1 h) and acetone (36 h), and the blue-green residue dried at 110 °C after washing with ether. Yield: (90%).

Synthesis of Rh(III)(Cl)Pc (Ref. 306)

A mixture of $RhCl_3$ (0.2 g), *o*-cyanobenzamide (1.8 g) and naphthalene (0.6 g) was heated for 1 h, and the resulting blue solid Soxhlet-extracted with glacial acetic acid until the washings were colourless, then dried at 100 °C for 12 h to remove traces of acetic acid. Yield: (60%).

Synthesis of Rh(III)(Cl)Pc$[OCH_2C(CH_3)_3]_8^{\beta}$ (Ref. 304)

4-Neopentoxyphthalonitrile (1.28 g) and $Rh_2(CO)_2C1_2$ (0.25 g) were heated in (dimethy1amino)ethanol (25 mL) at 165 °C for 48 h. The crude reaction mixture was poured into water, and the resulting precipitate filtered off, washed, and dried overnight, to yield 1.5 g of crude green solid. This solid was preabsorbed onto flash silica gel and eluted with 1:l hexane:chloroform, to give, upon evaporation, a bluish green solid. Yield: 0.12 g (7%) of a dark blue shining powder.

Synthesis of Rh(Cl)Pc$[OPh\text{-}4\text{-}CH_2(CH_3)_2]_8^{\beta\beta}$ (Ref. 308)

A suspension of rhodium(III) chloride trihydrate (0.3 g) in deoxygenated 2-ethoxyethanol (10 mL) was boiled under a stream of dry nitrogen gas until the water was removed, then added to a solution of 4,5-bis(4-isopropylphenoxy)phthalonitrile (1.5 g) and DBU (0.45 g) in deoxygenated 2-ethoxyethanol (10 mL). The mixture was refluxed for 22 h under dry nitrogen. A dark blue solution was formed, cooled to room temperature, and then poured into $MeOH:H_2O$ (3:1). The suspended solution was centrifuged, and the precipitate filtered off, washed with a methanol/water mixture (1/ 1, v/v), and dried under vacuum. The crude product (1.2 g) was subjected to column chromatography over silica gel, using a mixture of *n*-hexane/dichloromethane (5:2, v/ v). Yield: 1.02 g (59%) as a blue violet powder.

Synthesis of Ir(III)(Cl)Pc (Ref. 310)

Phthalonitrile (857 mg) was dissolved in dry 1-chloronaphtalene (5 mL) at about 150 °C, and iridium(III) chloride (500 mg) added. The reaction mixture was refluxed for 2 h, cooled to room temperature and filtered. The green filtrate was added to concentrated sulfuric acid (250 mL). The residue (374 mg) turned out to be unreacted iridium chloride. The sulfuric acid solution was diluted to a total volume of 2 L, filtered, washed with water, and dried for 7 h at 150 °C and 0.1 mbar. Yield: 45 mg (<16%, based on reacted amount of iridium chloride).

Synthesis of Ir(Cl)Pc(Py) (Ref. 311)

Under a nitrogen atmosphere, a mixture of tetra-*tert*-butyl-phthalocyanine (50.0 mg) and $[Ir(COD)Cl]_2$ (114 mg) was heated under reflux in pyridine (8 mL) for 3 days. After removal of the solvent under reduced pressure, the crude product was purified by silica gel column chromatography ($CH_2Cl_2 \rightarrow CH_2Cl_2$/MeOH = 1:20), to give the target complex. Yield: 9.5 mg (13%)

11. *Group 10: NiPc, PdPc, and PtPc*

The unsubstituted nickel phthalocyanine **Ni-1** was prepared for the first time by Linstead and coworkers in 1936.[21] Here, fusion of *o*-cyanobenzamide in the presence of nickel foil afforded **Ni-1** (Scheme 89, Table 50).[21] Later, Elvidge and Linstead published a procedure for the preparation of **Ni-1** in high yield by using 1,3-diiminoisoindoline as a starting material (Scheme 89, Table 50).[20] A simple method for the preparation of **Ni-1** was published by Tomoda *et al.* in 1983.[29] **Ni-1** was also obtained *via* direct template condensation of unsubstituted phthalonitrile and a nickel(II) salt (bromide or acetylacetonate) in the presence of a strong organic base, such as DBU (Scheme 89, Table 50).[29] In addition, **Ni-1** can be prepared by template condensation of phthalonitrile with nickel(II) salt in the absence of DBU, or by using reducing agents such as hydroquinone (Table 50).[41] Phthalimide can be used as a starting material for preparing **Ni-1** with good yield (Scheme 89, Table 50).[287] In addition, **Ni-1** can also be prepared by using high-pressure conditions,[42] microwave irradiation,[184] ring contraction reaction of $U(O_2)SPc$[152] or nickel insertion into free H_2Pc[41] (Schemes 89 and 90, Table 50). The yield of **Ni-1** depends on the nickel salt and reaction conditions (Table 50).

1,8(11),15(18),22(25)-Tetra-substituted NiPcs (**Ni-2**, **Ni-3**) have been obtained by the nickel metal insertion reaction into metal-free Pcs[60,76] (Scheme 91, Table 51). 2,9(10),16(17),23(24)-Tetra-substituted NiPcs (**Ni-4** – **Ni-11**) can be obtained by two approaches: (i) *via* direct template condensation of 4-substituted phthalonitriles with nickel(II) salts (**Ni-4** – **Ni-10**) and (ii) metal insertion into H_2Pcs (**Ni-11**) (Scheme 92, Table 51).[62–64,291–293,295]

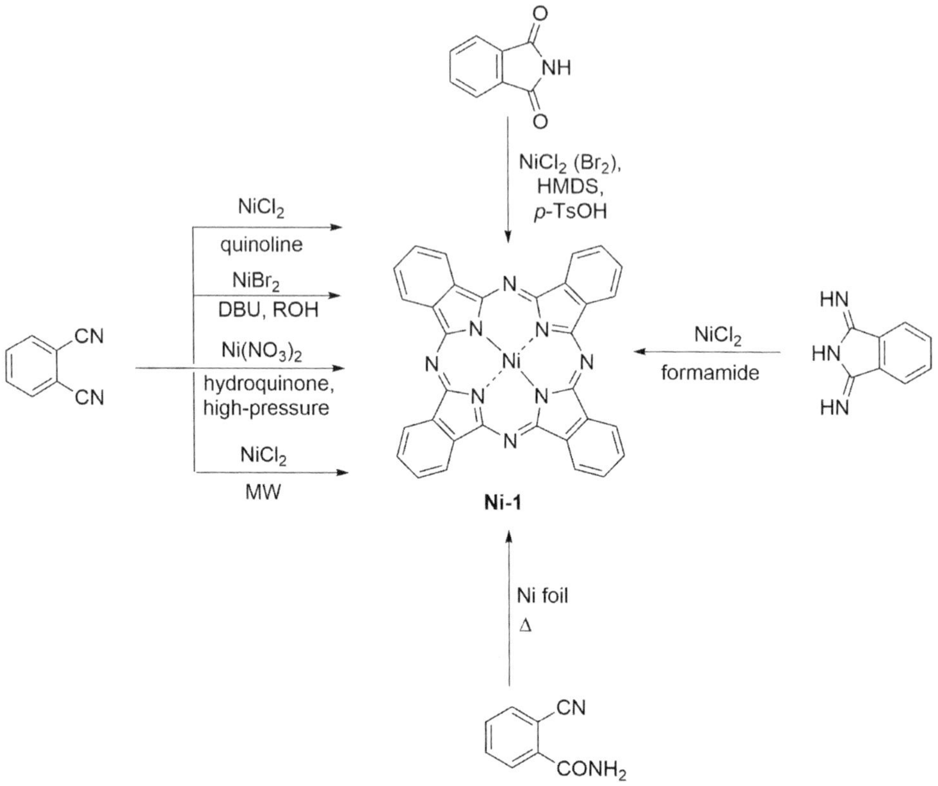

Ni-1

Scheme 89. The preparation of unsubsituted **Ni-1** *via* direct template condensation.[20,21,29,41,42,184,287]

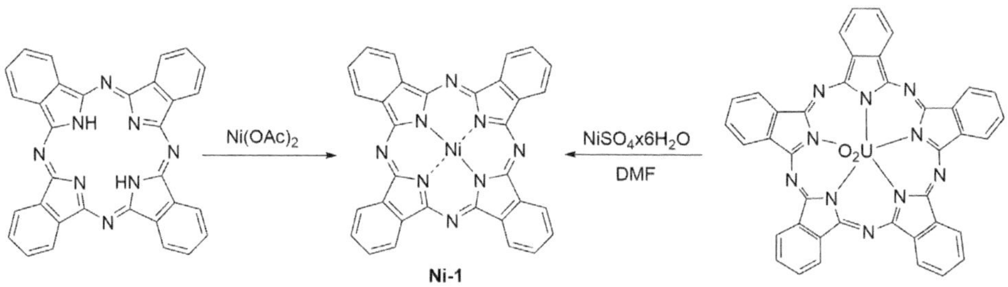

Ni-1

Scheme 90 The preparation of unsubsituted **Ni-1** by metalation into metal-free Pc or ring contraction reaction.[41,152]

1,4,8,11,15,18,22,25-Octa-substituted (**Ni-12** – **Ni-14**),[298,312,313] as well as 2,3,9,10,16,17,23,24-octasubstituted NiPcs (**Ni-15** – **Ni-23**)[153,195,251,302,314,315] were prepared by direct template condensation of the corresponding 3,6- or 4,5-disubstituted phthalonitriles (**Ni-12**, **Ni-13**, **Ni-15** – **Ni-22**) or 1,3-diiminoisoindolines (**Ni-14**, **Ni-23**) in the presence of nickel(II) chloride as the metal source (Schemes 93 and 94, Table 52).

Table 50. Preparation of unsubstituted **Ni-1**.

	Conditions	Yield (%)	Ref.
Ni-1	CBA, Ni foil, melting, 270 °C, 3 h	20	21
Ni-1	DI, $NiCl_2$, formamide, refluxung, 24 h	96	20
Ni-1	PN, $NiBr_2$, DBU, 2-methoxyethanol, refluxing, 6 h	75	29
Ni-1	PN, $Ni(acac)_2$, DBU, *n*-pentanol, refluxing, 6 h	59	29
Ni-1	PN, $Ni(OAc)_2$, hydroquinone, quinoline, refluxing, 5 h	21	41
Ni-1	PN, $NiCl_2$, quinoline, refluxing, 10 min	46	41
Ni-1	PN, $Ni(OAc)_2$, quinoline, refluxing, 3 h	23	41
Ni-1	H_2Pc, $NiCl_2$x$6H_2O$, quinoline-1-ClN (1:9, v/v), refluxing, 72 h	56	41
Ni-1	PA, $NiCl_2$, HMDS, *p*-TsOH, DMF, 150 °C, 4 h	53	287
Ni-1	PA, $NiBr_2$, HMDS, *p*-TsOH, DMF, 150 °C, 10 h	45	287
Ni-1	PN, $Ni(NO_3)_2$, hydroquinone, MeOH, P = 11.5 kbar, 100 °C, 1.75 h	46	42
Ni-1	PN, $NiCl_2$, MW (560 W), 10 min	78	184
Ni-1	$U(O_2)$SuperPc, $NiSO_4$x$6H_2O$, DMF, 140 °C, 50 h	82	152

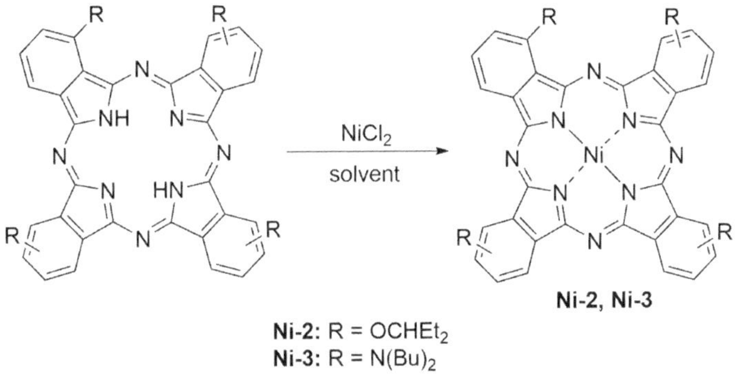

Scheme 91. Preparation of 1,8(11),15(18),22(25)-tetra-substituted **Ni-2**, **Ni-4**.[60,76]

1,3,8,10(9,11),15,17(16,18),22,24(23,25)-Octa-substituted NiPcs (**Ni-24**, **Ni-25**) were yielded *via* direct template condensation, using 3,5-disubstituted phthalonitrile (**Ni-24**) or 3,5-disubstituted phthalic acid (**Ni-25**) as the starting materials in the presence of nickel(II) chloride as the metal source (Scheme 95, Table 52).[296,316]

The hexadeca-alkoxy-substituted NiPcs **Ni-26** – **Ni-32** were produced by direct template condensation of tetraalkoxyphthalonitriles with nickel(II) chloride as a metal source (Scheme 96, Table 53).[317]

The first method for preparing an unsubstituted *palladium* phthalocyanine (PdPc) was published in 1938 by Linstead and coworkers (Scheme 97).[52] Here,

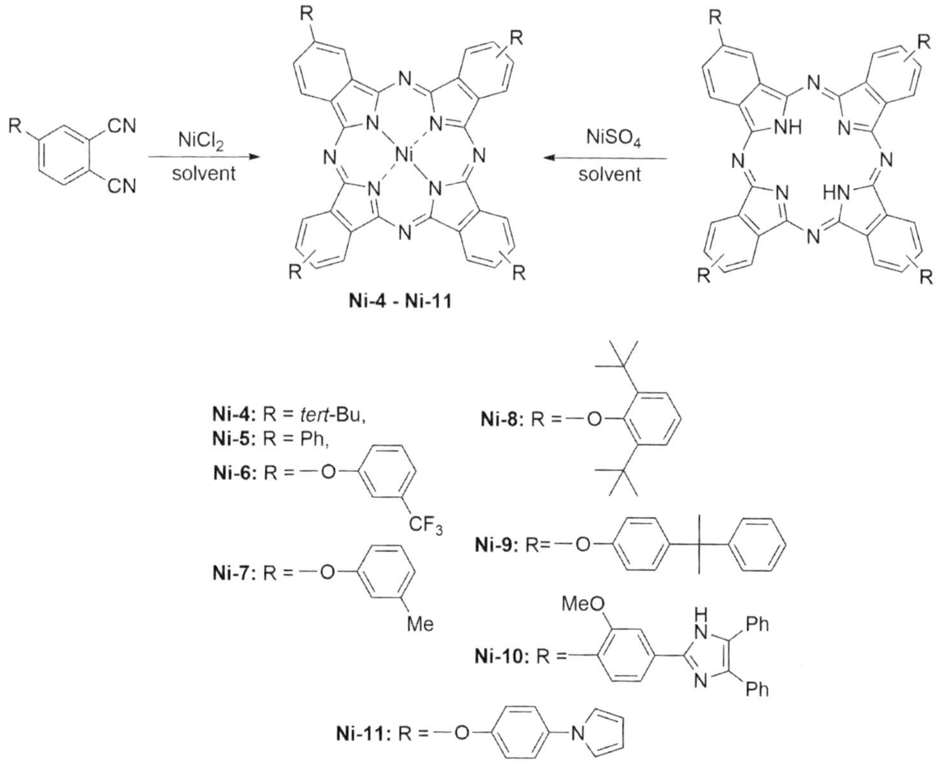

Scheme 92. Preparation of 2,9(10),16(17),23(24)-tetra-substituted **Ni-5** – **Ni-11**.[62–64,291–293,295]

Table 51. Preparation of tetra-substituted **Ni-2** – **Ni-11**.

	Conditions	Yield (%)	Ref.
Ni-2	H_2PcR_4, $NiCl_2 \times 6H_2O$, DMF, refluxing, 4 h	67	76
Ni-3	H_2PcR_4, DBU, $Ni(OAc)_2 \times 4H_2O$, *n*-pentanol, refluxing, 6 h	74.5	60
Ni-4	PN, $NiCl_2$, melting, 200–215 °C, 1 h	31	63
Ni-5	PN, $NiCl_2$, MOA, melting, 265–280 °C, 5 h	38	62
Ni-6	PN, $NiCl_2$, DBU, *n*-pentanol, refluxing, 40 h	71	292
Ni-7	PN, $NiCl_2$, DBU, *n*-pentanol, refluxing, 40 h	65	292
Ni-8	PN, $NiCl_2$, DBU, DMF, refluxing, 9 h	21.8	291
Ni-9	PN, Ni powder, melting, 280 °C, 44 h	76	64
Ni-10	PN, $NiCl_2$, DBU, DMAE, 160 °C, 24 h	31	293
Ni-11	H_2PcR_4, $NiSO_4$, *n*-pentanol, refluxing, 1.5 h	20	295

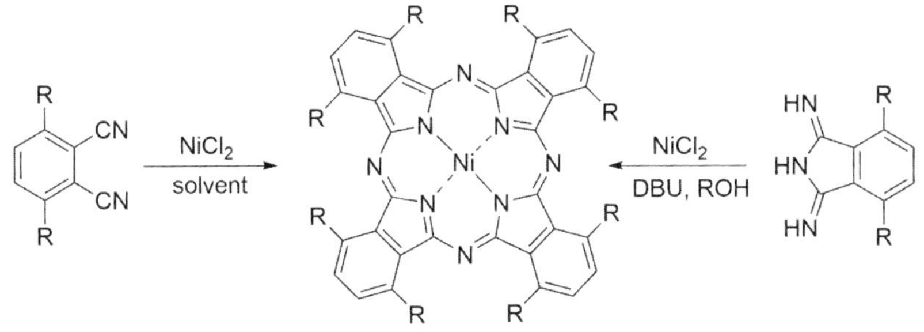

Ni-12 - Ni-14

Ni-12: R = C_7H_{15},
Ni-13: R = $O(CH_2)_2CH(Me_2)$,
Ni-14: R = OCH_2SiMe_3

Scheme 93. Preparation of 1,4,8,11,15,18,22,25-octa-substituted **Ni-12** – **Ni-14**.[298,312,313]

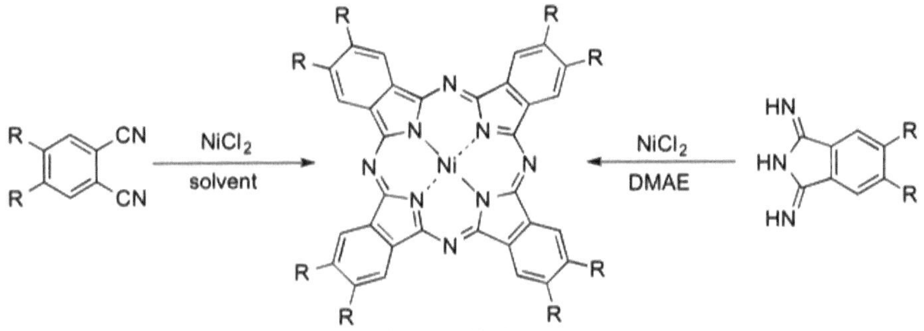

Ni-15 - Ni-23

Ni-15: R = Me,
Ni-16: R = Bu,
Ni-17: R = Ph,
Ni-18: R = SC_5H_{11},
Ni-19: R = SC_6H_{13},
Ni-20: R = SC_8H_{17},
Ni-21: R = $SC_{12}H_{25}$,
Ni-22: RR = 15-crown-5,
Ni-23: R = $CH_2OC_8H_{17}$

Scheme 94. Preparation of 2,3,9,10,16,17,23,24-octa-substituted (**Ni-15** – **Ni-23**).[153,195,251,302, 314,315]

Pd-1 was obtained by metal exchange of Li_2Pc with palladium(II) chloride in absolute ethanol (Table 54). In addition, for the preparation of **Pd-1**, other phthalogenes, such as phthalonitrile[28] or 1,3-diiminoisoindoline,[318] could be used (Scheme 97). In both cases, **Pd-1** was obtained in good yield (Table 54). In addition, **Pd-1** was also produced under microwave- assisted conditions (Scheme 97, Table 54).[184]

1,8(11),15(18),22(25)-Tetra-substituted PdPcs (**Pd-2** – **Pd-7**) have been obtained *via* direct template condensation of the corresponding 3-substituted phthalonitriles and palladium(II) chloride as the metal source in the presence of alcohols[313,319] (*e.g.* *n*-pentanol) as a reaction media (Scheme 98, Table 55).

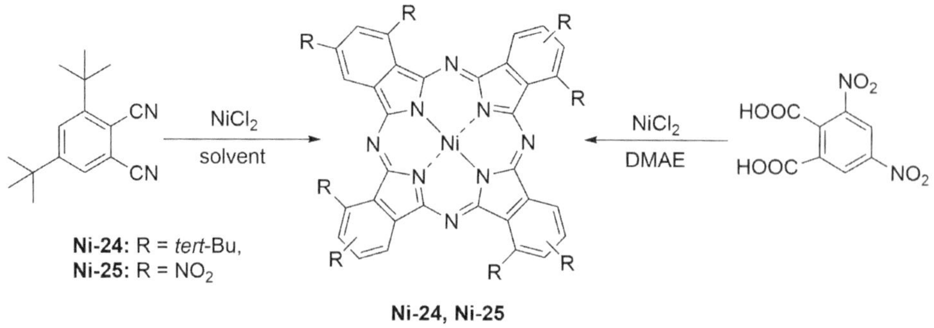

Scheme 95. Preparation of 1,3,8,10(9,11),15,17(16,18),22,24(23,25)-octa-substituted (**Ni-24**, **Ni-25**).[296,316]

Table 52. Preparation of octa-subsituted **Ni-12** – **Ni-25.**

	Conditions	Yield (%)	Ref.
Ni-12	PN, $Ni(OAc)_2$x$4H_2O$, DBU, *n*-pentanol, refluxing, 48 h	20	312
Ni-13	PN, $NiCl_2$, DBU, *n*-pentanol, 160 °C, 10 h	23	298
Ni-14	DI, $NiCl_2$, DBU, *n*-pentanol, refluxing, 60 h	8.4	313
Ni-15	PN, $NiCl_2$x$6H_2O$, quinoline, 200–230 °C, 2.5 h	57	153
Ni-16	PN, $NiCl_2$x$6H_2O$, quinoline, 200–230 °C, 2.5 h	26	153
Ni-17	PN, $NiCl_2$, MOA, 1-bromonaphthalene, 240–250 °C, 6 h	38	195
Ni-18	PN, $NiCl_2$, DBU, *n*-pentanol, refluxing, 1 h	80	314
Ni-19	PN, $NiCl_2$, quinoline, 200 °C, 5 h	38	302
Ni-20	PN, $NiCl_2$, DBU, *n*-pentanol, refluxing, 1 h	60	314
Ni-21	PN, $NiCl_2$, quinoline, 200 °C, 5 h	19	302
Ni-22	PN, $NiCl_2$, quinoline, 195–200 °C, 2.5 h	49.71	251
Ni-23	DI, $NiCl_2$, DMAE, refluxing, 7 h	33	315
Ni-24	PN, $Ni(OAc)_2$x$2H_2O$, DBU, *n*-pentanol, 130 °C, 4 d	86	316
Ni-25	3,5-R_2-Phthalic acid, $NiCl_2$, NH_4Cl, MOA, urea, $PhNO_2$, 180 °C, 30–45 min	50–60	296

2,9(10),16(17),23(24)-Tetra-substituted PdPcs (**Pd-8** – **Pd-10**) were obtained by melting the corresponding 4-substituted phthalonitriles and palladium(II) chloride in the presence of ammonium molybdate as a catalyst (Scheme 99, Table 56).[62–64]

1,4,8,11,15,18,22,25-Octa-substituted PdPcs (**Pd-11** – **Pd-15**) can be obtained by two synthetic approaches: (i) by direct template condensation of 3,6-disubstituted phthalonitriles (**Pd-11** – **Pd-13**)[320,321] or (ii) by metal insertion into H_2Pcs (**Pd-14**, **Pd-15**)[279,322] (Scheme 100, Table 57).

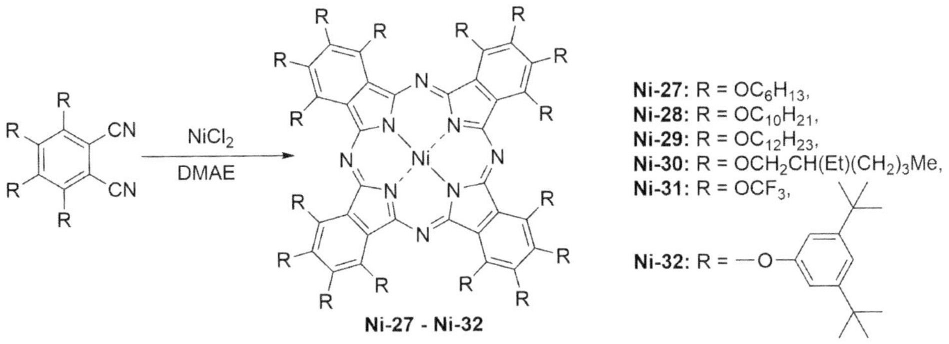

Scheme 96. Preparation of hexadeca-substituted (**Ni-26** – **Ni-32**).[317]

Table 53. Preparation of octa-subsituted **Ni-26** – **Ni-32**.

	Conditions	Yield (%)	Ref.
Ni-27	$NiCl_2$, DMAE, refluxing, 48 h	17	317
Ni-28	$NiCl_2$, DMAE, refluxing, 48 h	9	317
Ni-29	$NiCl_2$, DMAE, refluxing, 48 h	15	317
Ni-30	$NiCl_2$, DMAE, refluxing, 48 h	18	317
Ni-31	$NiCl_2$, DMAE, refluxing, 48 h	12	317
Ni-32	$NiCl_2$, DMAE, refluxing, 48 h	16	317

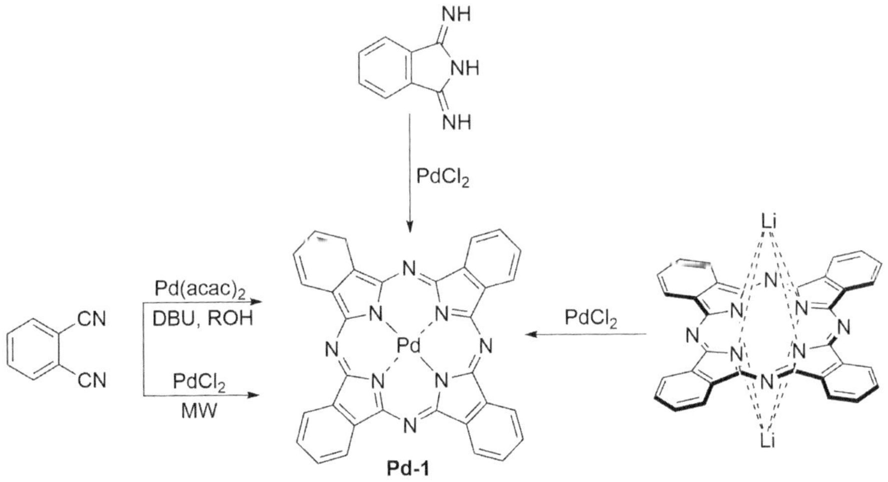

Scheme 97. Synthetic approaches for Synthesis of the unsubstituted **Pd-1**.[28,52,184,318]

Table 54. Preparation of unsubstituted **Pd-1**.

	Conditions	Yield (%)	Ref.
Pd-1	Li_2Pc, $PdCl_2$, EtOH	n.r.	52
Pd-1	DI, $PdCl_2$, urea, MOA, $PhNO_2$, refluxing, 4 h	68	28
Pd-1	PN, $Pd(acac)_2$, *n*-pentanol, refluxing, 24 h	52	318
Pd-1	PN, $PdCl_2$, MW (630 W), 10 min	65	184

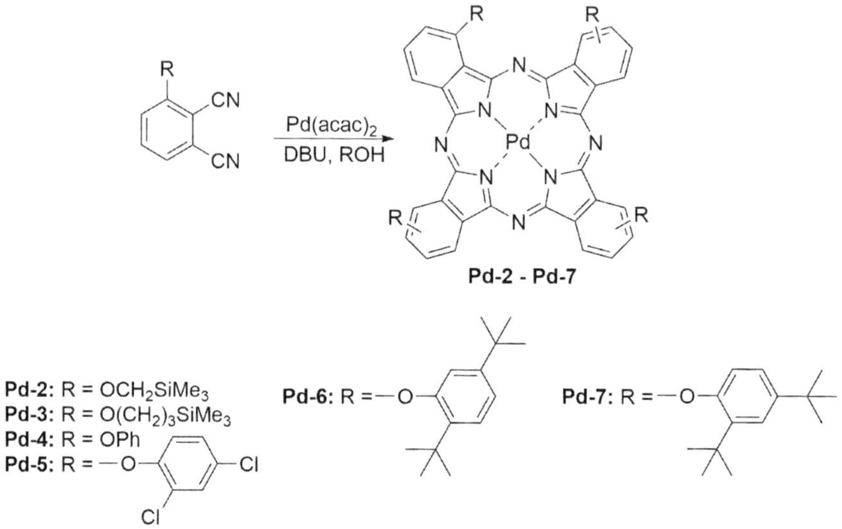

Scheme 98. Preparation of 1,8(11),15(18),22(25)-tetra-substituted **Pd-2** – **Pd-7**.[313,319]

Table 55. Preparation of 1,8(11),15(18),22(25)-tetra-substituted **Pd-2** – **Pd-7**.

	Conditions	Yield (%)	Ref.
Pd-2	$PdCl_2$, DBU, *n*-pentanol, 110–-120 °C, 24 h	32.6	313
Pd-3	$PdCl_2$, DBU, *n*-pentanol, 110–-120 °C, 30 h	38.1	313
Pd-4	$PdCl_2$, DBU, *n*-pentanol, refluxing, 5 h	20	319
Pd-5	$PdCl_2$, DBU, *n*-pentanol, refluxing, 5 h	9.1	319
Pd-6	$PdCl_2$, DBU, *n*-pentanol, refluxing, 5 h	54	319
Pd-7	$PdCl_2$, DBU, *n*-pentanol, refluxing, 5 h	70	319

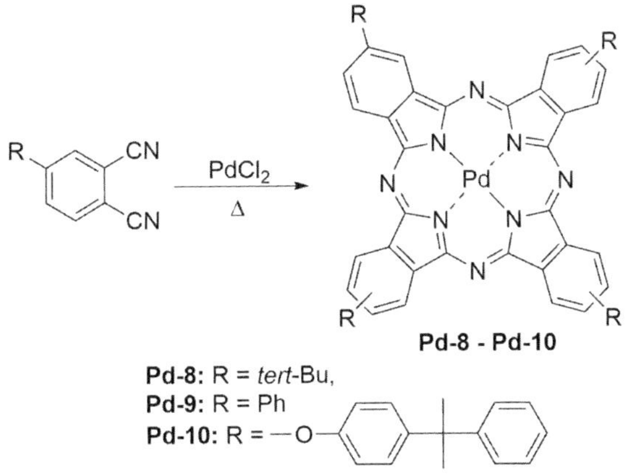

Scheme 99. Preparation of 2,9(10),16(17),23(24)-tetra-substituted **Pd-8** – **Pd-10**.[62–64]

Table 56. Preparation of 1,8(11),15(18),22(25)-tetra-substituted **Pd-9**, **Pd-10**.

	Conditions	Yield (%)	Ref.
Pd-8	$PdCl_2$, MOA, melting, 200–215 °C, 1 h	55	63
Pd-9	$PdCl_2$, MOA, melting, 265–280 °C, 5 h	64.5	62
Pd-10	$PdCl_2$, melting, 290 °C, 48 h	15	64

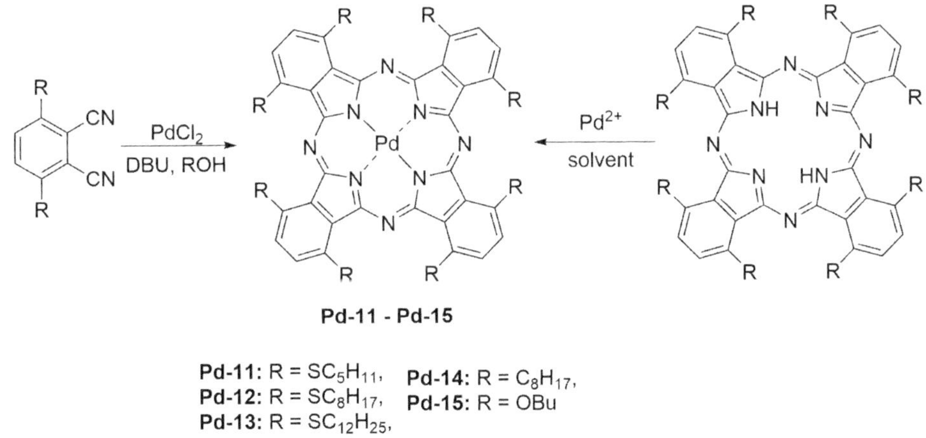

Scheme 100. Preparation of 1,4,8,11,15,18,22,25-octa-substituted **Pd-11** – **Pd-15**.[279,320–322]

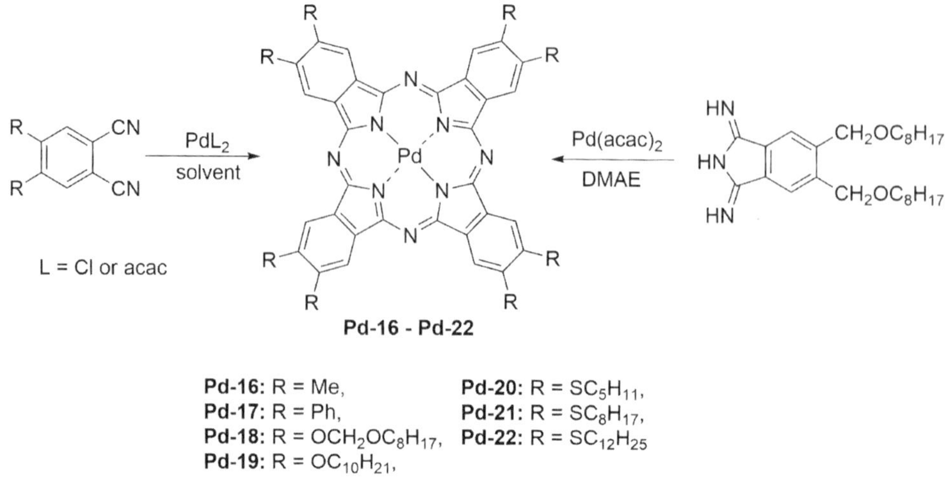

Scheme 101. Preparation of 2,3,9,10,16,17,23,24-octa-substituted **Pd-16** – **Pd-22**.[195,312,314,323,324]

Table 57. Preparation of octa-substituted **Pd-11** – **Pd-22**.

	Conditions	Yield (%)	Ref.
Pd-11	PN, $PdCl_2$, DBU, *n*-pentanol, refluxing, 18 h	40	320
Pd-12	PN, $PdCl_2$, DBU, *n*-pentanol, refluxing, 5 h	56	321
Pd-13	PN, $PdCl_2$, DBU, *n*-pentanol, refluxing, 5 h	53	321
Pd-14	H_2Pc, $Pd(OAc)_2$, *n*-pentanol, refluxed, 4 h	77	322
Pd-15	H_2Pc, $PdCl_2(PPh_3)_2$, DMF, refluxed, 1 h	45	279
Pd-16	PN, $PdCl_2$, DMF, NH_4Cl, 150 °C, 8 h	n.r.	323
Pd-17	PN, PdCl2, MOA, 1-bromonaphthalene, 240–250 °C, 6 h	25	195
Pd-18	DI, $Pd(acac)_2$, DMAE, refluxing, 4.5 h	47	324
Pd-19	PN, $Pd(acac)_2$, DBU, DMAE, refluxing, 12 h	30	312
Pd-20	PN, $PdCl_2$, DBU, *n*-pentanol, refluxing, 1 h	50	314
Pd-21	PN, $PdCl_2$, DBU, *n*-pentanol, refluxing, 1 h	60	314
Pd-22	PN, $PdCl_2$, DBU, *n*-pentanol, refluxing, 1 h	55	314

2,3,9,10,16,17,23,24-Octa-substituted PdPcs (**Pd-16** – **Pd-22**) were prepared using 4,5-disubstituted phthalonitriles (**Pd-16**, **Pd-17**, **Pd-19** – **Pd-22**), as well as 4,5-disubstituted-1,3-diiminoisoindoline (**Pd-18**) (Scheme 101, Table 57).[195,312,314,323,324]

Linstead and coworkers reported the preparation of *platinum* phthalocyanine (PtPc) for the first time in 1936.[21] Indeed, unsubstituted **Pt-1** was prepared by fusing unsubstituted phthalonitrile with platinous chloride (Scheme 102, Table 58).[21] Later, **Pt-1** was obtained in moderate yield *via* direct template condensation of

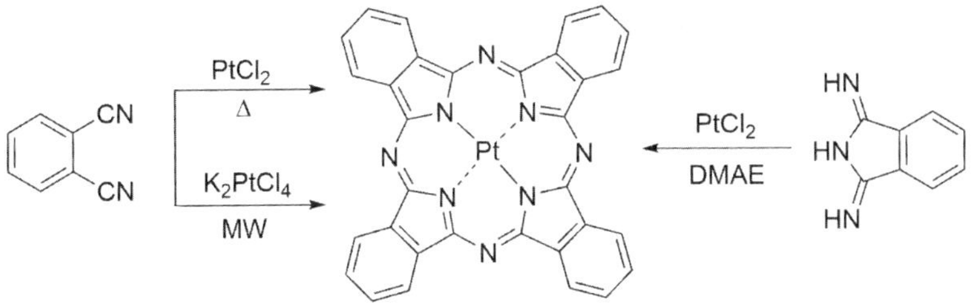

Pt-1

Scheme 102. Synthetic pathway of the preparation of **Pt-1**.[21,184,325]

Table 58. Preparation of unsubstituted **Pt-1**.

	Conditions	Yield (%)	Ref.
Pt-1	PN, $PtCl_2$, melting	99	21
Pt-1	DI, $PtCl_2$, DMAE, refluxing, 24 h	32	325
Pt-1	PN, K_2PtCl_4, MW (560 W), 10 min	78	184

unsubstituted 1,3-diiminoisoindoline and platinous chloride, using 2-dimethylaminoethanol as a solvent (Scheme 102, Table 58).[325] In addition, **Pt-1** could also be prepared under microwave-assisted conditions using phthalonitrile and K_2PtCl_4 as the starting materials (Scheme 102, Table 58).[184]

1,8(11),15(18),22(25)-Tetra-substituted PtPc (**Pt-2**) was obtained *via* direct template condensation of the 3-substituted-1,3-diiminoisoindoline and platinum(II) chloride as a metal source (Scheme 103, Table 59).[325]

2,9(10),16(17),23(24)-Tetra-substituted PtPcs (**Pt-3** – **Pt-5**) were produced by melting the corresponding 4-substituted phthalonitrile (**Pt-3**, **Pt-4**)[64,326] or 4-substituted-1,3-diiminoisoindoline (**Pt-5**)[325] and platinous chloride or potassium tetrachloroplatinate(II) as a metal source (Scheme 104, Table 59).

1,4,8,11,15,18,22,25-Octa-(**Pt-6**), as well as 2,3,9,10,16,17,23,24-octa-substituted PtPcs (**Pt-7** – **Pt-12**), were obtained by direct template condensation of 3,6-or 4,5-disubstituted phthalonitriles, as well as disubstituted 1,3-diiminoisoindolines (Schemes 105 and 106, Table 60).[312,324,325,327]

Synthesis of Ni(II)Pc (Ref. 20)

1,3-Diiminoisoindoline (0.2 g) and nickel chloride (0.2 g) were dissolved separately in hot formamide (2.5 mL and 12.5 mL, respectively), and the hot solutions mixed. Pigment separated almost immediately. After leaving overnight, the nickel phthalocyanine was collected, washed with ethanol, and extractively-crystallized from chloronaphthalene. Yield: 0.19 g (96%).

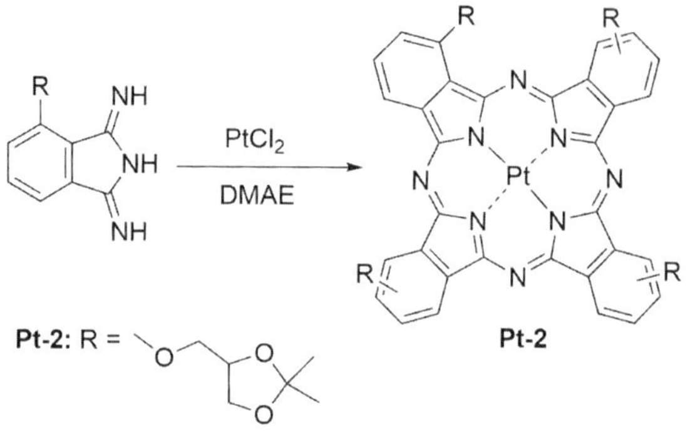

Scheme 103. Synthetic pathway of preparation of 1,8(11),15(18),22(25)-tetra-substituted **Pt-2**.[325]

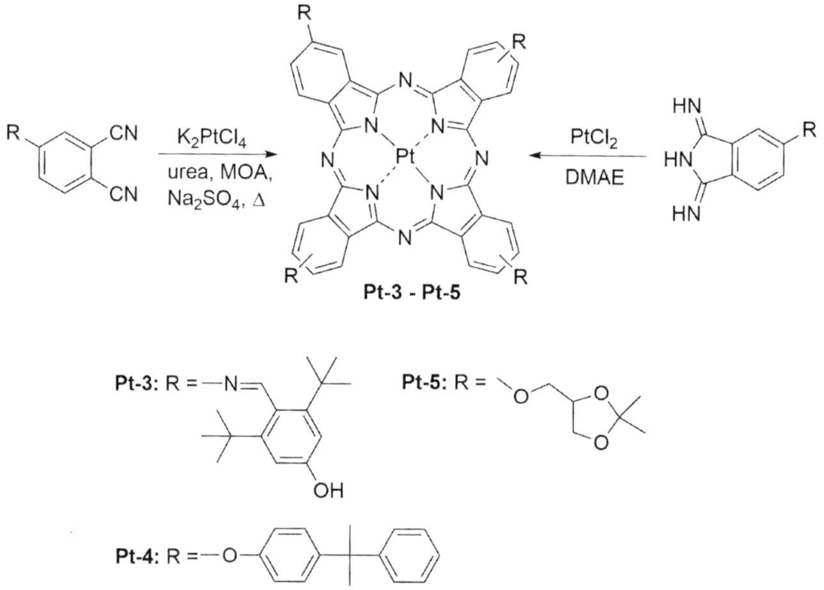

Scheme 104. Preparation of 2,9(10),16(17),23(24)-tetra-substituted **Pt-4** and **Pt-5**.[64,325,326]

Table 59. Preparation of tetra-substituted **Pt-2** – **Pt-5**.

	Conditions	Yield (%)	Ref.
Pt-2	DI, $PtCl_2$, DMAE, refluxing, 24 h	10	325
Pt-3	PN, K_2PtCl_4, urea, MOA, Na_2SO_4, melting	28	326
Pt-4	PN, $PtCl_2$, melting, 290 °C, 48 h	23	64
Pt-5	DI, $PtCl_2$, DMAE, refluxing, 24 h	35	325

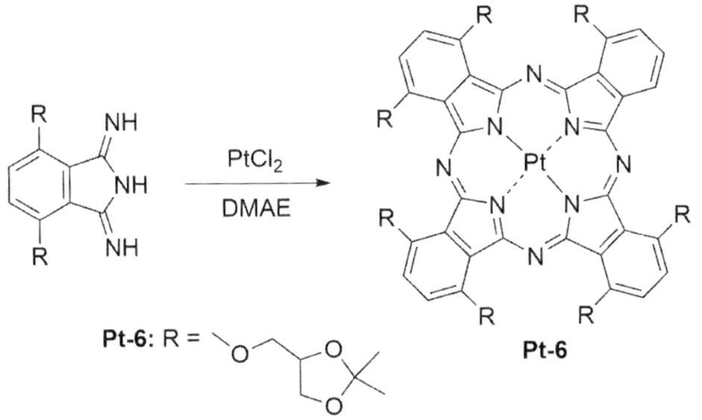

Pt-6

Scheme 105. Preparation of 1,4,8,11,15,18,22,25-octa-substituted **Pt-6**.[325]

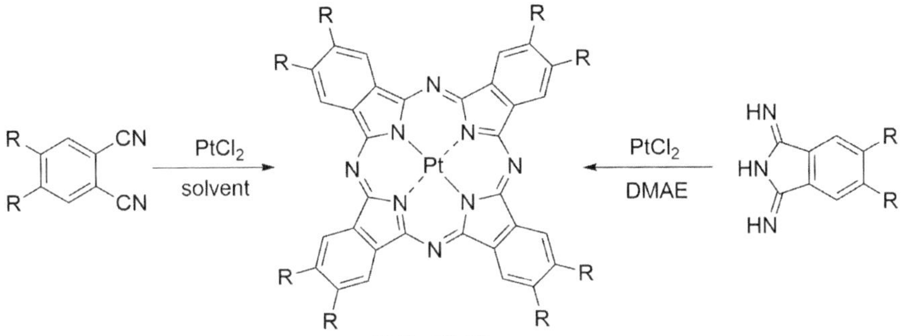

Pt-7 - Pt-12

Pt-7: R = $CH_2OC_8H_{17}$ **Pt-12:** R =
Pt-8: R = C_6H_{13},
Pt-9: R = OC_5H_{11},
Pt-10: R = $OC_{10}H_{21}$,
Pt-11: R = $OC_{12}H_{25}$,

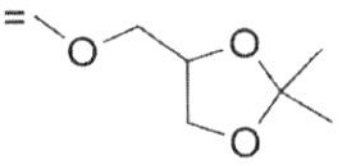

Scheme 106. Preparation of 2,3,9,10,16,17,23,24-octa-substituted **Pt-7** – **Pt-12**.[312,324,325,327]

Table 60. Preparation of octa-substituted **Pt-6** – **Pt-12**.

	Conditions	Yield (%)	Ref.
Pt-6	DI, $PtCl_2$, DMAE, refluxing, 24 h	8	325
Pt-7	DI, $PtCl_2$, DMAE, refluxing, 4.5 h	32	324
Pt-8	DI, $PtCl_2$, DMAE, refluxing, 48 h	32	324
Pt-9	DI, $PtCl_2$, DMAE, refluxing, 5 h	18	324
Pt-10	PN, $PtCl_2$, DBU, DMAE, refluxing, 24 h	11	312
Pt-11	DI, $PtCl_2$, DMAE, 20 h	26	327
Pt-12	DI, $PtCl_2$, DMAE, refluxing, 24 h	35	325

Synthesis of Ni(II)Pc (Ref. 41)

A sample of H_2Pc (1.00 g) was metalated with $NiCl_2 \times 6H_2O$ (0.92 g) in a 10% quinoline/l-chloronaphthalene solvent system (20 mL). The solution was refluxed in a quartz vessel for 72 h and then cooled in an ice bath. The purple $Ni^{II}Pc$ crystals which precipitated out were removed by filtration and washed with hot water and acetone. The progress of the metalation was followed by UV-vis spectroscopy, and the reaction was repeated until the peak at 696 nm in the UV-vis spectrum associated with H_2Pc had disappeared. The resulting material was sublimed twice under vacuum at 400 °C. Yield: 0.62 g (56%).

Synthesis of Ni(II)Pc (Ref. 41)

A phthalonitrile (2.0 g) was cyclized with anhydrous $NiC1_2$ (1.0 g) in quinoline (4 mL). The solution was refluxed for 10 min until the solution solidified, and the solid mixture was ground, washed with acetone and hot water, and sublimed twice. Yield: 1.02 g (46%).

Synthesis of Ni(II)Pc(OCH(C$_2$H$_5$)$_2$)$_4^{\alpha}$ (Ref. 76)

Nickel(II) chloride hexahydrate (70 mg) was added to a DMF solution (1–2 mL) of 1,8,15,22-tetrakis-(pentan-3′-yloxy)phthalocyanine (100 mg), and refluxed for 4 h. After the addition of adequate amounts of chloroform to the reaction mixture, the solution was poured onto a SiO_2 column. A green solution was eluted out with chloroform, and dark-green precipitates were obtained upon addition of hexane to the eluent. Yield: 71 mg (67%).

Synthesis of Ni(II)Pc(OPh-3-CF$_3$)$_4^{\beta}$ (Ref. 287)

A 3-fluoromethylphenoxyphthalonitrile (1.0 g), *n*-pentanol (20 mL) and DBU (1.5 mL) were mixed and homogenized. The mixture was stirred at 50 °C for 10 min in a nitrogen atmosphere and then anhydrous nickel chloride (0.1 g) added. The suspension was slowly brought to boiling over 2 h and then refluxed for 40 h. The reactant was cooled to 60 °C, 50 mL methanol added, and the obtained deep blue product filtered, and washed with hydrochloric acid (5%, 150 mL) and then methanol (2 × 100 mL). The crude product was extracted with methanol in a Soxhlet extractor, and purified by column chromatography (SiO_2, toluene). The blue fraction was collected, evaporated under vacuum, and dried *in vacuo* at 100 °C. Yield: 0.69 g (71%) as a deep blue powder.

Synthesis of Ni(II)Pc(C_7H_{15})$_8^{\alpha\alpha}$ (Ref. 312)

A mixture of 3,6-bis(heptyl)phthalonitrile (324 mg), $Ni(OAc)_2 \times 4H_2O$ (75 mg) and DBU (152 mg) was heated under nitrogen in *n*-pentanol (15 mL) for 48 h under reflux. The cooled solution was poured into methanol (50 mL), and the precipitate centrifuged, washed with ethyl acetate and ethanol, and dried. Further purification was performed by column chromatography (SiO_2, toluene) and recrystallization from ethyl acetate. Yield 68 mg (20%) as a green powder.

Synthesis of Ni(II)^{t}PcMe$_8^{\beta\beta}$ (Ref. 153)

A 100-mL one-neck flask was fitted with a magentic stirring bar and charged with a finely-ground mixture of 4,5-dimethylphthalonitrile (2.0 g), $NiCI_2 \times 6H_2O$ (0.76 g), and quinoline (5 mL). The reaction mixture was stirred at 200–230 °C for approximately 2.5 h, cooled, diluted with methanol, filtered, and then washed successively with more methanol, water, hexane, and acetone. Vacuum drying of the remaining dark blue-green powder for several hours afforded 2.2 g (57%) of product.

Synthesis of Ni(II)Pc(SC_6H_{13})$_8^{\beta\beta}$ (Ref. 302)

A mixture of 4,5-di(hexylsulfanyl)phthalonitrile (1.0 g), anhydrous $NiCl_2$ (0.104 g) and quinoline (1.5 mL) was heated and stirred at 200 °C for 5 h under argon. After cooling to room temperature, the reaction mixture was treated with ethanol (5 mL), and the product filtered off and washed with the same solvent. The dark green metallophthalocyanine was isolated on a SiO_2 column with $CHC1_3$ as the eluent. Yield: 0.390 g (38%).

Synthesis of Ni(II)Pc(tert-Bu)$_8$ (Ref. 316)

A mixture of 3,5-di(*tert*-butyl)phthalonitrile (200 mg), nickel acetate tetrahydrate (100 mg) and DBU (four drops) in *n*-pentanol (12 mL) was heated at 139 °C under nitrogen for 4 d. The cooled green solution was poured into methanol/water (5:l, v/v) mixture (70 mL), and the suspension centrifuged. The filtrate was removed, and the solid re-suspended in methanol, filtered and dried. The purification of the crude product was performed by column chromatography on silica gel using a hexane/dichloromethane (2:1, v/v) mixture as eluent. Further recrystallization from dichloromethane/methanol mixture gave the desired compound as a green powder. Yield: 192 mg (86%).

Synthesis of Pd(II)Pc (Ref. 28)

Phthalonitrile (0.51 g) was added under nitrogen and stirring to a three-necked flask containing *n*-pentanol (40 mL). The mixture developed into a colorless transparent solution after heating at 50 °C for 1 h. To this solution was added palladium acetylacetonate (0.3 g) and DBU (0.4 mL), upon which the solution immediately became bright orange, and turned black after refluxing for 0.5 h. The reflux was continued for 24 h under a nitrogen atmosphere until the initial compounds had completely disappeared by TLC detection, then the mixture was evaporated under reduced pressure, to produce a black powder. The crude powder was washed with ethanol, DMF, chloroform and *n*-hexane, respectively; and then dried. Yield: 0.32 g (52%) as a black powder.

Synthesis of Pd(II)Pc$(OCH_2SiMe_3)_4^{\alpha}$ (Ref. 313)

To a stirred solution of 3-(trimethylsilylmethoxy)phthalonitrile (2.5 g) and DBU (1.16 g) in dry *n*-pentanol (10 mL) was added under a nitrogen atmosphere at 110–120 °C, $PdCl_2$ (0.49 g), and the reaction continued for 24 h. After removing the *n*-pentanol under reduced pressure, the resultant solid was passed through a SiO_2 column using toluene as eluent. Evaporation of the solvent, followed by recrystallization from toluene–MeOH, provided a mixture of isomers as a green powder. Yield: 0.910 g (32.6%).

Synthesis of Pd(II)Pc$(SC_5H_{11})_8^{\alpha\alpha}$ (Ref. 320)

To refluxing *n*-pentanol (10 mL), 3,6-bis(pentylsulfanyl)phthalonitrile (0.5 g), $PdCl_2$ (0.12 g) and DBU (1.70 g) were added. Reflux was continued for a further 18 h, then the reaction mixture cooled, and the solvent removed under reduced pressure, followed by titration with cold methanol, to precipitate the product. The black precipitate was dissolved in CH_2Cl_2 and passed through a silica column using CH_2Cl_2 as the eluting solvent. Yield: 0.12 g (40%).

Synthesis of Pd(II)Pc$(C_8H_{17})_8^{\alpha\alpha}$ (Ref. 322)

Metal-free 1,4,8,11,15,18,22,25-octakis(octyl)phthalocyanine (320 mg) was dissolved in *n*-pentanol (10 mL) and heated to reflux. Excess metal salt, palladium(II) acetate (0.2 g), was added, and reflux continued for 4 h. After the reaction mixture had cooled, methanol (10 mL) was added, and the precipitate collected and chromatographed over SiO_2 (eluent: petroleum ether:THF, 2:1, v/v), collecting the green fraction, which was recrystallised from THF–methanol, to give the title product as a blue powder. Yield: 270 mg (77%).

Synthesis of Pt(II)Pc (Ref. 21)

A mixture of platinous chloride (1.8 g) and phthalonitrile (20 g) was heated to 280 °C over 20 min. Pigment was first formed at about 180 °C, and the melt began to solidify at 230 °C; some hydrogen chloride was then evolved. The cooled product was ground and thoroughly washed with alcohol. The recovered phthalonitrile was free from chlorine. Yield of residue, 4.2 g. Yield of pure PtPc: 2.75 g (99%) as blue needles with a bronze lustre from chloronaphthalene.

Synthesis of Pt(II)Pc (Ref. 325)

A mixture of 1,3-diiminoisoindoline (250 mg) and $PtCl_2$ (230 mg) was refluxed in *N,N*-dimethylaminoethanol under an ammonia atmosphere for 24 h. The cooled reaction mixture was then poured into hexane. The solid product was filtered, dissolved in a minimum amount of dichloromethane, and precipitated in hexane (several times). On silica gel column chromatography, initial elution by hexane/ethyl acetate (2:1, v/v) eliminated the remaining organic impurities. The platinum phthalocyanine was then eluted by tetrahydrofuran. Yield: 98 mg (32%).

Synthesis of Pt(II)Pc$(OC_{10}H_{21})_8^{\beta\beta}$ (Ref. 312)

4,5-Bis(decyloxy)phthalonitrile (880 mg), $PtCl_2$ (142 mg) and a catalytic amount of DBU were added to DMAE (8 mL) and refluxed under nitrogen for 24 h. The cooled solution was poured into a mixture of methanol:water 1:l (v/v) (50 mL), and the precipitate centrifuged, washed with acetone, and dried. The crude product was purified by column chromatography (Al_2O_3, eluent: $CHCl_3$). Yield: 120 mg (11%).

12. *Group 11: CuPc, AgPc, and AuPc*

Copper phthalocyanine (CuPc) was also first prepared by Linstead and coworkers in 1934.[328] Today, CuPc (as trade name *Phthalocyanine Blue BN*, *Monastral blue*, *Winsor blue, C.I. 74160* and numerous other trade names) and its derivatives, in particular, hexadecachlorinated $CuPcCl_{16}$ (as trade name *Phthalocyanine Green Blue, pigment green 7, C.I. 74260* and numerous other trade names) are some of the most important colorants in the world.[5,8,329,330]

Unsubstituted *copper* Pc (**Cu-1**) was prepared by direct template condensation of phthalic acid, as well as its derivatives, such as an anhydride, imide and dinitrile, with copper metal or copper salts as templates (Scheme 107, Table 61).[57,328] **Cu-1** could also be produced using 1,3-diiminoisoindoline and copper powder or copper(II) acetate as starting materials[20,331] or *via* metal exchange of Li_2Pc with copper(I) chloride (Scheme 108, Table 61).[52] A simple method of synthesis of **Cu-1** was published by Tomoda *et al.* (Scheme 107).[29] Here, **Cu-1** was obtained in high yield using phthalonitrile and metal salt as starting materials in the presence of strong organic bases, such as DBU and DBN (Table 61). In this method, the yield of **Cu-1** depended on the nature of the base and copper salt.[29] Later, a simple preparative method for preparing **Cu-1** was published by Toru and coworkers.[287] Here, **Cu-1** was obtained by condensation of phthalimide (without urea as N source) in the presence of hexamethyldisilazane and *p*-toluenesulfonic acid (Scheme107, Table 61).[287] In addition, **Cu-1** could be obtained as a ring-contraction product of uranyl superphthalocyanine in the presence of copper(II) chloride (Scheme 108,

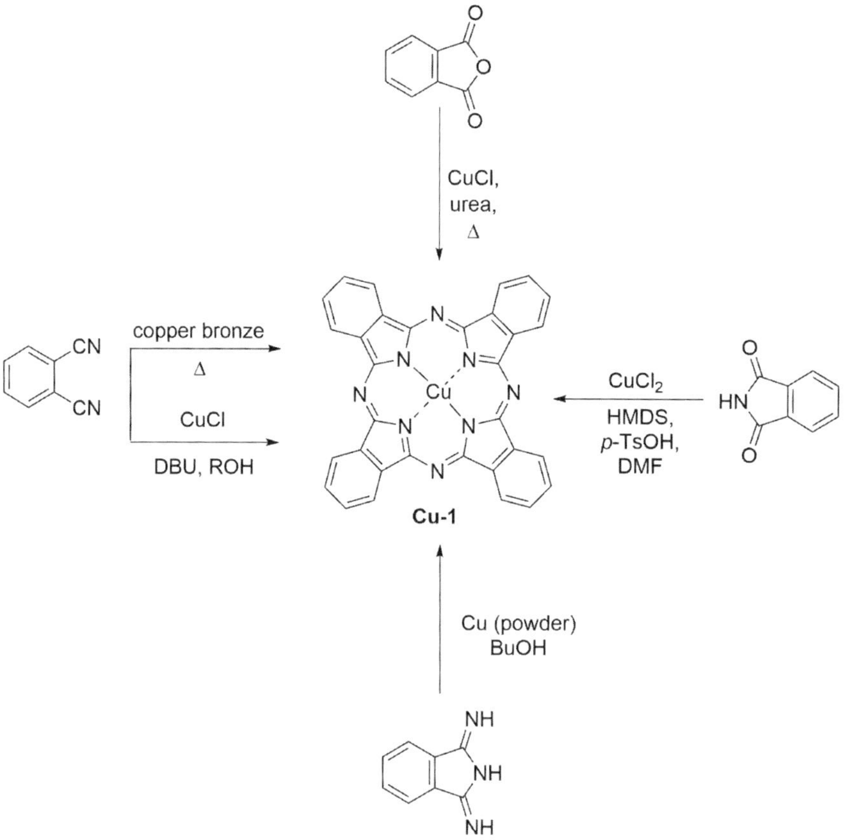

Scheme 107. Synthetic approaches for preparation of the unsubstituted CuPs (**Cu-1**).[20,52,287,328,331]

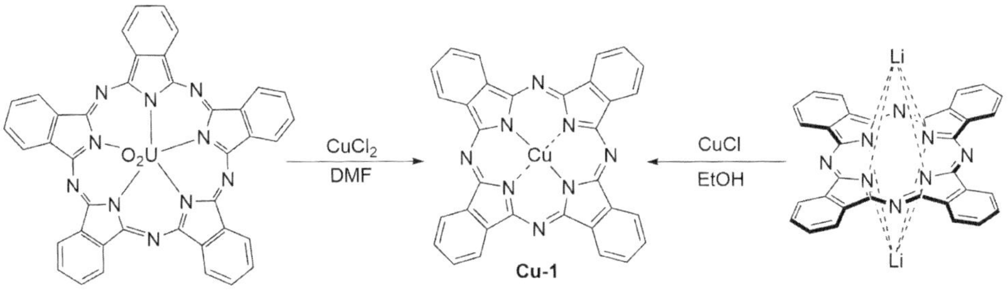

Scheme 108. Synthetic approaches for preparation of the **Cu-1** *via* ring contraction and transmetalation reactions.[29,152,244]

Table 61. Preparation of unsubstituted **Cu-1**.

	Conditions	Yield (%)	Ref.
Cu-1	PN, copper bronze, melting	85	328
Cu-1	DI, Cu powder, *n*-butanol, refluxing	24	20
Cu-1	DI, $Cu(OAc)_2$, quinoline, 270 °C, 8 h	51.3	331
Cu-1	Li_2Pc, CuCl, EtOH, 3 h	75	52
Cu-1	PN, CuCl, DBU, *n*-pentanol, refluxing, 6 h	85	29
Cu-1	PN, CuCl, DBN, *n*-pentanol, refluxing, 6 h	79	29
Cu-1	PI, $CuCl_2$, HMDS, *p*-TsOH, DMF, 100 °C 10 h	72	287
Cu-1	$U(O_2)$SuperPc, $CuCl_2$, DMF, 120 °C, 3 h	95	152
Cu-1	$CuCl_2$, MeOH, 9 kbar, 90 °C, 70 h	51	42
Cu-1	$CuCl_2$, (560 W), 10 min	92–98	184

Table 61).[152,244] A method for preparing **Cu-1** in good yield under high-pressure conditions appeared in 1990 (Table 61).[42] In addition, **Cu-1**, as well as substituted CuPcs, were prepared *via* a microwave-assisted approach or by using ionic liquids (*e.g.* tetrabutylammonium bromide) as reaction media (Table 61).[184,332] Copper bronze and many copper salts (*e.g.* CuCl, $CuCl_2$, $Cu(OAc)_2$, $Cu(acac)_2$) were successfully used as the metal source.[20,29,42,52,57,152,244,287,328,331]

1,8(11),15(18),22(25)-Tetra- (**Cu-2** – **Cu-7**), as well as 2,9(10),16(17),23(24)-tetra-substituted CuPcs (**Cu-8** – **Cu-19**), were obtained by two synthetic approaches: (i) *via* direct template condensation of 3- as well as 4-substituted phthalogenes (anhydrides, imides and dinitriles) (**Cu-2** – **Cu-6**, **Cu-8** – **Cu-18**)[63,64,66,76,189,247,289–291,332–334,336] and 4-substituted 1,2-dibromobenzene (**Cu-19**)[333] or 4-substituted phthalic acid (**Cu-8**)[335] and (ii) by metal insertion into H_2Pcs (**Cu-7**, **Cu-8**)[76,290] (Schemes 109–111, Table 62).

1,4,8,11,15,18,22,25-Octa-substituted CuPcs (**Cu-20** – **Cu-33**) were also obtained by direct template condensation of 3,6-disubstituted phthalonitriles or

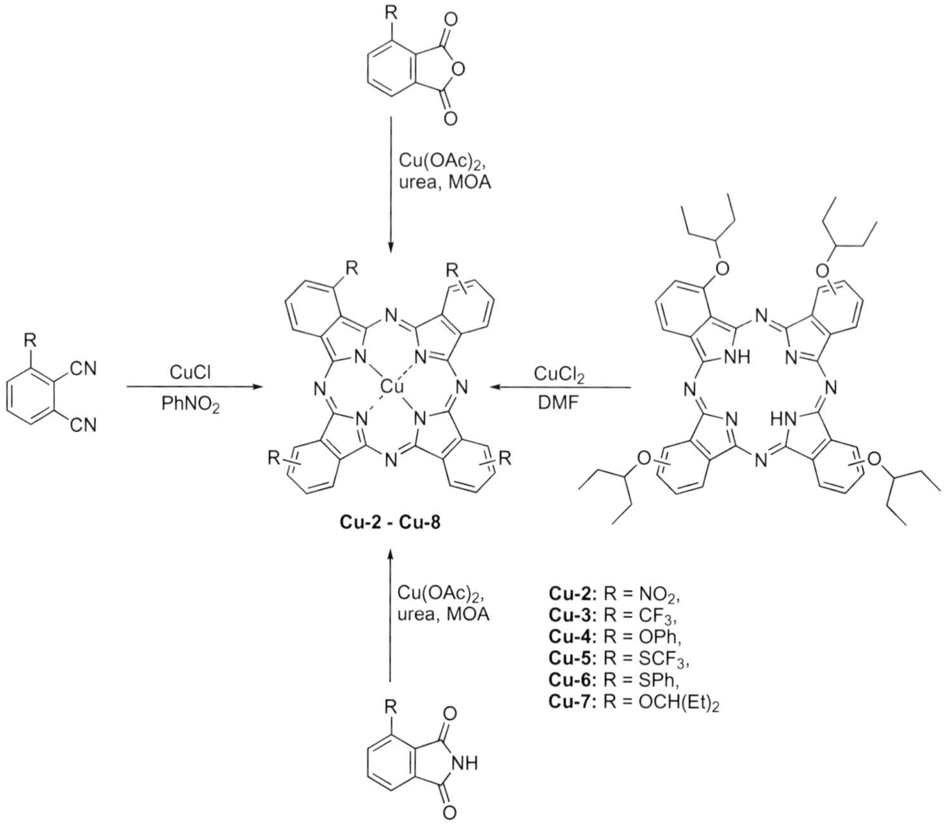

Scheme 109. The synthetic routes of preparation 1,8(11),15(18),22(25)-tetra-substituted **Cu-2** – **Cu-7**.[76,289,332–334]

phthalimides (**Cu-20**, **Cu-21**),[298,333] as well as by copper insertion into 1,4,8,11,15,18,22,25-octa-substituted H_2Pcs (**Cu-22** – **Cu-33**)[337,338] (Scheme 112, Table 63).

2,3,9,10,16,17,23,24-Octa-substituted, as well as tetra-$\beta\beta$-fused CuPcs (**Cu-34** – **Cu-39**), could be obtained by the following synthetic approaches: (i) by the direct template condensation of 4,5-disubstituted phthalonitriles (**Cu-34** – **Cu-37**) in the presence of copper(I) salts,[76,189,195,251] and (ii) by the reaction of 4,5-disubstituted-1,2-dibromobenzenes with copper(I) cyanide (**Cu-37**, **Cu-38**, **Cu-39**)[198,251,333] (Scheme 113, Table 63).

Hexadecasubstituted CuPcs (**Cu-40**, **Cu-41**) were synthesized by direct template condensation of tetra-substituted phthalonitriles with copper salts (**Cu-40**),[339] or *via* microwave-assisted conditions from tetra-substituted phthalic anhydride (**Cu-41**)[184] (Scheme 114, Table 64). The yield of **Cu-40** depended on the nature of the copper salt and the reaction conditions (Table 64).

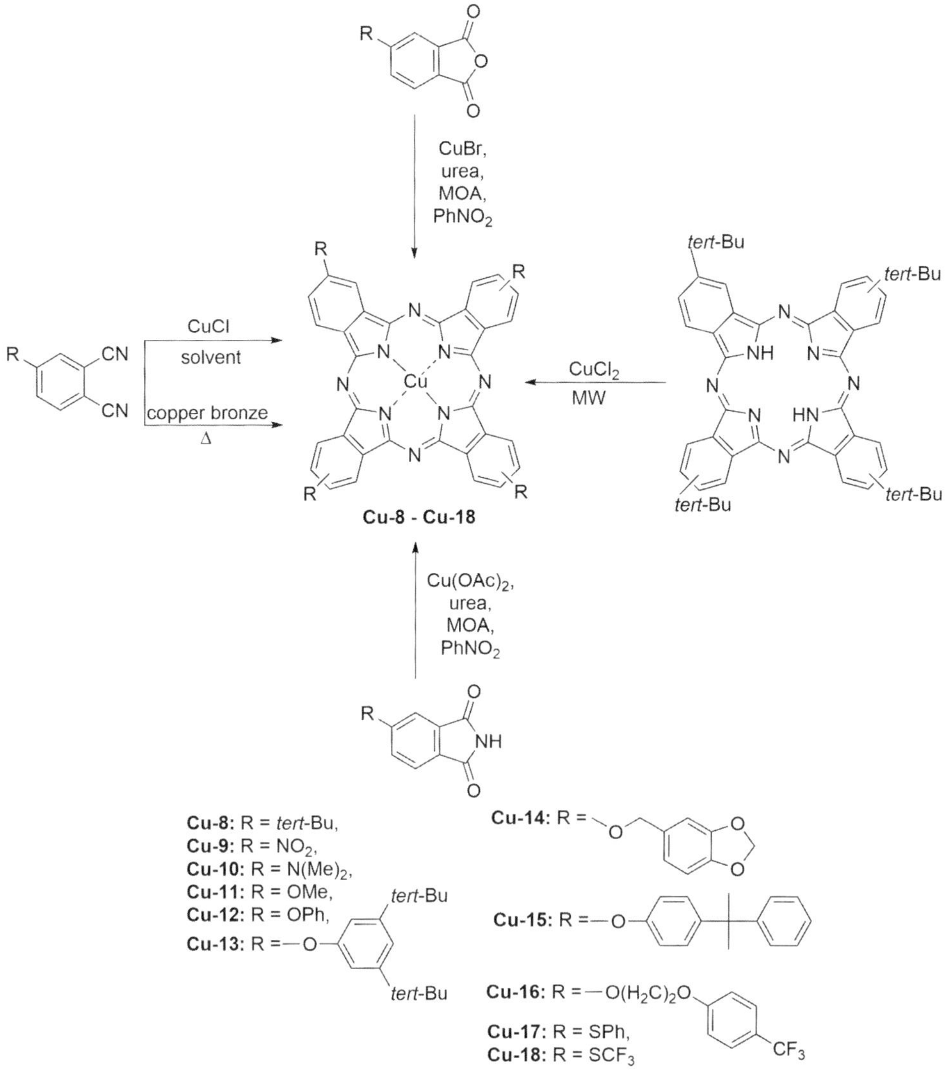

Scheme 110. The synthetic routes of preparation 2,9(10),16(17),23(24)-tetra-substituted **Cu-8** – **Cu-19**.[63,64,66,76,189,247,290,291,332–336]

At present, *silver* phthalocyanines can be prepared using two approaches: (i) metal exchange reaction between MPc and a silver salt[52] and (ii) metal insertion into H_2Pcs.[203,340] The first successful attempt at the preparation of AgPc was also published by Linstead and coworkers in 1938.[52] At that time, the desired AgPc (**Ag-1**) was prepared *via* metal exchange reaction between Li_2Pc (or PbPc) and a silver salt (nitrate or sulfate).[52] Unfortunately, in both cases the yields of the desired compounds were not reported (Table 65). Later, two examples of tetra-substituted

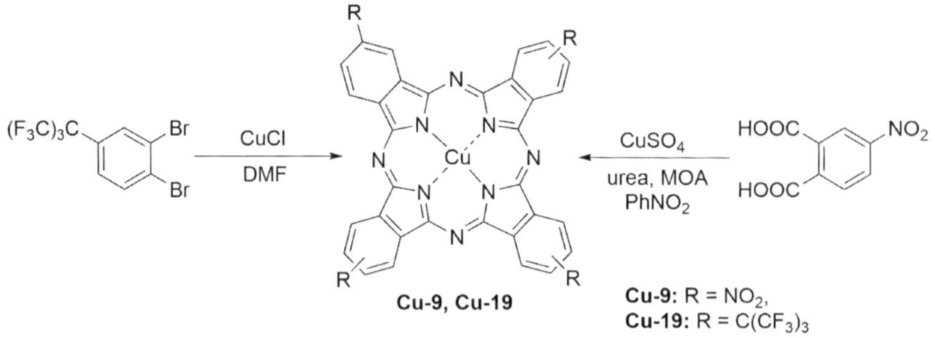

Scheme 111. Synthetic routes of preparation 2,9(10),16(17),23(24)-tetra-substituted **Cu-9**, **Cu-19**.[333,335]

Table 62. Preparation of tetra-substituted **Cu-2** – **Cu-19**.

	Conditions	Yield (%)	Ref.
Cu-2	PI, $Cu(OAc)_2 \times 2H_2O$, urea, MOA, $PhNO_2$, 180–190 °C, 4 h	51	332
Cu-2	PA, $Cu(OAc)_2 \times 2H_2O$, urea, MOA, $PhNO_2$, 180–190 °C, 4 h	41	332
Cu-3	PN, CuCl, MOA, $PhNO_2$, refluxing, 3 h	59.5	333
Cu-4	PA, $Cu(OAc)_2$, urea, Na_2SO_4, MOA, 160–170 °C 1 h; 200–210 °C, 3 h	32	289
Cu-5	PN, CuCl, MOA, $PhNO_2$, refluxing, 3 h	56	334
Cu-6	PA, $Cu(OAc)_2$, urea, Na_2SO_4, MOA, 160–170 °C 1 h; 200–210 °C, 3 h	26	289
Cu-7	H_2PcR_4, $CuCl_2$, DMF, refluxing, 4 h	56	76
Cu-8	PA, CuCl, urea, MOA, 210–215 °C, 1 h	35	63
Cu-8	PN, CuCl, MOA, $PhNO_2$, refluxing, 2 h	58	63
Cu-8	H_2PcR_4, CuCl, MW, 440 W, 10 min	31	290
Cu-9	PI, $Cu(OAc)_2 \times 2H_2O$, urea, MOA, $PhNO_2$, 180–190 °C, 4 h	55	332
Cu-9	4-nitrophthalic acid, $CuSO_4 \times 5H_2O$, NH_4Cl, urea, MOA, $PhNO_2$,180–190 °C, 4.5 h	n.r.	335
Cu-10	PN, CuCl, urea, MOA, quinoline, 200–205 °C, 2 h	55	189
Cu-11	PN, $Cu(OAc)_2$, urea, MOA, Na_2SO_4, 180–190 °C 1.5 h	50	336
Cu-12	PA, CuBr, urea, melting, 160–170 °C, 1 h; 180–190 °C 3 h	26	334
Cu-13	PN, $CuCl_2$, DBU, EtOH, refluxing, 24 h	22.8	291
Cu-14	PN, $CuCl_2$, DBU, *n*-hexanol, 160 °C, 24 h	23.8	66
Cu-15	PN, copper bronze, melting, 270 °C, 12 h	43	64
Cu-16	PN, $CuCl_2$, DBU, *n*-pentanol, 160 °C, 24 h	16	247
Cu-17	PA, CuBr, urea, melting, 160–170 °C, 1 h; 180–190 °C 3 h	48	334
Cu-18	PN, CuCl, MOA, $PhNO_2$, refluxing, 3 h	72	76
Cu-19	4-trifluoromethyl-1,2-dibromobenzene, CuCl, DMF, refluxing, 4 h	57.5	333

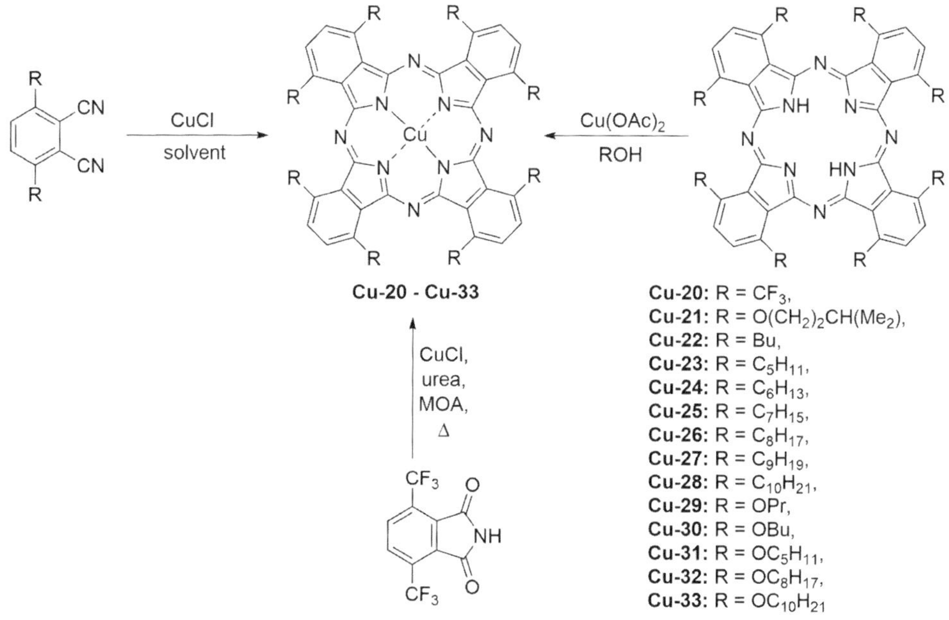

Scheme 112. The synthetic routes of preparation 1,4,8,11,15,18,22,25-octa-substituted **Cu-20** – **Cu-33**).[298,333,337,338]

AgPcs (**Ag-2** and **Ag-3**) prepared by metal insertion into the corresponding H_2Pcs were described (Scheme 115, Table 65).

Gold phthalocyanine (**Au-1**) was synthesized for the first time in 1965 by melting 1,3-diiminoisoindoline and gold(I) bromide (Table 66).[341] Until recently, this was the only procedure for the synthesis of AuPc. **Au-1** could also be prepared by template condensation of phthalonitrile with gold(III) chloride hydrate as the metal source (Table 66).[342] However, Kobayashi, Leznoff and coworkers re-examined the data of MacCragh and Koski by using modern analytical techniques, and observed formation of only Au(III)Pc instead of Au(II)Pc.[343] The A(III)Pcs (**Au-2** – **Au-4**) were prepared with moderate yields by template condensation of the (un)substituted 1,3-diiminoisoindolines using gold(I) chloride, potassium gold(III) chloride or gold(III) chloride hydrate as the metal source (Scheme 116, Table 66).[343,344] In addition, Kobayashi, Leznoff and coworkers observed the formation of the expected **Au-2** – **Au-4** complexes along with the unusual triphyrin-like macrocycles **Au-5** – **Au-7**, which can be converted into Au(III)Pcs under specific conditions (Scheme 116, Table 66).[343] The yield of complexes **Au-2** – **Au-4** depended on the type of gold salt (Table 66).

Tetra(substituted) complexes **Au-8a,8b** were prepared by gold insertion into the corresponding H_2PcR_4 with potassium gold(III) chloride (Scheme 117, Table 66).[345]

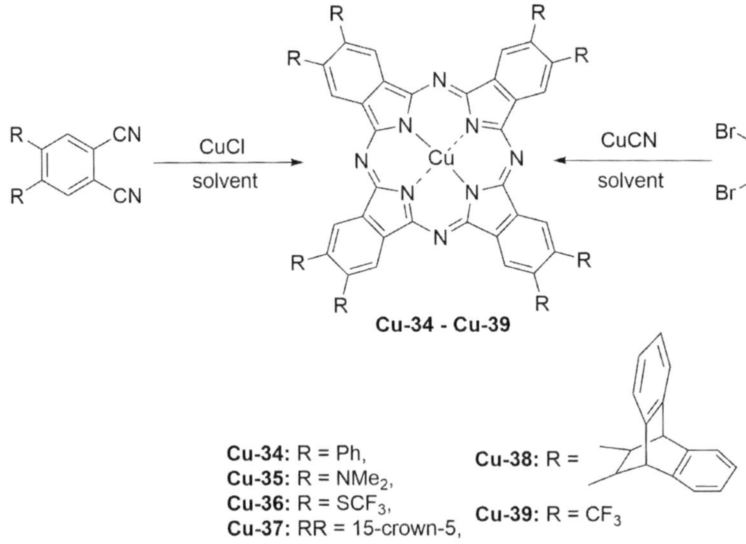

Scheme 113. Synthetic approaches of preparation of 2,3,9,10,16,17,23,24-octa-substituted and tetra-$\beta\beta$-fused **Cu-34** – **Cu-39**.[76,189,195,198,251,333]

Synthesis of Cu(II)Pc (Ref. 328)

A mixture of phthalonitrile (4 equiv) and copper bronze (1 equiv) was heated with stirring. A green color was first observed at 190 °C, the mass became pasty at 220 °C, and was too stiff to be stirred at 270 °C. At a bath temperature of about 220 °C, the internal temperature began to increase rapidly and at times exceeded that of the oil bath by 45 °C. After 5 min, the reaction mass was cooled, and ground with EtOH. The powdered product was repeatedly boiled with EtOH and then dried. The pigment was added to concentrated H_2SO_4 (10 times the mass of the product), and after an hour the solution was filtered and poured as a thin stream onto ice (100 times the mass of product). The flocculent blue precipitate was allowed to coalesce for 2 h, collected by filtration, washed with boiling H_2O, then boiling EtOH, and dried at 100 °C. Yield: 70–85%.

Synthesis of Cu(II)Pc (Ref. 52)

A solution of anhydrous cupric chloride (1.0 g) in absolute alcohol (15 mL) was added gradually with shaking to dilithium phthalocyanine (2.0 g) dissolved in 20 mL of the same solvent. A bright blue precipitate of copper phthalocyanine separated immediately. This was collected after 3 h, washed with alcohol until free of cupric chloride, dissolved in sulphuric acid, and precipitated with water. Yield: 1.8 g (75%).

Table 63. Preparation of octa-substituted **Cu-20** – **Cu-39**.

	Conditions	Yield (%)	Ref.
Cu-20	PI, CuCl, urea, MOA, 200 °C, 45 min; 200–210 °C, 15 min; 210–215 °C, 30 min	30	333
Cu-20	PN, CuCl, MOA, $PhNO_2$, refluxing, 3 h	60	333
Cu-21	PN, CuCl, DBU, *n*-pentanol, 160 °C, 10 h	57	298
Cu-22	H_2PcR_8, $Cu(OAc)_2$, *n*-pentanol, refluxing, 1 h	78	337
Cu-23	H_2PcR_8, $Cu(OAc)_2$, *n*-pentanol, refluxing, 1 h	80	337
Cu-24	H_2PcR_8, $Cu(OAc)_2$, *n*-pentanol, refluxing, 1 h	85	337
Cu-25	H_2PcR_8, $Cu(OAc)_2$, *n*-pentanol, refluxing, 1 h	85	337
Cu-26	H_2PcR_8, $Cu(OAc)_2$, *n*-pentanol, refluxing, 1 h	80	337
Cu-27	H_2PcR_8, $Cu(OAc)_2$, *n*-pentanol, refluxing, 1 h	82	337
Cu-28	H_2PcR_8, $Cu(OAc)_2$, *n*-pentanol, refluxing, 1 h	72	337
Cu-29	H_2PcR_8, $Cu(OAc)_2$, BuOH, refluxing, 0.5 h	81	338
Cu-30	H_2PcR_8, $Cu(OAc)_2$, BuOH, refluxing, 0.5 h	77	338
Cu-31	H_2PcR_8, $Cu(OAc)_2$, *n*-pentanol, refluxing, 0.5 h	85	338
Cu-32	H_2PcR_8, $Cu(OAc)_2$, *n*-pentanol, refluxing, 0.5 h	81	338
Cu-33	H_2PcR_8, $Cu(OAc)_2$, *n*-pentanol, refluxing, 0.5 h	79	338
Cu-34	PN, CuBr, MOA, 1-bromonaphthalene, 240–250 °C, 6 h	51	195
Cu-35	PN, $Cu(OAc)_2$, urea, MOA, quinoline, 200–205 °C, 2 h	47	189
Cu-36	PN, CuCl, MOA, $PhNO_2$, refluxing, 3 h	75	76
Cu-37	PN, CuCl, urea, melting, 160 °C, 10 min	38.1	251
Cu-37	1,2-Dibromobenzo-15-crown-5, CuCN, quinoline, 220 °C, 16 h	38	251
Cu-38	2,3-Dibromotriptycene, CuCN, DMF, refluxing 3 h	25	198
Cu-39	CuCN, *N*-methyl-2-pyrrolidone, 100 °C, 7 h	28	333

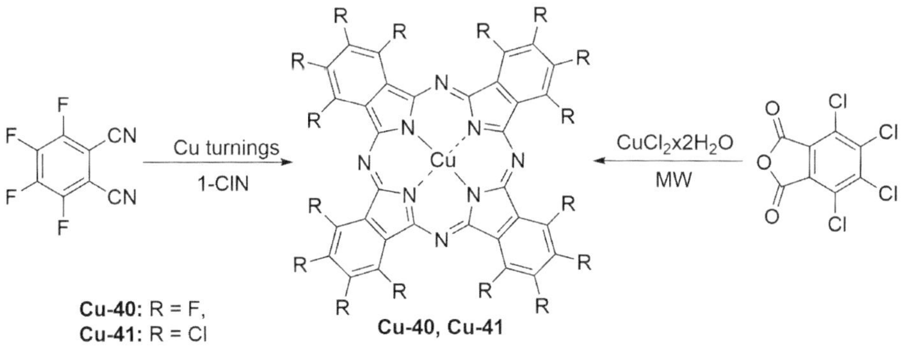

Scheme 114. Synthetic approaches of preparation of hexadeca-substituted **Cu-40**, **Cu-41**.[184,339]

Table 64. Preparation of hexadeca-substituted **Cu-40**, **Cu-41**.

	Conditions	Yield (%)	Ref.
Cu-40	PN, Cu turnings, 1-ClN, 48 h	78	339
Cu-40	PN, CuCl, melting, 220 °C, 2 h	87	339
Cu-40	PN, $CuCl_2$, melting, 250 °C, 7 h	62	339
Cu-41	PA, $CuCl_2 \times 2H_2O$, urea, MOA, MW, 620 W, 4 min	93	184

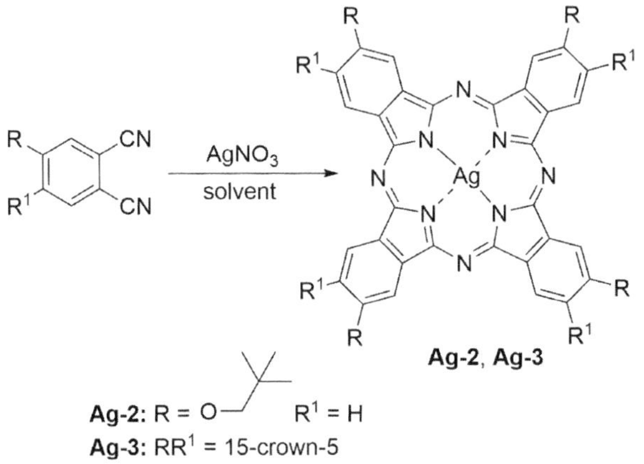

Scheme 115. Preparation of tetra-substituted $Ag^{II}Pcs$.[203,340]

Table 65. Preparation of **Ag-1** – **Ag-3**.

	Conditions	Yield (%)	Ref.
Ag-1	Li_2Pc, $AgNO_3$, EtOH(abs)	n.r.	52
Ag-1	PbPc, $AgSO_4$, 1-ClN	n.r.	52
Ag-2	$AgNO_3$, DMF, 75 °C, 6–7 h	93	340
Ag-3	$AgNO_3$, NaOAc, 1,2-dichloroethane: AcOH (3:4, v/v), refluxing, 2.5 h	40	203

Synthesis of Cu(II)Pc (Ref. 29)

A mixture of phthalonitrile (1.28 g), CuCl (0.25 g), and DBU (1.78 g) in *n*-pentanol (40 mL) was refluxed for 6 h. As the reaction proceeded, a blue powder was precipitated. On cooling, the product was collected by filtration and washed with aq HCl (3%), H_2O, and EtOH. Yield: 1.2 g (83%).

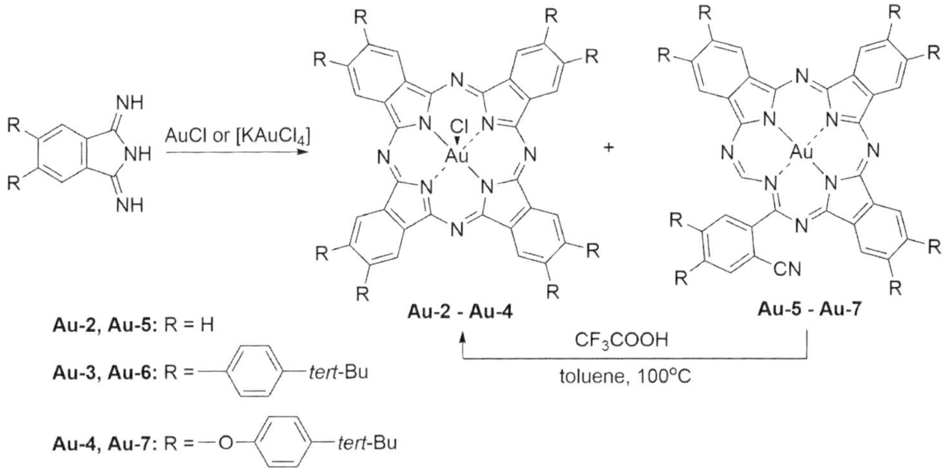

Scheme 116. The synthesis of $Au^{III}Pcs$.[343]

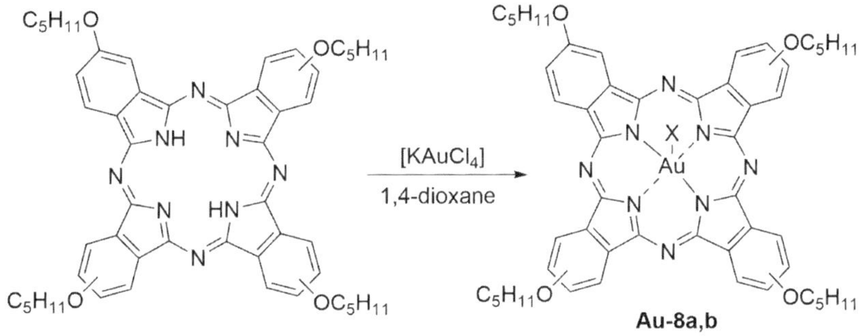

Scheme 117. Synthesis of the tetra-substituted Au(III)Pcs *via* metalation of the metal-free Pc.[345]

Synthesis of Cu(II)Pc (Ref. 331)

Mixtures of 1,3-diiminoisoindoline (0.058 g) and $Cu(OAc)_2 \times 2H_2O$ (0.0217 g), using quinoline (10 mL) as solvent, were heated in an autoclave (50 mL) under 270 °C for 8 h, then the mixture cooled to room temperature. Purple crystals of CuPc were obtained after removal of the solvent. Yield: 0.02954 g (51.3%).

Synthesis of Cu(II)Pc (Ref. 287)

A glass tube was charged with phthalimide (100 mg), $CuCl_2$ (23 mg), *p*-toluenesulfonic acid monohydrate (13 mg), and HMDS (438 mg), and the mixture

Table 66. Preparation of **Au-1 – Au-4, Au-8**.

	Conditions	Yield (%)	Ref.
Au-1	DI, AuBr, melting, 240–250 °C, 5 min	n.r.	341
Au-1	PN, $HAuCl_4 \times H_2O$, nitrobenzene, refluxing, 10 min	35	342
Au-2	AuCl, melting, 230 °C, 5 min	19	343
Au-2	$HAuCl_4 \times 4H_2O$, melting, 230 °C, 30 min	11	344
Au-3	AuCl, 1-ClN, 200 °C	12	343
Au-3	$[KAuCl_4]$, 1-ClN, 200 °C	24	343
Au-4	AuCl, 1-ClN, 200 °C	14	343
Au-4	$[KAuCl_4]$, 1-ClN, 200 °C	14	343
Au-8a,b	$[KAuCl_4]$, 1,4-dioxane, AcOH/NaOAc buffer, refluxing, 24 h	n.r.	345

stirred at 100 °C for 1 h under an argon atmosphere. The mixture was cooled, and DMF (50 μL) was added, then the tube was sealed, and the mixture heated at 150 °C. A dark purple solid immediately appeared. After heating for 10 h, the mixture was cooled and filtered. The solid was washed with methanol and then dissolved in concentrated H_2SO_4 (5 mL). The solution was poured into water (100 mL), and the resulting blue precipitates filtered and washed successively with dilute H_2SO_4, water, and methanol. The solid was further purified by extraction with methanol by Soxhlet extractor. Yield: 62 mg (63%) as a blue solid.

Synthesis of Cu(II)Pc (Ref. 152)

To anhydrous $CuCl_2$ (0.030 g) dissolved in dry DMF (2.0 mL) was added $U(O_2)$ SuperPc (0.10 g), and the mixture heated under nitrogen for 3 h at 120 °C with stirring. After this time, the solvent was removed *in vacuo* and the resulting solid washed with water, acetone, and ether, and then vacuum dried. Yield: 0.060 g (95%) as a violet, microcrystalline powder.

Synthesis of Cu(II)Pc$(C_7H_{15})_8^{\alpha\alpha}$ (Ref. 337)

Metal-free 1,4,8,11,15,18,22,25-octaheptylphthalocyanine (70 mg) and copper(II) acetate (500 mg) were heated in refluxing *n*-pentanol (7 mL) for 1 h. On cooling, the resulting slurry was dissolved in toluene (20 mL) and filtered to remove excess copper(II) acetate. The crude product was purified by elution through a silica column with toluene, to yield the desired compound. Yield: 62 mg (85%).

Synthesis of Cu(II)Pc(OPr)$_8^{\alpha\alpha}$ (Ref. 338)

A copper(II) acetate monohydrate (0.75 g) was added to a refluxing solution of metal-free 1,4,8,11,15,18,22,25-octapropyloxyphthalocyanine (0.1 g) in *n*-butanol (5 mL). The solution was maintained at reflux for 0.5 h, then cooled, and chromatographed (SiO_2, CH_2Cl_2 and ether as eluents). The green-blue fraction which eluted first was crystallized from a slowly-evaporating solution in CH_2Cl_2-pyridine, to afford the target compound. Yield: 0.086 g (81%).

Synthesis of Cu(II)Pc(15-crown-5)$_4^{\beta\beta}$ (Ref. 251)

A mixture of dicyanobenzo-15-crown-5 (0.25 g), CuCl (0.02 g) and urea (0.02 g) was well ground in a ball-mill, and placed in a flask. The reaction temperature was brought to 160 °C, and held for 10 min, then cooled to room temperature. The crude product was dissolved in CH_2Cl_2 and added dropwise into ethyl acetate (*ca.* 150 mL), then stirred for 10 min. The precipitate was filtered off, washed with ethyl acetate, and dried with diethyl ether. Yield: 0.29 g (38.1%).

Synthesis of Cu(II)Pc(15-crown-5)$_4^{\beta\beta}$ (Ref. 251)

A mixture of dibromobenzo-15-crown-5 (3.5 g), CuCN (1.93 g), and quinoline (6 mL) was heated and stirred at 220 °C for 16 h under N_2. After cooling to room temperature, the reaction mixture was treated with ethanol, and the crude product precipitated. This was filtered off and washed with ethanol and diethyl ether to remove unreacted organic material. The excess CuCN was dissolved by treating the precipitate with a solution of NaCN in a water-ethanol (1:2, v/v) mixture. This solution was extracted with chloroform (5 × 20 mL), and the chloroform solution was dried over $MgSO_2$ and evaporated to dryness, to afford the pure product. Yield: 1.03 g (38%).

Synthesis of Cu(II)PcF$_{16}$ (Ref. 339)

Tetrafluorophthalonitrile (5.01 g) and fine copper turnings (12.7 g) were heated in refluxing 1-chloronaphthalene (50 mL) for 48 h with agitation. The cold mixture was added to light petroleum (b.p. 100–120 °C; 400 mL), and the insoluble product subjected to flotation in light petroleum to remove the excess copper. The crude material was dried, dissolved in concentrated sulphuric acid (100 mL) at 80 °C, and the solution filtered through a glass sinter, diluted with cold water (1 L), and kept for 24 h. The resulting purple crystals were identified as copper(II) hexadecafluorophthalocyanine. Yield: 4.20 g (78%).

Synthesis of Ag(II)Pc(OCH$_2$C(CH$_3$)$_3$)$_4^\beta$ (Ref. 340)

Metal-free tetra(neopentoxy)phthalocyanine (50 mg) and silver nitrate (62 mg) were heated in dimethylformamide (10 mL) at about 75 °C for 6–7 h. Drops of the reaction mixture were removed from time to time, to monitor the electronic absorption at 707 nm due to the starting metal-free Pc, relative to that at 685 nm due to the desired AgPc. The reaction was stopped when the 707 nm peak disappeared, and solution stored overnight at room temperature. The resulting green precipitate was washed with water and acetone, and recrystallized from toluene. Yield: 52 mg (93%).

Synthesis of Ag(II)Pc(15-crown-5)$_4$ (Ref. 203)

Metal-free tetra(15-crown-5)Pc (20 mg), silver nitrate (20 mg) and sodium acetate (20 mg) were taken in a mixture of 1,2-dichloroethane (15 mL) and acetic acid (20 mL), and the mixture refluxed for 2.5 h. The reaction mixture was filtered, and the resulting solid washed with excess water. The crude product was dissolved in chloroform and subjected to column chromatography over basic alumina. An initial light green band eluted in $CHCl_3$: CH_3OH (98:2, v/v), to give unreacted initial Pc. A bluish-green band eluted in $CHCl_3$: CH_3OH (95:5, v/v) gave the pure Ag(II) Pc(15-crown-5)$_4$. Yield: 8 mg (40%).

Synthesis of Au(III)(Cl)Pc (Ref. 343)

Gold(I) chloride (40 mg) and 1,3-diiminoisoindoline (50 mg) were heated at 230 °C for 5 min. The dark green product was then allowed to cool to room temperature and ground into a fine powder. The powder was extracted with a mixture of $CHCl_3$ and MeOH (v:v = 4:1). The solvent was removed under reduced pressure, and the crude product purified by silica gel column chromatography ($CHCl_3$/ MeOH (v:v = 10:1), followed by $CHCl_3$/ MeOH (v:v = 4:1)), to give a blue-green solid. Yield: 12 mg (19%).

Synthesis of Au(III)(Cl)Pc(Ph-p-tert-Bu)$_8^{\beta\beta}$ (Ref. 343)

To 5,6-di(*p-tert*-butylphenyl)-1,3-diiminoisoindoline (88 mg) and [$KAuCl_4$] (41 mg) in a Schlenk tube was added 1-chloronaphthalene (0.3–1.0 mL). The mixture was deoxygenated *via* three freeze-pump-thaw degassing cycles, and the resulting mixture stirred for 1.5 h at 200 °C under argon. After the reaction, 1-chloronaphthalene was removed by silica-gel and alumina-gel column chromatography using hexane as eluent. The resultant mixture was purified into fractions containing target gold complex (**Au-3**) and by-product (**Au-6**) by silica-gel column chromatography using a mixture of $CHCl_3$-MeOH (10:1 v/v) or $CHCl_3$ as eluent, respectively. Yield: 20 mg (21%).

13. *Group 12: ZnPc, CdPc, and HgPc*

Zinc Pc (ZnPc) is one of the most studied phthalocyanines spectroscopically, because the Zn(II) ion has a d^{10} configuration, so that LMCT and d–d transitions should not appear in the UV-region. Many syntheses of unsubstituted ZnPc (**Zn-1**) have been reported (Scheme 118).[21,29,42,339,346] Zinc metal successfully reacts with phthalonitrile in the melt to provide ZnPc (Table 67).[21] When a zinc salt was used, the reaction in the melt with zinc(II) chloride and zinc(II) chloride gave chlorinated or brominated ZnPc, respectively (Table 67).[21,339] On the other hand, the corresponding reaction with zinc iodide is faster, and produces only non-halogenated ZnPc (Table 67).[339] **Zn-1** can also be synthesized using the Tomoda *et al.* method.[29] Refluxing phthalonitrile and $ZnCl_2$ in *n*-pentanol in the presence of DBU gives **Zn-1** in good yield (Table 67).[29] Furthermore, the effects of high pressure and microwave irradiation on the synthesis of ZnPc has been studied (Table 67).[42,184,346] *Cadmium* Pc (CdPc) can be prepared by similar methods to ZnPc, by reaction of phthalonitrile with cadmium metal,[21] the Tomoda *et al.* method,[29] and using microwave irradiation (Scheme 118. Table 67).[184] To date, there are only a few reports on the synthesis of *mercury* Pc (HgPc). HgPc can be synthesized from Li_2Pc[52] and H_2Pc[347] using $HgCl_2$ and $Hg(Me)N(SiMe_3)$, respectively (Scheme 118, Table 67).

Synthesis of ZnPc (Ref. 29)

A mixture of phthalonitrile (1.28 g, 10 mmol) and zinc chloride ($ZnCl_2$) (2.5 mmol) in the presence of DBU (10 mmol) was refluxed for 6 h in *n*-pentanol. As the reaction proceeded, a blue or greenish-blue powder gradually precipitated. The precipitate was collected by filtration, washed with 3% HCl solution, water, and then ethanol, and purified by an extraction technique with chloroform. Yield: 74%.

Synthesis of CdPc (Ref. 29)

A mixture of phthalonitrile (1.28 g, 10 mmol) and cadmium chloride ($CdCl_2$) (2.5 mmol) in the presence of DBU (10 mmol) was refluxed for 6 h in *n*-pentanol. As the reaction proceeded, a blue or greenish-blue powder gradually precipitated. The precipitate was collected by filtration, washed with 3% HCl solution, water, and then ethanol, and purified by an extraction technique with chloroform. Yield: 64%.

Synthesis of HgPc (Ref. 52)

A solution of mercuric chloride ($HgCl_2$) in EtOH (15 ml) was added gradually with shaking to Li_2Pc (2 g) dissolved in 20 ml of EtOH. A dull green precipitate of mercuric phthalocyanine separated immediately. and was collected by filtration. The yield is not reported in the original article.

14. *Group 13: B-subPc, AlPc, GaPc, InPc, and TlPc*

Unlike other metallic elements, boron forms a subphthalocyanine (B-SubPc), in which the boron is surrounded by three isoindole units linked by *meso*-nitrogen atoms. In 1972, Meller and Ossko reported the first synthesis of B-SubPc.[348] They accidentally found that this unique reaction of phthalonitrile with boron complexes such as BCl_3 provided macrocyclic boron complexes. Of the various boron reagents (BCl_3, BBr_3, BF_3, BPh_3, $BPhCl_2$) employed to date in SubPc synthesis, BCl_3 is the most commonly used, since it provides the best compromise between macrocyclization yield and stability of the formed chloro-SubPcs.[349–352] The SubPc is typically synthesized by the reaction of phthalonitrile with BCl_3 in refluxing *p*-xylene (Scheme 119, Table 68). There are a few reports on the synthesis of SubPcs from an 1,3-diiminoisoindoline.[353] Axial ligand exchange reactions of SubPc can introduce various axial ligands to chloro-SubPc. As shown in Scheme 120, the chloro-SubPc can be transformed into fluoro-, hydroxy-, or phenoxy-SubPcs, by reaction with excess $EtOBF_3$, SiO_2 and H_2O, or phenol, respectively.[354–360] In addition, a wide range of nucleophiles can be reacted with the chlorine atom of chloro-SubPc.[361]

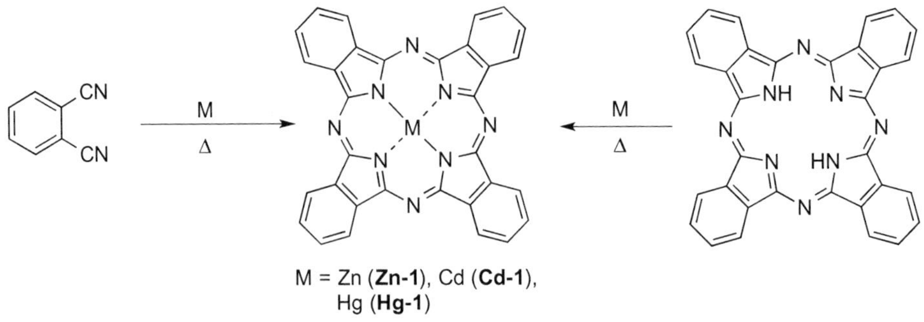

Scheme118. Synthesis of ZnPc, CdPc, and HgPc.[21,29,347]

Table 67 Synthesis of unsubstituted **ZnPc, CdPc,** and **HgPc**.

	Conditions	Yield (%)	Ref.
Zn-1	PN, Zn metal, 245 °C, 0.5 h	80	21
Zn-1	PN, ZnI_2, 280 °C, 0.5 h	87	339
Zn-1	PN, $ZnCl_2$, urea, Na_2SO_4, microwave, 4 min	80	346
Zn-1	PN, $Zn(NO_3)_2$, hydroquinone, MeOH, 90 °C, 12 kbar, 16 h	28	42
Zn-1	PN, $Zn(acac)_2$, DBU, *n*-pentanol, reflux, 6 h	58	29
Zn-1	PN, $ZnCl_2$, DBU, *n*-pentanol, reflux, 6 h	74	29
Zn-1	PN, $ZnCl_2$, microwave (490 W), 5 min	87	184
Cd-1	PN, Cd metal, 290 °C, 3h	n.r.	21
Cd-1	PN, $CdCl_2$, DBU, petanol, reflux, 6 h	64	29
Cd-1	PN, $CdCl_2$, MW (630 W), 10 min	88	184
Hg-1	Li_2Pc, $HgCl_2$, EtOH, 3 h	n.r.	52
Hg-1	H_2Pc, $Hg(Me)N(SiMe_3)$, benzene, reflux, 48 h	57	347

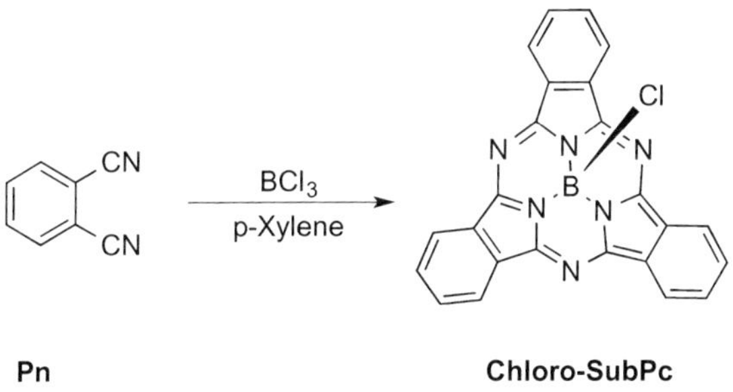

Scheme 119. Synthesis of SubPc.

The *aluminum* [Al(III)], *gallium* [Ga(III)], and *indium* [In(III)] ions in AlPc, GaPc, and InPc are trivalent; thus neutral AlPc, GaPc, and InPc have an axial ligand. AlPcs having Cl or Br as their axial ligand can be synthesized by reacting aluminum trichloride with phthalonitrile in refluxing quinoline or 1-chloronaphthalene. The chloro-AlPc can be transformed to a hydroxy derivative by reaction with sulfuric acid (Scheme 121, Table 69).[21,372] The hydroxy-AlPc can also be prepared from phthalonitrile, tributoxyaluminum ($Al(OBu)_3$), and urea in refluxing butanol, followed by hydrolysis in aqueous sulfuric acid.[165] Moreover, AlPc can be prepared by metalation reaction.[279,372] Similarly to AlPc, GaPc can be synthesized from phthalonitrile or 1,3-diiminoisoinndoline by reaction with gallium trichloride ($GaCl_3$) in refluxing quinoline (Table 69).[362,365] In addition, the metalation reaction of H_2Pc with $GaCl_3$ in DMF is possible.[279] InPc can be prepared

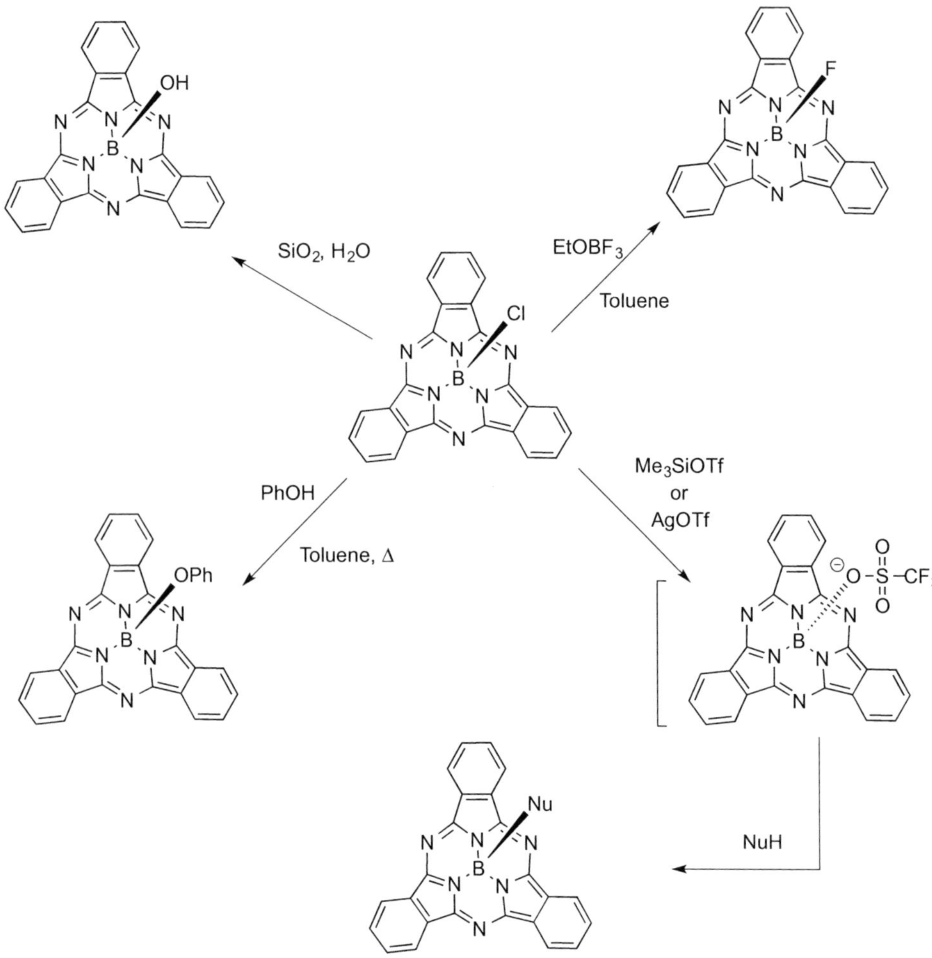

Scheme 120. Ligand exchange reaction of Chloro-SubPc.

Table 68 Synthesis of unsubstituted **B-subPc**.

	Conditions	Yield (%)	Ref.
B-1	PN, BCl_3, *p*-xylene	82	349
B-1	PN, BCl_3, 1,2-dichlorobenzene, reflux, 1.5 h	59–72	350
B-1	PN, BCl_3, *p*-xylene, reflux, 0.5 h	75	351
B-1	PN, BCl_3, 1-chloronaphthalene, microwave (540 W), 4 min	90	352

similarly to AlPc and GaPc (Table 69). The reaction of phthalonitrile with indium trichloride ($InCl_3$), indium tribromide ($InBr_3$), and iodine (I_2) provides chloro-InPc,[362] bromo-InPc,[365] and iodo-InPc,[366] respectively (Table 69). Janczak and Kubiak reported *thallium* Pcs (TlPc). They found that the reaction of phthalonitrile with thallium metal gives bicyclic TlPc (**Tl-1**),[368] while the reaction of phthalonitrile with InTl alloy provides Tl_2Pc (**Tl-2**) (Scheme 122. Table 69).[369,370]

Synthesis of B(Cl)SubPc (Ref. 349)

BCl_3 (2 ml, 1 M solution in *p*-xylene) was added to dry phthalonitrile (2 mmol) under an argon atmosphere, and the reaction mixture stirred under reflux for 20 min. The resulting purple solution was then flushed with argon, the solvent evaporated, and the solid extracted with toluene (100 ml). The solution was evaporated, and the resulting purple solid thoroughly washed with methanol (50 ml) and hexane (50 ml). Yield: 235 mg (82%).

Synthesis of Al(Cl)Pc (Ref. 362)

Phthalonitrile (20 g), aluminum trichloride ($AlCl_3$) (5. 0 g), and quinoline (distilled twice and then deoxygenated with nitrogen) (100 ml) were refluxed for 30 min. The reaction mixture was then cooled to approximately 0 °C and filtered. The solid thus isolated was washed with benzene, carbon tetrachloride and acetone, and dried at ~110 °C. Yield: 10.7 g (50%).

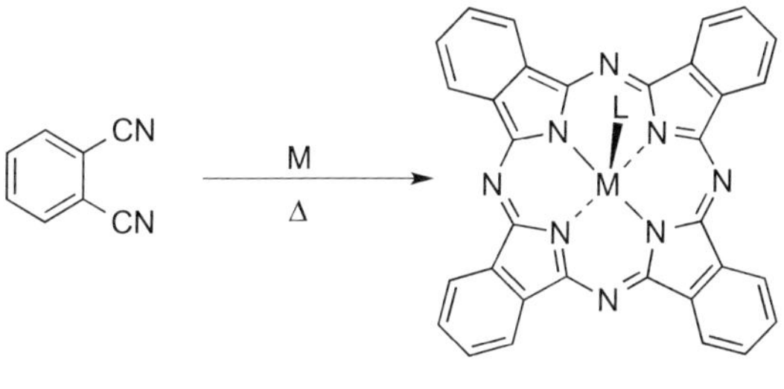

M = **Al-1** (L: -OH), **Al-2** (L: -Cl),
Ga-1 (L: -Cl), **In-1** (L: -Cl),
In-2 (L: -Br), **In-3** (L: -I),

Scheme 121. Synthesis of AlPc, GaPc, and InPc.[21,165,362–367]

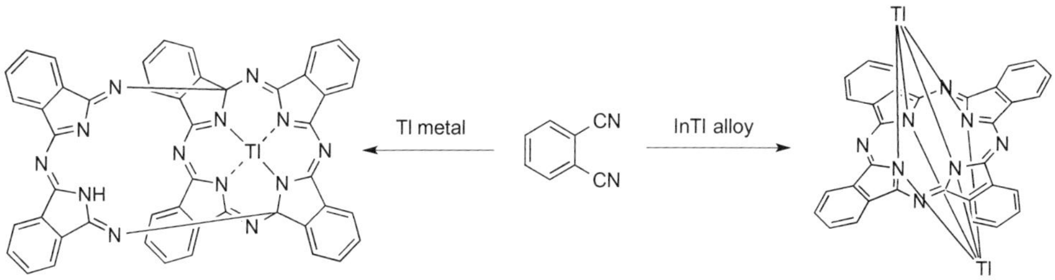

Tl1 **Tl-2**

Scheme 122. Synthesis of TlPc.[368–370]

Table 69. Synthesis of unsubstituted **AlPc, GaPc, InPc,** and **TlPc**.

	Conditions	Yield (%)	Ref.
Al-1	1. PN, $Al(OBu)_3$, urea, 1-butanol, 140 °C, 6 h; 2. H_2SO_4, 100 °C, 2 h	65	165
Al-2	PN, $AlCl_3$, quinoline, reflux, 0.5 h	70	21
Al-2	PN, $AlCl_3$, quinoline, reflux, 1 h	50	362
Al-2	PN, $AlCl_3$, quinoline, reflux, 1 h	n.r	363–365
Ga-1	PN, $GaCl_3$, quinoline, reflux, 1 h	45	362
Ga-1	DI, $GaCl_3$, quinoline, reflux	n.r.	365
In-1	PN, $InCl_3$, quinoline, reflux, 1 h	60	362
In-2	PN, $InBr_3$, quinoline, reflux, 1 h	70	365
In-3	PN, In, I_2, 200 °C, 3 h	n.r.	366
In	PN, InMg, 210 °C, 24 h	25	367
Tl-1	PN, Tl metal, 210 °C, 48 h	n.r.	368
Tl-2	PN, InTl alloy, 200 °C	n.r.	369,370

Synthesis of In(Br)Pc (Ref. 365)

In(Br)Pc was prepared by refluxing a mixture of 20 g phthalonitrile, 12.5 g of $In(Br)Pc_3$, and 100 ml quinoline (doubly distilled and deoxygenated under nitrogen) for 2 h. The reaction mixture was then cooled to 0 °C and filtered. The obtained blue/black solid was isolated, washed with methanol and acetone, and dried at 70 °C. Yield: 10.8 g (70%).

15. *Group 14: SiPc, GePc, SnPc, $SnPc_2$, and PbPc*

Silicon Pc (SiPc), *germanium* Pc (GePc), and *tin* Pc (SnPc) have two axial ligands, one above and the other below the Pc plane, because silicon, germanium and tin are tetravalent. Although $Si(Cl_2)Pc$ can be prepared from phthalonitrile, the yield of this reaction is poor (8%) (Scheme 123, Table 70).[373] The reaction using 1,3-diiminoisoindoline or 2-cyanobenzamide as a precursor usually provides a better yield, because these compounds have higher reactivity than phthalonitrile (Scheme 123, Table 70).[52,374,380,381] GePc can be synthesized by reacting phthalonitrile, 2-cyanobenzamide, or 1,3-diiminoisoindoline with $GeCl_4$ in quinoline (Table 70).[375–378] In addition, metalation of H_2Pc by $GeCl_4$ in quinoline provides GePc (Table 70).[378] In the case of SnPcs, Sn(II)Pc and Sn)IV)Pc species ($Sn(Cl_2)Pc$) have been reported. These can be synthesized by the reaction of phthalonitrile with metallic tin or tin(II) chloride ($SnCl_2$), respectively

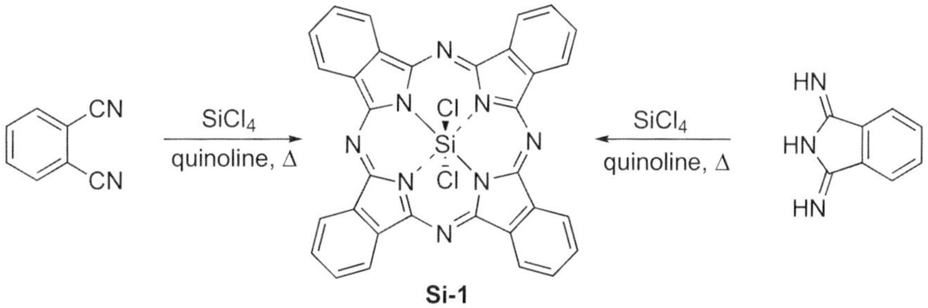

Scheme 123. Synthesis of SiPcs.

Table 70. Synthesis of unsubstituted **SiPc, GePc, SnPc** and **PbPc**.

	Conditions	Yield (%)	Ref.
Si-1	PN, $SiCl_4$, quinoline	8	373
Si-1	DI, $SiCl_4$, quinoline, reflux, 0.5 h	71	374
Si-1	CBA, $SiCl_4$, quinoline, 205 °C, 0.1 h	35	374
Si-1	DI, $SiCl_4$, quinoline, reflux	80	380
Si-1	DI, $SiCl_4$, Bu_3N, tetraline,	70	381
Si-1	DI, $SiCl_4$, quinoline, reflux, 0.5 h	71	382
Ge	PN, Ge powder, I_2, 200 °C, 3 h	n.r	375,376
Ge	PN, $GeCl_4$, quinoline, reflux, 4 h	n.r.	377
Ge	CBA, $GeCl_4$, quinoline, reflux, 4 h	20	378
Ge	H_2Pc, $GeCl_4$, quinoline, reflux, 4 h	50	378
Sn	PN, $SnCl_2$, DBU, 2-ethoxyethanol, reflux, 6 h	50	29
Sn	PN, Sn metal, 300 °C, 3 h	70	21
Sn	PN, $SnCl_2$, 210 °C	90	21
Sn	PN, $SnCl_2$, 1-chloronaphthalene, reflux, 3 h	83	365
Sn	PN, Sn powder, I_2, 200 °C, 3 h	n.r.	375
Pb	PN, PbO, 1-chloronaphthalene	16	379
Pb	PN, $PbCl_2$, DBU, *n*-pentanol, reflux, 6 h	77	29
Pb	Pn, $PbCl_2$, DBN, *n*-pentanol, reflux, 36 h	69	37
Pb	Li_2Pc, $Pb(OAc)_2$, EtOH, 3 h	n.r.	52

(Table 70).[21,365,375] SnPc can also be synthesized by the reaction of phthalonitrile, $SnCl_2$, and a catalytic amount of DBU in refluxing *n*-pentanol.[29] In addition, SnPc can be synthesized by metal insertion into H_2Pc, *via* reaction with tin(IV) chloride in DMF.[279] It has been reported that reacting Na_2Pc with $SnCl_2Pc$ in 1-chloronaphthalene at reflux provides a sandwich-like bis-phthalocyanine ($SnPc_2$).[379]

Unlike SiPc, GePc, and SnPc, only Pb(III)Pc species have been isolated to date. PbPc can be synthesized from the reaction of lead oxide with phthalonitrile in refluxing 1-chloronaphthalene solution (Table 70).[379] The Tomoda *et al.* method is also useful for the synthesis of PbPc. It can be obtained in good yield from a reaction mixture of phthalonitrile, tin(II) chloride, and DBU as catalyst in *n*-pentanol (Table 70).[29,37] PbPc can also be prepared by reacting Li_2Pc with $Pb(OAc)_2$ in EtOH (Table 70).[52]

Synthesis of Si(Cl_2)Pc (Ref. 374)

In a flask equipped with a water condenser, a mechanically-stirred mixture of 36.5 g (0.25 mol) 1,3-diiminoisoindoline, 41.5 ml (0.36 mol) silicon tetrarchloride, and 415 ml quinoline was brought slowly to reflux (219 °C). This temperature was maintained for 30 min and then cooled to 184 °C and filtered. The purple crystalline product was washed with quinoline, benzene, methanol, and acetone, and dried at 110 °C. Yield: 27.4 g (71%).

Synthesis of Ge(Cl_2)Pc (Ref. 378)

A mixture of 2.6 g (0.0051 mol) H_2Pc, 0.60 ml (0.0052 mol) $GeCl_4$ and 30 ml quinoline was heated to reflux in 10 min, and held at this temperature for 110 min. The crystalline product was separated from the hot reaction mixture by filtration, and washed with quinoline, benzene and acetone. Yield: 2.7 g (50%).

Synthesis of SnPc (Ref. 29)

A mixture of phthalonitrile (1.28 g, 10 mmol) and tin chloride ($SnCl_2$) (2.5 mmol) in the presence of DBU (10 mmol) was refluxed for 6 h in 2-ethoxyethanol. As the reaction proceeded, a blue or greenish-blue powder gradually precipitated. The precipitate was collected by filtration, washed with water and then ethanol, and purified by an extraction technique with chloroform. Yield: (50%).

Synthesis of PbPc (Ref. 29)

A mixture of phthalonitrile (1.28 g, 10 mmol) and lead chloride ($PbCl_2$) (2.5 mmol) in the presence of DBU (10 mmol) was refluxed for 6 h in *n*-pentanol. As the reaction proceeded, a blue or greenish-blue powder gradually precipitated. The precipitate was collected by filtration, washed with 3% HCl solution, water and then ethanol, and purified by an extraction technique with chloroform. Yield: (77%).

16. *Group 15: PPc, AsPc, SbPc, BiPc, BiPc$_2$, and Bi$_2$Pc$_3$*

Although the first synthesis of *phosphorus* Pc (PPc) was reported in 1981,[396,397] a later report pointed out that it did not have a phthalocyanine structure.[398] PPcs can be synthesized by the reaction of H_2Pc with PCl_3 or $POBr_3$ in dichloromethane or pyridine (Table 71).[384,394] Chloro *arsenic* Pc (As(Cl)Pc) can be prepared by metal exchange reaction of Li_2Pc with arsenic(III) chloride$_3$ in DFM.[231] The reaction of phthalonitrile with powdered arsenic provides iodine-doped arsenic phthalocyanine ([(AsPc)(Pc)]I_2) (Table 71).[399] The first synthesis of As(V)Pc was reported by Isago and Kagaya.[400] This was synthesized by the oxidative addition of XeF_2 to As(III)Pc. *Antimony* Pc (SbPc) can be synthesized by the reaction of phthalonitrile with Sb powder or $SbCl_3$.[52,387,388] SbPc can also be prepared by the metalation reaction of H_2Pc with SbF_3 or $SbCl_3$ in chloronaphthalene (Table 71).[52,389] Examples of unsubstituted PPcs, AsPcs, and SbPcs are still rare. α-Substituted PPc, AsPc, and SbPc have been studied because of their interesting spectroscopic properties. In 2011 and 2014, Furuyama, Kobayashi *et al.*[401,402] reported that α-*octa*-substituted P(V)Pcs, As(V)Pcs, and Sb(V)Pcs show the Q band beyond 1000 nm. The P(V)Pc was synthesized by the reaction of H_2Pc containing Group16 (S, Se, and Te) elements at α positions with $POBr_3$ in pyridine, following treatment with KPF_6.[402] *Bismuth* (III) Pc (BiPc) can be synthesized by the reaction of Li_2Pc

Table 71. Synthesis of unsubstituted **PPc, AsPc, SbPc, BiPc, BiPc$_2$** and **Bi$_2$Pc$_3$**.

	Conditions	Yield (%)	Ref.
P	H_2Pc, PCl_3, (*i*-Pr)$_2$EtN, CH_2Cl_2, r.t., 1 h	39	384
P	H_2Pc, $POBr_3$, pyridine, 80–90 °C, overnight	65	394
AsPc(Pc)I2	PN, powdered arsenic, I_2, 200 °C, 3 h	n.r.	399
As-Cl	Li_2Pc, $AsCl_3$, DMF, r.t., 0.5 h	n.r.	231
Sb	PN, Sb_4O_6, I_2, 220 °C, 3 h	n.r.	385,386
Sb	PN, Sb powder, 300 °C, 4 h	75	52
Sb	PN, $SbCl_3$, 150 °C, 10 h	1	387,388
Sb	H_2Pc, SbF_3, chloronaphthalene, reflux	n.r.	389
Sb	H_2Pc, $SbCl_3$, chloronaphthalene, reflux	90	52
Sb	PN, SbI_3, 150 °C, 24 h	?	395
Bi	Li_2Pc, $BiCl_3$, acetonitrile	70	390
Bi	Li_2Pc, $BiBr_3$, acetonitrile	40	390
BiPc$_2$	PN, Bi, I_2, 200 °C, 24 h	n.r.	391
Bi$_2$Pc$_3$	PN, Bi_2Se_3, 220 °C, 24 h	n.r.	392
Bi$_2$Pc$_3$	PN, $Bi(OAc)_3$, 310 °C, 0.5 h	n.r.	393

with bismuth(III) chloride ($BiCl_3$) or bismuth(III) bromide ($BiBr_3$) in acetonitrile.[390] The sandwich-type $BiPc_2$ can be prepared by reacting phthalonitrile with Bi metal and iodine at 200 °C. The triple-decker Bi_2Pc_3 can also be obtained by reacting phthalonitrile with bismuth acetate[393] or Bi_2Se_3[392].

Synthesis of $P(OH)_2Pc$ (Ref. 394)

$POBr_3$ (1.0 g) was added to a suspension of H_2Pc (50 mg) in pyridine (30 mL) and the stired mixture was heated to 80–90 °C overnight. After cooling to room temperature, the mixture was poured into water (100 mL), and the resulting precipitate was collected by vacuum filtration and washed thoroughly with water. The crude product thus obtained was taked up in pyridine and purified by chromatography on silica gel using a mixture of pydidine and methanol (1:1) as eluent. The blue fraction cantaining the bulk of the product was collected and the solvents were removed under reduced pressure to give 38 mg (65%) of a blue solid.

Synthesis of $[P(OME)_2PcSPh_8^{\alpha\alpha}]^+PF_{6-}$ (Ref. 401)

$POBr_3$ (150 mg) was added a solution of $H_2PcSPh_8^{\alpha\alpha}$ (8.3 mg) in 2 mL of pyridine and stirred for 1.5 h under reflux. After the solvent was removed *in vacuo*, the residue was dissolved in a solution of CH_2Cl_2/MeOH (50/50, v/v) and stirred for 30 min at room temperature. The organic layer was collected, washed wish water, and the solvent was removed to yield a dark green product (P(V)Pc), which was dissolved in CH_2Cl_2 and then KPF_6 (4.0 mg) was added. After the mixture was stirred for 12 h at room temperature, the solvent was removed, and the residue was recrystallized from CH_2Cl_2/ *n*-hexane. Yield: 3.8 mg (40%) as a dark bluish green powder.

Synthesis of Sb(Cl)Pc (Ref. 52)

A mixture of 5 g H_2Pc, 5 g antimony trichloride ($SbCl_3$), and 5 ml chloronaphthalene was heated in a sealed tube at 300 °C for 6 h. The liquid was filtered and the dark green residue washed with ether, stirred with water, heated to boiling and allowed to settle. The crystalline phthalocyanine derivative sank to the bottom, while basic antimony compounds formed by the hydrolysis of the excess of reagent remained in suspension. This was cleared by the dropwise addition of hydrochloric acid. The product was then collected and washed with water and alcohol. Yield: 5 g (90%).

B. Water-Soluble Metallophthalocyanines

Pcs and MPcs are also used in biomedical applications, such as photodynamic therapy due to their high thermal and chemical stabilities and strong π conjugative structures. Hydrophilic and unaggregated Pcs and MPcs are important and potentially useful biomaterials. Therefore, many water-soluble Pcs and MPcs derivatives have been synthesized over recent years. A number of phthalocyanines bearing hydrophilic moieties, such as carboxylates, sulfonates, glucose, phosphonates, polyoxyethylene, amino and carboranyl groups have been reported. In this chapter, the synthetic methods of water-soluble Pcs are described. The water-soluble Pcs can be classified into three types: anionic water-soluble Pcs, cationic water-soluble Pcs and non-ionic water-soluble Pcs.

1. *Anionic Water-Soluble Phthalocyanines*

a. *Sulfonyl group*

Tetramerization of phthalic acid containing sulfonyl groups

Fukada, Bernauer and Baumann reported a method for synthesizing tetrasulfonated Pcs and MPcs.[403–406] In this classical synthetic method, tetrasulfonated Pcs were directly prepared from sulfonated phthalic acid and its corresponding sulfonated phthalic anhydride. However, it was pointed out that these methods generate significant amounts of uncomplexed metal ion as persistent impurities.[407] In 1965, Weber and Busch reported the preparation of highly purified tetra-sulfoMPcs by modifications of Fukuda and Baumann's procedures (Scheme 124).[407] The procedure for synthesizing tetra-sulfo CoPc[407] is as follows: 4-sulfophthalic acid (43.2 g, 0.162 mol), ammonium chloride (4.7 g, 0.09 mol), urea (58 g, 0.97 mol), ammonium molybdate (0.68 g, 0.6 mmol) and cobalt(II) sulfate 7-hydrate (13.6 g, 0.048 mol) are ground together until homogeneous.

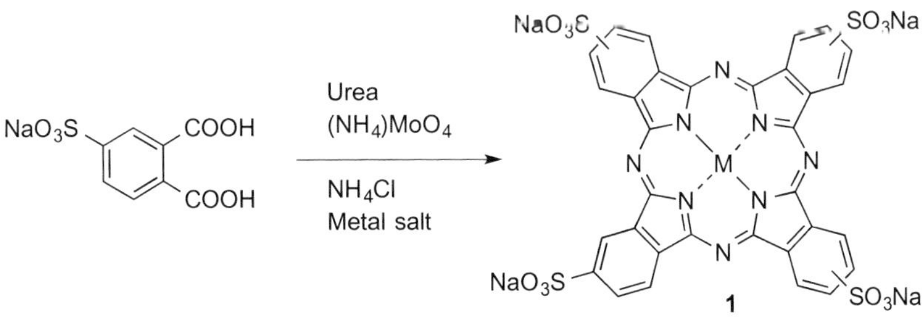

1

Scheme 124. Synthesis of tetrasulfonated MPc.[407]

Nitrobenzene (40 ml) was added to a three-necked flask, and heated to 180 °C. The solid mixture was added slowly, while keeping the temperature between 160–190 °C. The reaction mixture was heated for 6 h at 180 °C, then the crude product ground and washed with methanol until the odor of nitrobenzene could no longer be detected. The resulting solid was added to 1100 mL of 1 N hydrochloric acid saturated with sodium chloride. This treatment is crucial for the removal of excess cobalt(II) from the product. The solution and accompanying undissolved material were briefly heated to boiling, cooled to room temperature and filtered. The resulting solid was dissolved in 700 mL 0.1 N sodium hydroxide. The solution was heated to 80 °C, and insoluble impurities immediately separated by filtration. Sodium chloride (270 g) was then added to the solution. At this point, some of the solid product precipitated. The slurry was heated and stirred at 80 °C until ammonia evolution stopped. The product was obtained by filtration and the initial reprecipitation process was repeated twice. The solid was separated and washed with 80% aqueous ethanol until the filtrate was chloride-free. This product was refluxed for 4 h in 200 mL absolute ethanol, and the blue product filtered and dried overnight *in vacuo*. Yield: (80%).

Sulfonylation of phthalocyanine

Sulfonation of the phthalocyanine core is also widely used for preparing sulfonated Pcs. The sulfonation of the Pc core is usually carried out by oleum (fuming H_2SO_4 containing 30% free SO_3) at 75–145 °C (Scheme 125).[408–414] This procedure provides a mixture of mono-, di- tri- and tetra-sulfonated Pcs which can be separated by high performance liquid chromatography (HPLC).[410,413,414] Attempts to control the sulfonation process to obtain particular sulfo derivatives as pure products have been carried out. Kim *et al.* reported that by varying the oleum concentration, reaction temperature and time, up to four sulfonate groups could be introduced in the 4-position of the benzene ring.[409]

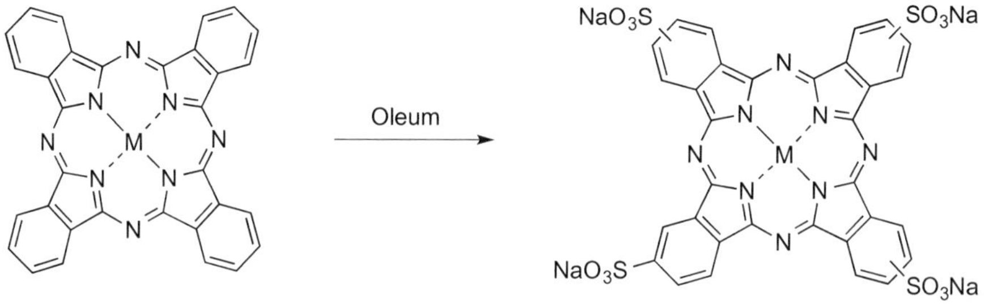

1

Scheme 125. Sulfonation of MPcs.[408–414]

The resulting products were mixtures of different regio-isomers. The general procedure of sulfonylation is as follows: [409] a total of 5 g MPc (M = Co, Cu, Ni or Zn) with 20 mL of oleum (20% SO_3) are refluxed at 145 °C for 3 h, and the obtained solution added to 500 mL distilled water. After adding sodium chloride (80 g) to the solution, the formed precipitate was filtered, washed with ethanol, and dried in a vacuum oven at 40 °C. The crude product was refluxed in ethanol for 2 h, hot filtered, washed with hot ethanol and subsequently dried in a vacuum oven.

b. *Carboxyl group*

Carboxylated Pcs and MPcs are also popular as anionic water-soluble Pcs. The carboxylated Pcs and MPcs show good solubility in alkaline aqueous media in which they are not protonated. Usually, carboxylated Pcs are prepared from the corresponding phthalic acid and phthalic anhydride as starting materials.

Synthesis of tetra-carboxylated MPcs

Tetra-carboxylated MPcs are typically prepared from trimellitic acid or trimellitic anhydride in two steps (Scheme 126).[415–420] The cyclo-tetramerization of these starting materials gives tetra-amide derivatives (**2**), and the following hydrolysis of amide groups under harsh alkaline or acidic conditions provides the desired tetra-carboxylated MPcs.

Synthesis of tetra-carboxylated H_2Pc, CoPc, CuPc and FePc[417]

Trimellitic anhydride (50 mol), urea (0.5 mol), and ammonium molybdate (0.1 mol) as a catalyst were placed in a three-necked flask having a reflux condenser. For MPc (CoPc, CuPc and FePc), the corresponding anhydrous metal salts

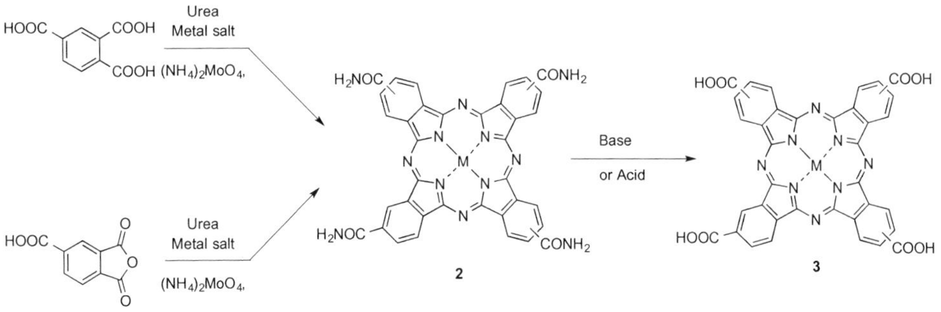

Scheme 126. Synthesis of a tetra-carboxylated MPc.[415–420]

($CoCl_2$, $CuSO_4$ and $FeCl_3$, 50 mmol) were added to the reaction mixture. After grinding and blending, the reaction mixture was stirred at 190 °C for 2–6 h, cooled to room temperature, then boiled in 5 mol/L HCl and filtered. The resulting mixture was washed with anhydrous ether and absolute methanol until the filtrate became colorless. The obtained tetraformamido-Pcs were dissolved in 50% sulfuric acid solution (170 mL) and added to 2 mol/L sodium nitrite solution (14 mL) dropwise while stirring at 5 °C. The reaction mixture was maintained at 30 °C for 1 h with constant stirring. The solvent was removed through centrifugal separation and filtration, and the resulting solid dissolved in deionized water. The obtained solution was neutralized with 2 mol/L H_2SO_4 solution to give a pH of 3. The solution was then filtered after standing for 12 h, and the resulting solid dried in a vacuum oven after washing. Yield: (9.8 – 26.9%).

Synthesis of octa-carboxylated MPcs

Octa-carboxylated MPcs are generally prepared from pyromellitic anhydride[421,422] or tetracyanobenzene[423] in two steps (Scheme 127). As shown in Scheme 127, the tetraphthalimide MPc (**4**) can be prepared by the reaction of pyromellitic anhydride with urea and a metal salt in DBU, or by the reaction of tetracyanobenzene with urea, metal salt and ammonium molybdate. The hydrolyzation of the tetraphthalimide MPc (**4**) gives the desired octa-carboxylated MPc (**5**). The alternative pathway for preparing octa-carboxylated MPcs is by hydrolysis of the ester groups.[424] 1,2,3,5-Benzentetracarboxylic acid can also be used as the starting material of octa-carboxy Pcs. Reaction of the 1,2,3,5-benzenetetracarboxylic acid with a metal salt, urea and $(NH_4)_2MnO_4$ provides octa-carboxylated MPcs **6** as a mixture of isomers (Scheme 128).[423]

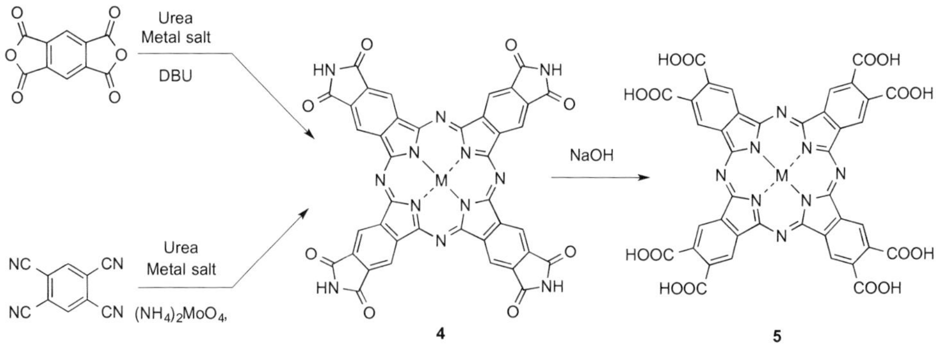

Scheme 127. Synthesis of an octa-carboxylated MPc.[421–423]

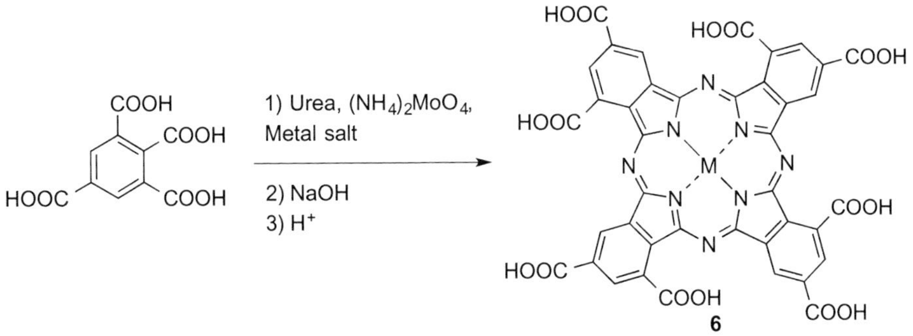

Scheme 128. Synthesis of an octa-carboxylated MPc **6**.[423]

Synthesis of octa-carboxylated CoPc[422]

Pyromellitic dianhydride (2.50 g, 11.5 mmol), urea (13.0 g, 0.22 mol), $CoCl_2$ (23.5 mmol), and DBU (0.1 g) are added to a two-necked flask. The flask is heated at 250 °C until the reaction mixture fused, then the reaction product washed with water, acetone and 6 N hydrochloric acid. After being dried, the obtained solid was hydrolyzed, and the crude product (30 g) and potassium hydroxide (KOH) (30 g) added to water (90 mL). The reaction mixture was heated at 100 °C for 8 h, then diluted with 200 mL water and filtered. The product precipitated as a blue solid at this point. The blue product was separated from the solution by centrifuging, and the obtained solid washed three times with water and dried. Yield: (30%).

2. *Cationic Water-Soluble Phthalocyanines*

a. *Quaternary amine*

Quaternary amine is the most popular functional group for cationic water-soluble Pcs. The main synthetic pathway is to first build a Pc structure bearing amine groups, and then quaternize the amine groups by multi-methylation. Several quaternization agents have been reported so far, namely, methyl bromide, methyl iodide and dimethylsulfate. Leznoff and coworkers reported the first cationic ZnPc in 1989[425] using a Pd-catalyzed coupling reaction of 4-iodophthalonitrile with 3-*N,N*-diethylaminoprop-1-yene to provide 4-(*N,N*-diethylaminoprop-1-ynyl) phthalonitrile (**7**) in high yield (89%). Catalytic hydrogenation of **7** gave 4-(3-*N,N*-diethylaminopropyl)phthalonitrile (**8**) in 78% yield. After conversion of **8** into 1,3-diminoisoindoline derivative **9** by bubbling ammonia, cyclo-tetramerization was carried out in DMAE, followed by metalation with $Zn(OAc)_2$ to give ZnPc

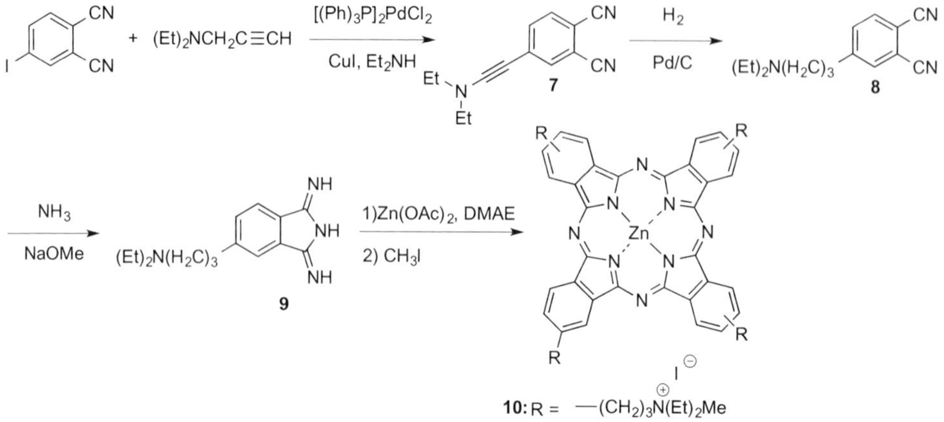

Scheme 129. Synthesis of quaternary amine-substituted ZnPc **10**.[425]

containing 3-*N,N*-diethylaminopropyl groups. Quaternization of the four amino groups of the ZnPc derivative was carried out by treatment with excess iodomethane in toluene at room temperature, to give the cationic water-soluble ZnPc **10** (Scheme 129). The authors also reported quaternary amine substituted CoPc and CuPc using the same synthetic method.[426] The Pd-catalyzed coupling reactions on halogenated phthalonitriles are a useful strategy for introducing amino functional groups. Tetra- and octa-trimethylaniliniumyl MgPcs can be synthesized using a Suzuki-Miyaura coupling reaction.[427] The coupling reaction of 4-iodo- or 4,5-dibromophthalonitrile with 4-*N,N*-methylaminophenyl boronic acid gives the phthalonitrile **11** or **12** (Scheme 130).[427] MgPcs substituted with dimethylaminophenyl were prepared by the reaction of the corresponding phthalonitriles and magnesium propoxide. The resulting MgPc **13** was subsequently treated with iodomethane in chloroform, to produce cationic MgPc **14**.

Sesalan *et al.* reported the synthesis of a MPc having 12 tertiary amino groups, which shows high water solubility (Scheme 131).[428] The reaction of 4-nitorophthalonitrile with 2,4,6-tris(*N,N*-dimethylaminomethyl)phenol provides the phthalonitrile derivative **15**. After cyclo-tetramerization of the resulting phthalonitrile **1** in the presence of anhydrous metal salt and DBU in *n*-pentanol, the obtained MPc (**16**) was reacted with excess iodomethane in dichloromethane for 4 h, giving a water-soluble MPc (**17**) containing 12 tertiary amino groups in 34% yield.

b. *Pyridine*

There are two main routes to access pyridine-substituted phthalocyanines: firstly, by synthesis from pyridine-substituted phthalonitrile; and secondly, by the introduction of pyridine into the phthalocyanine structure.

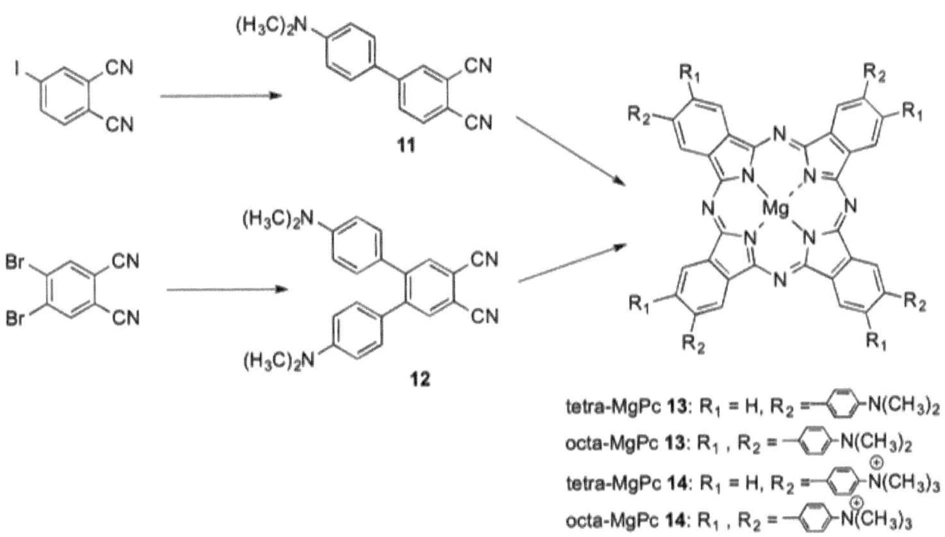

Scheme 130. Synthesis of MgPcs **13** and **14**.[427]

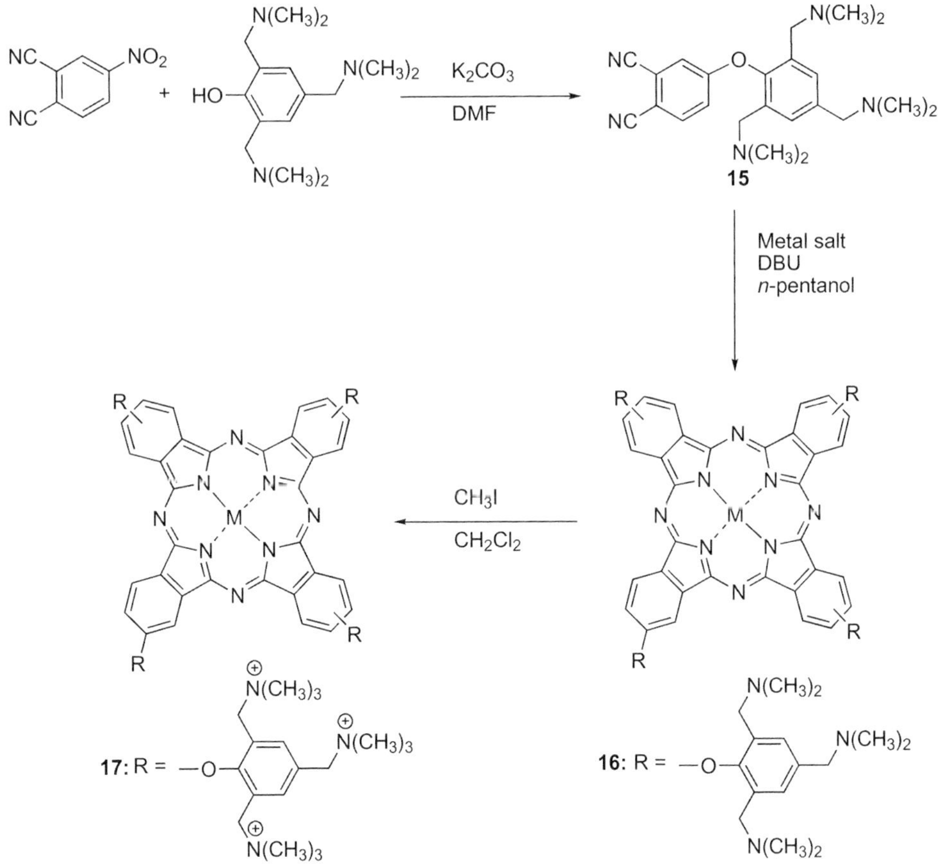

Scheme 131. Synthesis of MPc **17**.[428]

Synthesis from pyridine-substituted phthalonitrile

3- or 4-Pyridine-substituted phthalonitrile can be synthesized from 3- or 4-nitrophthalonitrile reacted with 4-hydroxypyridine, 3-hydroxyphridine, 2-hydrooxpyridine, and 2-mercaptopyridine[422,429–432,433, 437] by base catalysis nucleophilic aromatic displacement (Scheme 132).[422, 429–433] The tetramerization of the resulting phthalonitrile containing pyridine units provides pyridine-substituted phthalocyanines.

Efimov and coworkers reported a direct C–C link between the α positions of the phthalocyanine core (Scheme 133).[436] 3-(Pyridin-4-yl) phthalonitrile

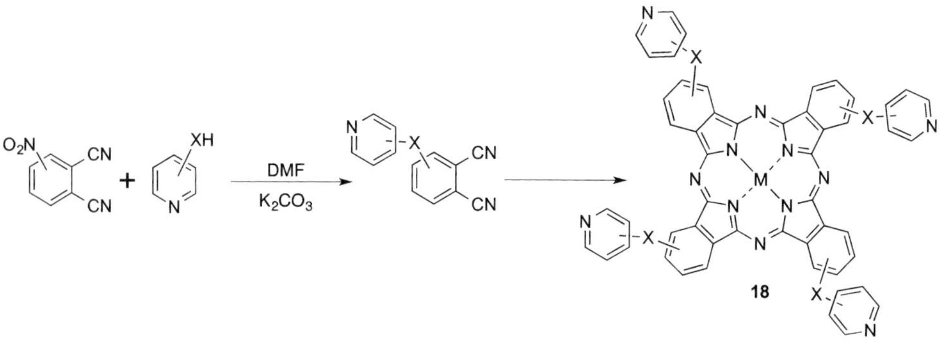

Scheme 132. Synthesis of pyridine-substituted Pc **18**.[422,429–433]

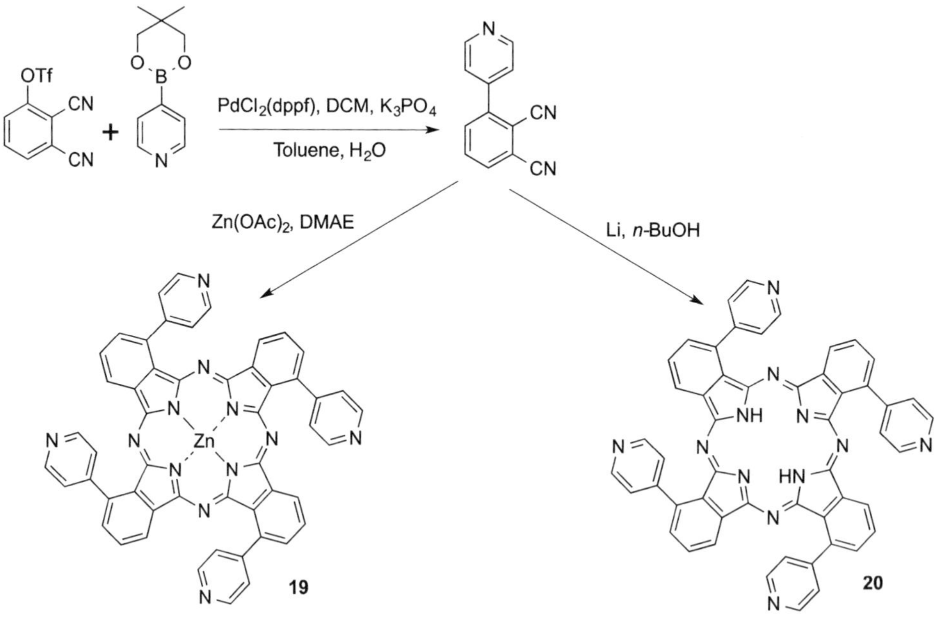

Scheme 133. Synthesis of pyridine-substituted Pcs **19** and **20**.[436]

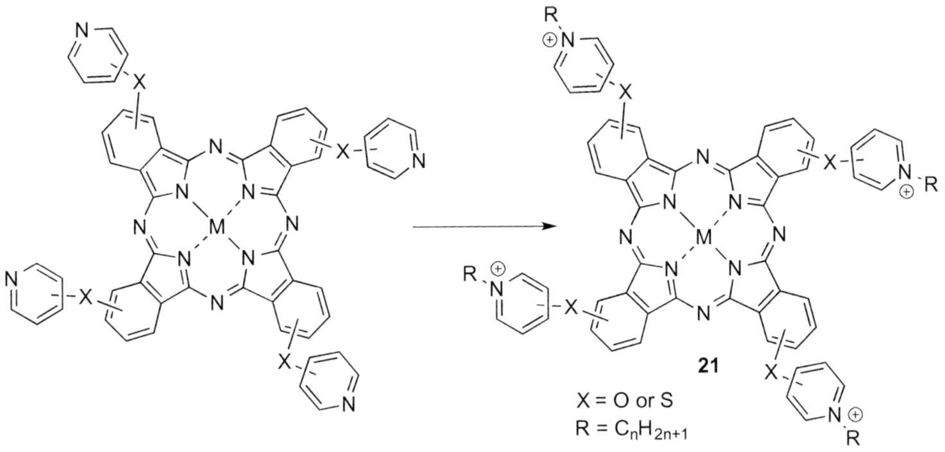

Scheme 134. Quaternization of the pyridine nitrogen.

was synthesized by Pd-catalyzed coupling reaction between 2,3-dicyanophenyl trifluoromethanesulfonate and 4-(5,5-dimethyl-1,3,2- dioxaborinan-2-yl) pyridine. The zinc phthalocyanine **19** and metal-free phthalocyanine **20** were synthesized by reaction with $Zn(OAc)_2$ in DMAE, or reaction with lithium in *n*-BuOH, respectively.

After cyclo-tetramerization, the quaternization of the pyridine nitrogen enhances the water-solubility and reduces aggregation (Scheme 134). The water-solubility of the phthalocyanine increases with the number of quaternized functional groups.

Introduction of pyridine unit into phthalocyanine core

The pyridine unit can be introduced into the phthalocyanine structure using aromatic substitution[434], esterification[435] and palladium-catalyzed coupling reaction.[438]

Aromatic substitution

Tolbin and Tomilova introduced pyridine groups into H_2Pc or ZnPc by nucleophilic substitution (Scheme 135).[434] The reaction between 4-bromopyridine and a Pc having phenol moieties (**22**) in DMSO produced the target compound **23** containing pyridine segments in high yield (88–91%). Subsequently, the corresponding water-soluble Pc bearing pyridinium hydrochloride segments was obtained by the reaction of MPc **24** with HCl.

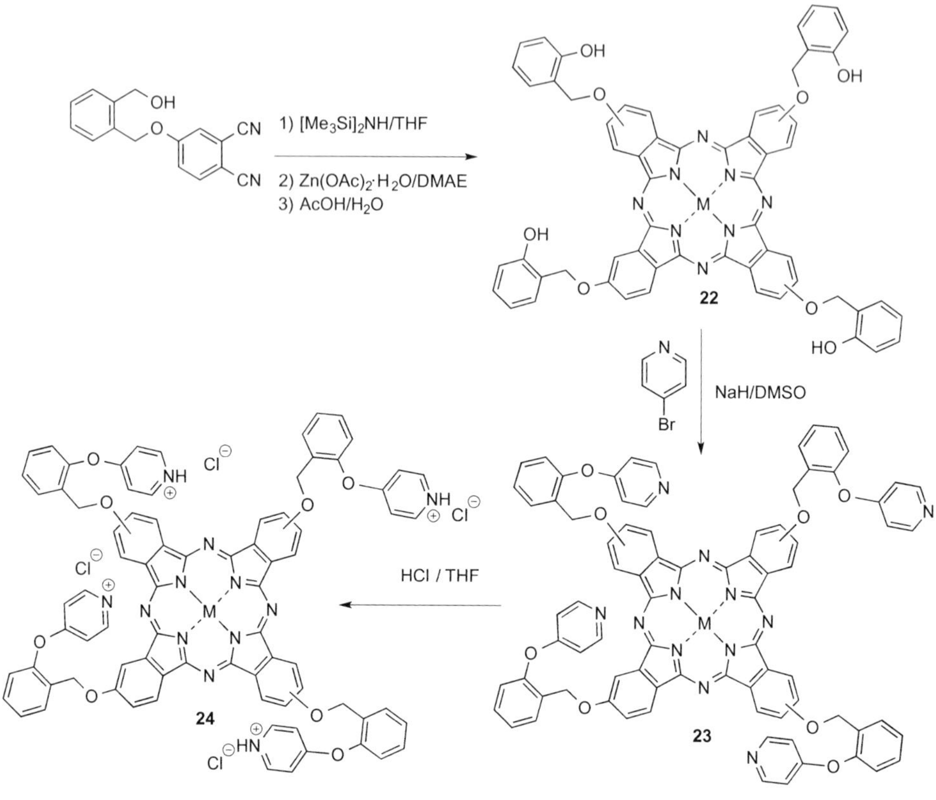

Scheme 135. Synthesis of pyridine-substituted Pc **24**.[434]

Esterification

Özçesmeci *et al.* reported an H_2Pc with four pyridyl groups, bound thorough ethylthio ester bridges on the peripheral positions (Scheme 136).[435] These pyridyl groups were introduced onto the Pc core by esterification. The H_2Pc bearing hydroxy groups (**25**) were reacted with pyridine-4-carboxylic acid in the presence of dicyclohexylcarbodiimide (DCCI) and *p*-toluenesulfonic acid (*p*-TSA) as catalysts for 2 days at ambient temperature. The desired tetra-pyridyl substituted H_2Pc **26** was obtained with a relatively high yield of 70%. Subsequently, the nitrogen atoms of pyridine were quaternized with iodomethane, to give the water-soluble tetra-cationic H_2Pc **27**.

Palladium-catalyzed coupling reaction

The palladium-catalyzed coupling reaction is also useful for the introduction of functional groups into the phthalocyanine core.[439] Sharman and van Lier reported

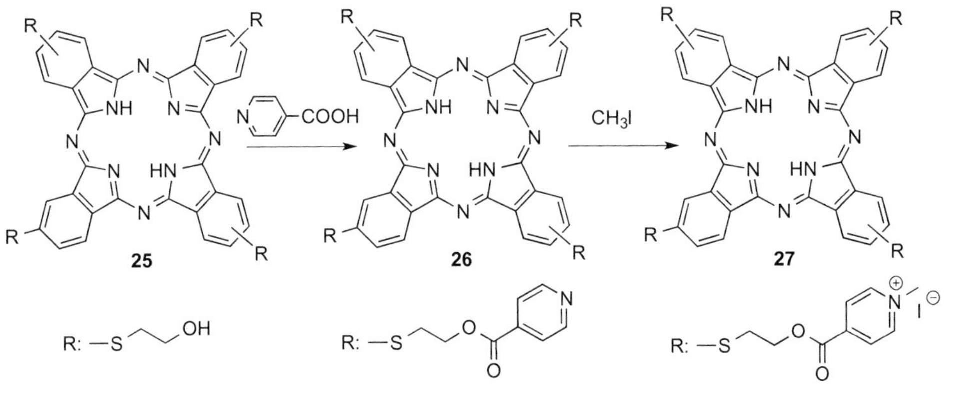

Scheme 136. Synthesis of pyridine-substituted Pc **27**.[435]

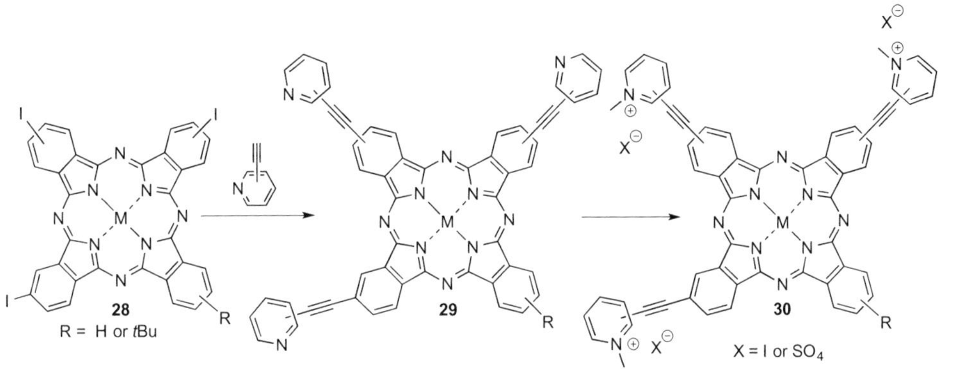

Scheme 137. Synthesis of pyridine-substituted Pc **30**.[438]

the Sonogashira coupling reaction of iodinated phthalocyanines with 2-ethynylpyridine, followed by *N*-methylation, to give the water-soluble tri-cationic substituted phthalocyanine **30** (Scheme 137).[438] The Sonogashira coupling reaction was carried out in the presence of dichlorobis(triphenylphosphine)palladium(II) and copper iodide in DMF/Et_3N.

3. *Non-Ionic Water-Soluble Phthalocyanines*

Ethylene glycol

Polyethylene glycol (PEG) substitution is a classical strategy for increasing the water-solubility of phthalocyanine. A number of PEG-substituted phthalocyanines have been reported to date, with many variations of the chain number and/or various ethylene glycol units. However, some of the MPcs with short chains such as ethylene glycol chains, are highly soluble in highly polar solvents but not in water. In this section, the synthetic pathways of tetra-PEG, octa-PEG, branched PEG, and axially PEG substituted water-soluble MPcs are described.

Tetra-PEG substituted water-soluble Pcs

Tetra-PEG substituted Pcs are usually prepared from 4-nitrophthalonitrile in two steps. McKeown and coworkers reported the synthesis of tetra-[oligo(ethyleneoxy)]-substituted phthalocyanine, and found that water-soluble Pcs having only four methyl-terminated oligoethyleneoxy groups showed lyotropic and thermotropic mesophases (Scheme 138).[440] The water-soluble Pcs are prepared from 4-nitrophthalonitrile in two steps. Although H_2Pc **30** is insoluble in water and anhydrous ethanol, H_2Pc **31** shows high water solubility. Non-peripherally PEG substituted MPcs were also synthesized from 3-ntitrophthalonitrile using the same synthetic strategy as for H_2Pc **30** and **31**.[441]

Octa-PEG substituted water-soluble Pcs

Octa-PEG substituted MPc can be accessed from catechol in four steps (Scheme 139)[442,443]: bromination of catechol using molecular bromine, alkylation of dibromocatechol in the presence of a base, a Rosenmund-von Braun reaction yielding the dicyanide, and the conversion of the phthalonitrile into MPc **32** by refluxing in absolute ethanol in the presence of DBN.

Branched PEG substituted water-soluble Pcs

Usually, increasing the number of side chains provides further solubility in solvents and decreases the probability of aggregate formation, because the

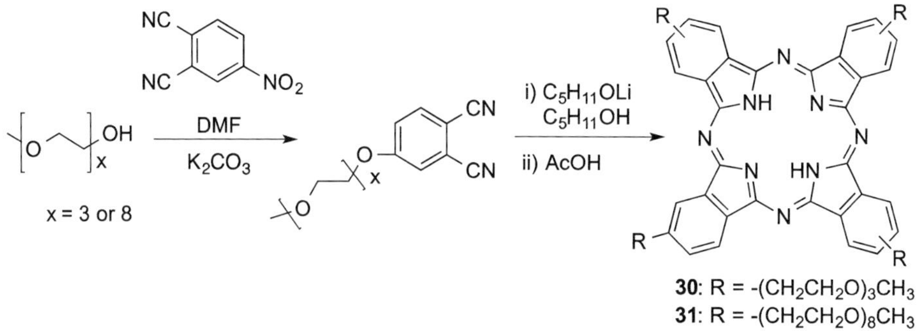

Scheme 138. Tetra-PEG substituted H_2Pcs.[440]

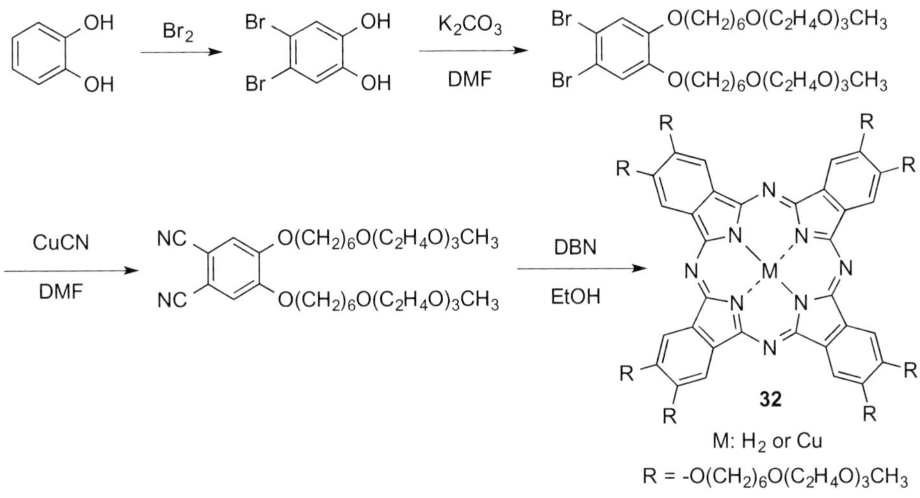

Scheme 139. Octa-PEG substituted H_2Pcs and CuPc.[443]

substituents can prevent interactions between the Pcs. Thus, Pcs having branched PEG chains show better solubility compared to those with linear PEG chains. Vacus and Simon reported the water-soluble ZnPc **37** and CuPc **38** having branched PEG, which show much better water-solubility than the unbranched octa-substituted MPcs (Scheme 140).[444] The synthetic pathway used for obtaining ZnPc **37** and CuPc **38** is shown in Scheme 140. 2-(2-Methoxyethoxy)ethanol opens epichlorohydrin under basic conditions and the resulting branched alcohol (**33**) achieves a nucleophilic substitution on **34**, followed by a dinitrilation step, leading to phthalonitrile **36**.

Ahsen and coworkers. reported thia-bridged branched PEG-substituted Pcs (**45** and **47**) (Scheme 141).[445–447] Tosylation of the hydroxy group was carried out in dry pyridine at room temperature with *p*-toluenesulfonyl chloride. Conversion of the tosyl compound **41** into the thiol derivative **42** was accomplished by boiling in dry ethanol in the presence of thiourea for 48 h. The obtained **42** was reacted with dichlorophthalonitrile, giving each of the desired phthalonitriles (**44** and **46**). Cyclo-tetramerization of phthalonitrile derivatives **44** and **46** provided the desired tetra- and octa-branched PEG-substituted MPcs **45** and **47**, respectively.

Axially PEG-substituted water-soluble MPcs

PEG can be introduced into the axial position of SiPcs[381,448] or AlPcs,[449] which can affect the solubility in aqueous media even more by improved steric shielding of the Pc core. As shown in Scheme 142, $Si(PEG)_2Pc$ was synthesized by the reaction

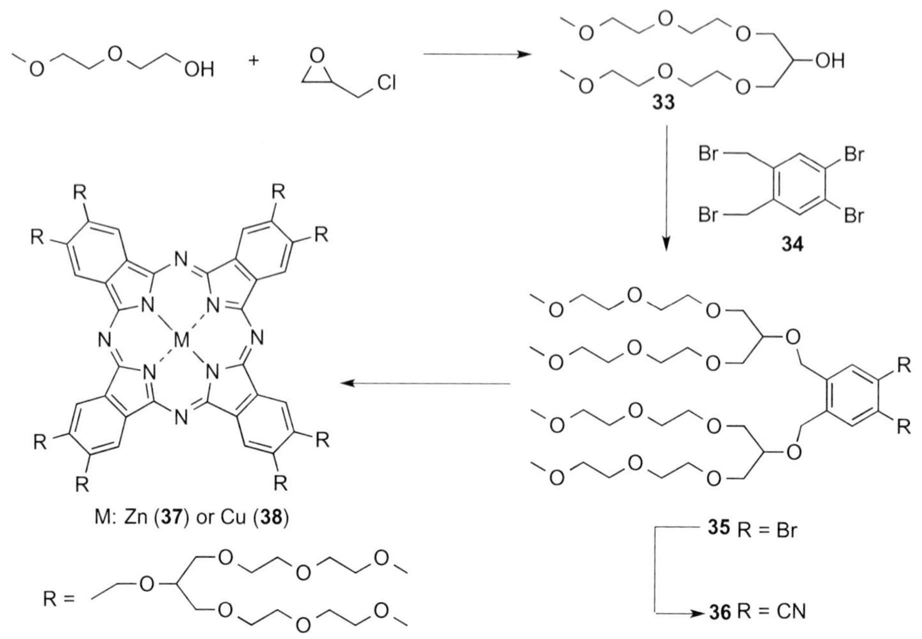

Scheme 140. Synthesis of branched-PEG substituted ZnPcs (**37**) and CuPc (**38**).[444]

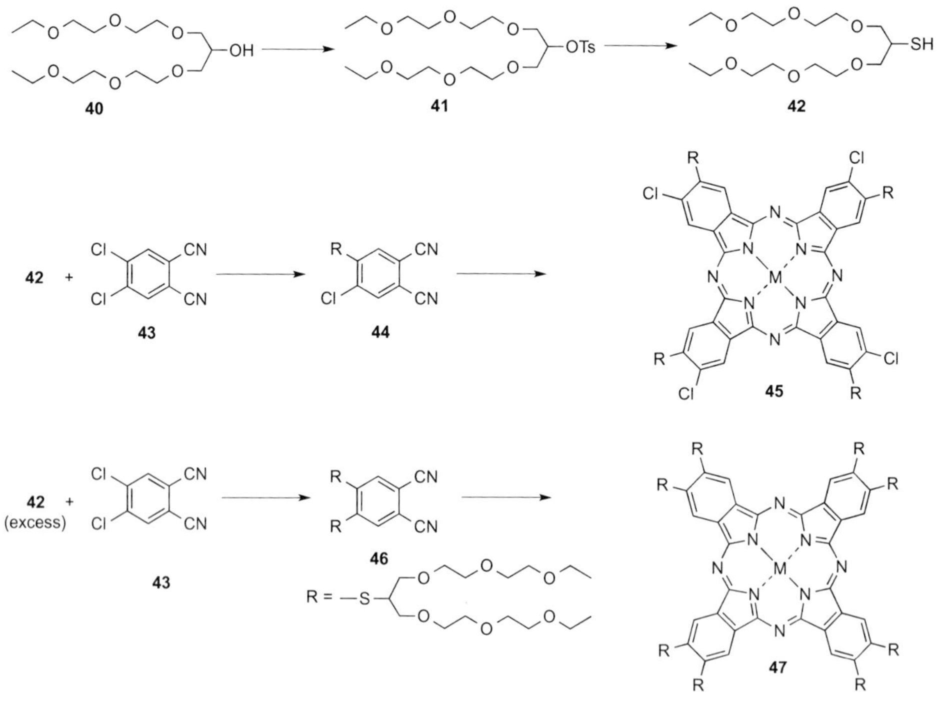

Scheme 141. Synthesis of branched-PEG-substituted MPcs **45** and **47**.[445–447]

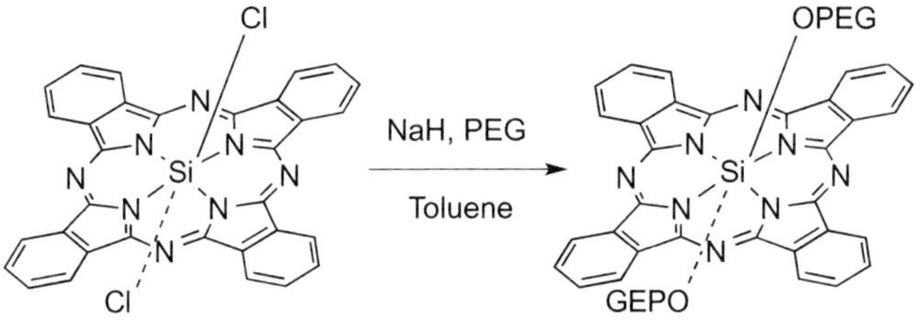

48

Scheme 142. Synthesis of axially PEG-substituted SiPc.[381,448]

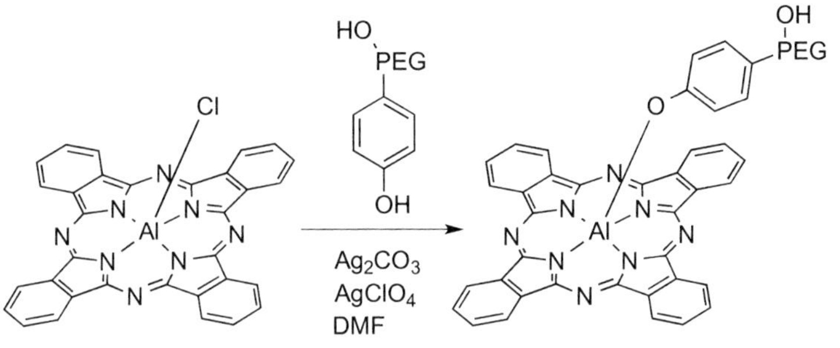

49

Scheme 143. Synthesis of axially PEG-substituted AlPc.[449]

of $Si(Cl)_2Pc$ with monomeroxy PEG and NaH in toluene. Hofman *et al.* found the hydrophilic PEG chains of $Si(PEG750)_2Pc$ to be very effective in shielding the hydrophobic Pc core, making the Pc much more hydrophilic and highly soluble in water.[381] Al(PEG)Pcs can also be accessed by the reaction of phenoxy-PEG with Al(Cl)Pc, silver carbonate, and silver perchlorate under Koenigs–Knorr conditions in DMF, to afford the axially substituted Al(PEG)Pc (Scheme 143).

Morpholine

Morpholine and its derivatives are important compounds as a result of their biological and pharmacological properties, such as anti-cancer,[450] anti-inflammatory,[451] anti-oxidant [452] and tyrosinase inhibitory activities[453], as well as in photodynamic therapy [454]. Bart and his coworkers reported tetra-morpholine-substituted ZnPc [455] (Scheme 144), synthesized by cyclotetramerization reaction of 4-morpholinophenol phthalonitriles. After cyclotetramerization, the authors quaternized the tetra-morpholinophenol-substituted ZnPc by mixing with iodomethane at room temperature, to increase the water-solubility (Scheme 144).

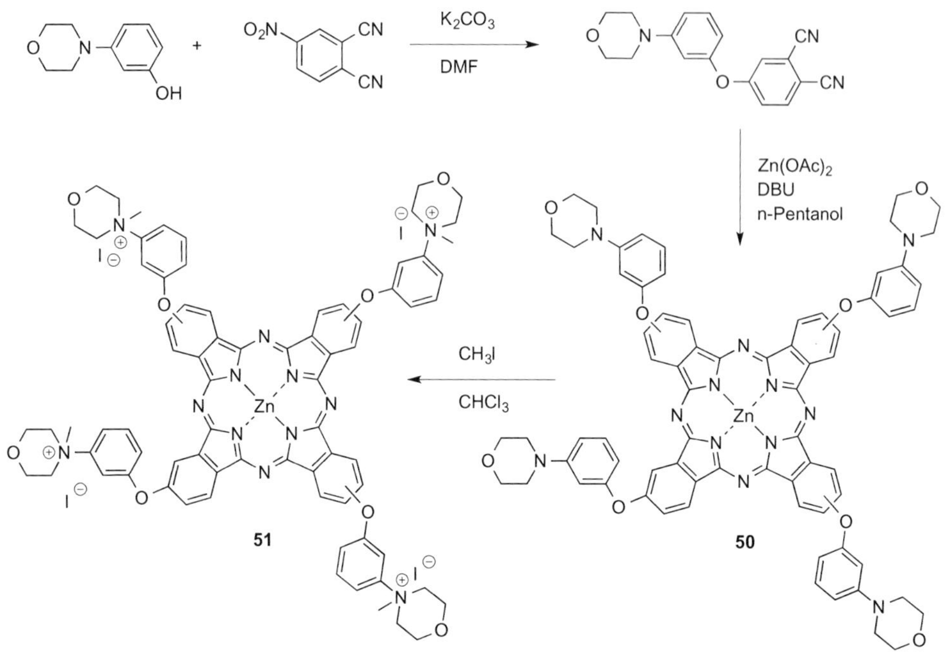

Scheme 144. Synthesis of a tetra-morpholine-substituted ZnPc.[455]

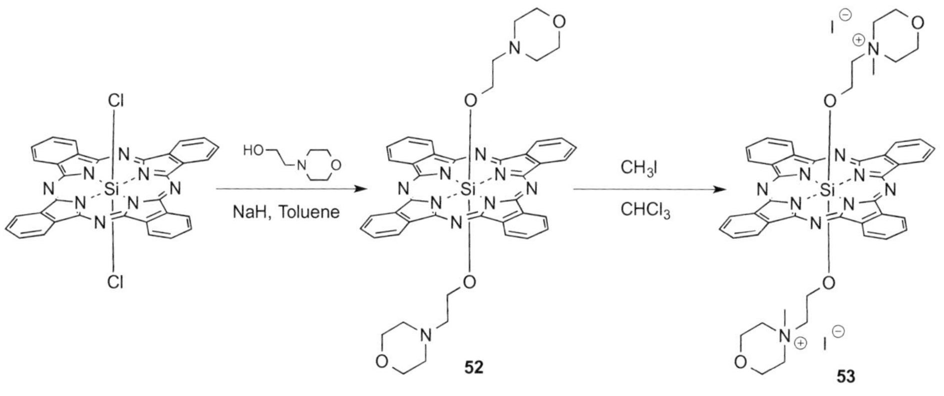

Scheme 145. Synthesis of axially morpholine-substituted SiPc.[456]

A water-soluble phthalocyanine axially modified by morpholine has also been reported.[456] Here, an axially morpholine-substituted SiPc was prepared from commercially-available SiPc dichloride (**1**) (Scheme 145). Reacting **1** with *N*-(2-hydroxy-ethyl)morpholine in the presence of NaH in toluene led to the axially morpholine-substituted SiPc (**2**). The quaternized Pc was obtained in 46% yield by treatment of **2** with an excess of iodomethane in $CHCl_3$.

V. Conclusions

The first synthesis of a Pc was achieved at the beginning of the 20th century, although the authors could not understand what they had obtained, except for the compound's intense blue color. Around 1927–1934, the Pc structure was confirmed. Since then, CuPcs have been mainly produced as blue-green dyes and pigments, while CoPcs have helped in removing sulfur from crude oil. The production of other metalloPcs was much lower as compared with CuPcs and CoPcs, but researchers in universities and research institutes began to work on other transition metalloPcs. However, very little modification of the Pc periphery had been attempted until around 1970, and it is only over the last 50 years that researchers have reported on substituent-linked Pcs. Because of the notoriously low solubility of Pcs and the lack of convenient purification methods, most pure, substituent-free Pcs were obtained only by repetitive sublimation. With the spread of column chromatography, the purification of Pcs with introduced *tert*-butyl groups became much easier. In particular, the use of gel-permeation chromatography from around 1990 improved the purity of the target Pc compounds. Accordingly, when Pcs purified by silica and/or alumina-chromatography were further subjected to gel-permeation chromatography, yellow unidentified impurities could almost always be separated. Although high purity may not be required as long as Pcs are used, for example, as dyes and pigments, highly purified compounds are indispensable in some studies, such as in the liquid crystalline field. In this area, the phase-transition temperature is examined, so that if impurities are included, the temperature always appears in the lower direction, since the melting point temperatures of chemicals are lower when impurities are included, making the data of little value.

Today, Pcs are often synthesized using substituted phthalonitriles as starting materials, because of the relative ease of purification (note, however, that not all metalloPcs can be synthesized using phthalonitrile: for example, high-yield synthesis of SiPc can only be achieved by using 1,3-diiminoisoindoline). However, even in this case, great care must be taken when characterizing the target compounds. Here, we confine ourselves to mentioning only the case of FePc as an example. FePcs can be prepared by template reactions using Fe(II) or Fe(III) salts and $Fe(CO)_5$, or iron insertion into H_2Pcs. Some researchers who are unfamiliar with this field believe that Fe(II)Pcs and Fe(III)Pcs can be obtained by the use of Fe(II) and Fe(III) salts, respectively, and insist, for example, that the catalytic activity for some reactions is higher for Fe(III)Pcs than Fe(II)Pcs. They even report the magnetic susceptibility data of the FePcs they obtained. However, the data are often unreliable, especially for Fe(III)Pcs. In particular, the compounds are not carefully purified, and in order to obtain precise magnetic susceptibility data, even a trace of unreacted Fe(III) salts must be excluded. Therefore, in addition to silica- or alumina-chromatography, further purification using gel-permeation chromatography or sublimation is required. A trace amount of Fe(III) salt in the sample makes the magnetic susceptibility data unreliable. In addition, researchers should take note that the oxidation state of iron in all *organic solvent-soluble* ***monomeric*** FePcs has been assigned as divalent, even if Fe(III) salts are used as a template. This situation is similar to CuPcs. Here, whether Cu(I) or Cu(II) salts are used as a template in the CuPc formation reaction, the oxidation state of Cu in the obtained CuPcs is always +2 (divalent). This conclusion that Fe is divalent is consistent with data from EPR and Mossbauer spectroscopies, and magnetic susceptibility data. Indeed, no-one has yet succeeded in finding an axial ligand of ***monomeric*** Fe(III)Pcs, although in the case of iron porphyrins, this is very easy. For example, in iron(III) octaethylporphyrin and tetraphenylporphyrin, we can find Cl^- as a counter anion in mass spectroscopy and elemental analysis. As organic solvent-soluble Fe(III)Pcs, only *dimeric* species such as μ-oxo[457] and μ-nitrido[458] derivatives are known. Their structures have been confirmed by X-ray crystallography,[457,458] and they show absorption bands at shorter wavelength than those of monomers,[459] in accordance with exciton theory.[460]

The iron atom of some *monomeric water-soluble* FePcs is judged to be trivalent under some circumstances, although an axial anionic ligand has not been identified. Thus, tetra-sulfonic and carboxylic, and octa-carboxylic FePcs show EPR signals assignable to Fe(III)Pcs in water or as powder.[461] When carboxylated Fe(III)Pcs are linked to polymers, the content of high-spin species increases through attaining a monomeric environment. Thus, for some reactions, the catalytic activities of Fe(III)Pcs increases with increasing high-spin content.[462] Although water-soluble FePcs can also be prepared by introducing pyridinium type ligands onto their periphery, the oxidation state of iron in this type of species has not been confirmed by EPR, magnetic susceptibility, or Mossbauer spectroscopies.

We have stressed the importance of purity in characterizing Pc compounds. To date, some researchers have asked us to characterize their compounds spectroscopically, although the samples are often not pure enough. We would like to ask synthetic researchers to purify their compounds as far as possible, since various suitable methods have been developed over the past 50 years. Some incorrectly written Pc papers have been published due to the lack of knowledge of the authors, and/or their use of samples having insufficient purity. These situations must be avoided in order to accumulate correct data on Pcs.

VI. Acknowledgments

S.V.D. is thankful to the Russian Science Foundation (17-73-20369) and the Russian Foundation for Basic Research (16-03-00368) for the partial financial support of this work. This work has been partially supported by Grants-in-Aid for Scientific Research (A) (No. 15H02172) and (C) (No. 18K05076) from the Japan Society for the Promotion of Science (JSPS).

VII. References

1. McKeown, N.B. *Phthalocyanine, Materials: Synthesis, Structure and Function*, Cambridge University Press: Cambridge, UK, **1998**.
2. Lezoff, C.C.; Lever, A. B. P. *Phthalocyanines: Properties and Applications*, VCH: New York, **1989**, Vol. 1.
3. Braun, A.; Tcherniac, J. *Ber. Deut. Chem. Ges.*, **1907**, *40*, 2709–2714.
4. Linstead, R.P. *J. Chem. Soc.*, **1934**, 1016–1017.
5. Erk, P.; Hengelsberg, H. *The Porphyrin Handbook*; Kadish, K. M.; Smith, K. M.; Guilard, R., Eds.; Academic Press: San Diego, **2003**, Vol. 19, pp. 105–149.
6. Herbst, W.; Hunter, K. *Industrial Organic Pigments Production, Properties, Applications*. Wiley-VCH, **2004**.
7. Berezin, B.D. *Coordination Compounds of Porphyrins and Phthalocyanines*, Wiley: Chichester, **1981**.
8. Gregory, P. *J. Porphyrins and Phthalocyanines*, **1999**, *3*, 468–476.
9. Sorokin, A.B. *Chem. Rev.* **2013**, *113*, 8152–8191.
10. Melville, O.A.; Lessard, B.H.; Bender, T.P. *ACS Appl. Mater. Interfaces*, **2015**, *7*, 13105–13118.
11. Gu, D.; Chen, Q.; Tang, X.; Gan, F.; Shen, S.; Liu, K.; Xu, H. *Opt. Commun.* **1995**, *121*, 125–129.
12. Bohrer, F.I.; Colesniuc, C.N.; Park, J.; Ruidiaz, M.E.; Schuller, I.K.; Kummel, A.C.; Trogler, W. C. *J. Am. Chem. Soc.* **2009**, *131*, 478–485.
13. de la Torre, G.; Vázquez, P.; Agulló-López, F.; Torres, T. *Chem. Rev.* **2004**, *104*, 3723–3750.
14. de la Torre, G.; Claessens, C.G.; Torres, T. *Chem. Commun.* **2007**, 2000–2015.
15. Ishii, K.; Kubo, K.; Sakurada, T.; Komori, K.; Sakai, Y. *Chem. Commun.* **2011**, *47*, 4932–4934.
16. Allen, C. M.; Sharman, W. M.; van Lier, J. E. *J. Porphyrins Phthalocyanines*, **2001**, *5*, 161–169.
17. Byrne, G.T.; Linstead, R.P.; Lowe, A.R. *J. Chem. Soc.* **1934**, 1017–1022.
18. Enokida, T.; Ehashi, S. *Chem. Lett.* **1988**, *17*, 179–182.
19. Brach, P.J.; Grammatica, S.J.; Ossanna, O.A.; Weinberger, L. *J. Heterocycl. Chem.* **1970**, *7*, 1403–1405.
20. Elvidge, J.A.; Linstead, R.P. *J. Chem. Soc.* **1955**, 3536–3541.
21. Barrett, P.A.; Dent, C.E.; Linstead, R.P. *J. Chem. Soc.* **1936**, 1719–1736.
22. Cong, F.-D.; Ning, B.; Du, X.-G.; Ma, C.-Y.; Yu, H.-F.; Chen, B. *Dyes Pigm.* **2005**, *66*, 149–154.
23. Fox, M.R. *A History of British Dye-Makers*, ICI Publications: Manchester **1987**, pp. 1856–1976.
24. Metz, J.; Schneider, O.; Hanack, M. *Inorg. Chem.* **1984**, *23*, 1065–1071.
25. Kharisov, B.I.; Méndez, U.O.; Rivera de la Rosa, J. *J. Russ. J. Coord. Chem.,* **2006**, *32*, 617–631.

26. Kharisov, B.I.; Méndez, U.O.; Almaraz Garza, J.L.; Almaguer Rodríguez, J.R. *New J. Chem.,* **2005**, *29*, 686–692.
27. Chauhan, S.M.S.; Srinivas, K.A.; Srivastave, P.K.; Sahoo, B. *J. Porphyrins Phthalocyanines,* **2003**, *7*, 548–550.
28. Kempa A.; Dobrowolski, J. *Can. J. Chem.* **1988**, *66*, 2553–2555.
29. Tomoda, H.; Saito, S.; Shiraishi, S. *Chem. Lett.* **1983**, *12*, 313–316.
30. Elvidge, J.A.; Linstead, R.P *J. Chem. Soc.* **1952**, 5000–5007.
31. Lo, P.-C.; Wang, S.; Zeug, A.; Meyer, M.; Röder, B.; Ng, D.K.P. *Tetrahedron Lett.* **2003**, *44*, 1967–1970.
32. Tamgho, I.-S.; Engle, J.T.; Ziegler, C.J. *J. Org. Chem.* **2012**, *77*, 11372–11376.
33. Baumann, F.; Bienert, B.; Rösch, G.; Vollmann, H.; Wolf, W. *Angew. Chem.* **1956**, *68*, 133–150.
34. Borodkin, V.F. *Zh. Prikl. Khim.* **1958**, 31, 813–816.
35. Oliver, S.W.; Smith, T.D. *J. Chem. Perkin Trans. II* **1987**, 1579–1582.
36. Baker, S.; Petty, M.C.; Roberts, G.G.; Twigg, M.V. *Thin Solid Films.* **1983**, *99*, 53–59.
37. Wöhrle, D.; Schnurpheil, G.; Knoteh, G. *Dyes Pigm.* **1992**, *18*, 91–102.
38. Canmenzind M.J.; Hill, C.L. *J. Heterocycl. Chem.* **1985**, *22*, 575–576.
39. Brewis, M.; Clarkson, G.J.; Humberstone, P.; Makhseed, S.; McKeown, N. B. *Chem. Eur. J.* **1998**, *4*, 1633–1640.
40. Lever, A.B.P. *Adv. Inorg. Radiochem.* **1965**, *7*, 27–105.
41. Thompson, J.A; Murata, K.; Miller, D.C.; Stanton, J.L.; Broderick, W.E.; Hoffman, B.M.; Ibers, J.A. *Inorg. Chem.* **1993**, *32*, 3546–3553.
42. Edmondson, S.J.; Hill, J.S.; Isaacs, N.S.; Mitchell, P.C.H. *J. Chem. Soc. Dalton Trans.* **1990**, 1115–1118.
43. Kharisov, B.I.; Blanco, L.M.; Torres-Martinez, L.M.; García-Luna, A. *Ind. Eng. Chem. Res.* **1999**, *38*, 2880–2887.
44. Alzeer, J.; Roth, P.J.C.; Luedtke, N.W. *Chem. Commun.* **2009**, 1970–1971.
45. Tomoda, H.; Saito, S.; Ogawa, S.; Shiraishi, S. *Chem. Lett.* **1980**, *9*, 1277–1280.
46. Beck A.; Hanack, M. *Chem. Ber.* **1993**, *126*, 1493–1494.
47. Ağirtaş, M.S.; Bekaroğlu, Ö. *J. Porphyrins Phthalocyanines* **2001**, *5*, 717–720.
48. Marcuccio, S.M.; Svirskaya, P.I.; Greenberg, S.; Lever, A.B.P.; Leznoff, C.C.; Tomer, K.B. *Can. J. Chem.* **1985**, *63*, 3057–3069.
49. Lee C.-H.; Ng, D.K.P. *Tetrahedron Lett.* **2002**, *43*, 4211–4214.
50. Chauhan, S.M.S.; Agarwal, S.; Kumari, P. *Synth. Commun.* **2007**, *37*, 2917–2925.
51. Terekhov, D.S.; Nolan, K.J.M.; McArthur, C.R.; Leznoff, C.C. *J. Org. Chem.* **1996**, *61*, 3034–3040.
52. Barrett, P. A.; Frye, D. A.; Linstead, R. P. *J. Chem. Soc.* **1938**, 1157–1163.
53. Dolphin, D.; Sams, J.R.; Tsin, T. *Inorg. Synth.* **1980**, *20*, 155–161.
54. Cadiou, C.; Pondaven, A.; L'Her, M.; Jèhan, P.; Guenot, P. *J. Org. Chem.* **1999**, *64*, 9046–9050.
55. Ueda, Y.; Yanagani, H.; Hayashi, S.; Ashida, M. *J. Electron. Microscopy* **1989**, *38*, 101–110.
56. Kubiak, R.; Waskowska, A.; Sledz, M.; Jezierski, A. *Inorg. Chim. Acta* **2006**, *359*, 1344–1350.
57. Moser, F.H.; Thomas, A.L. *Phthalocyanine Compounds*; Reinhold Publ. Corp.: New York, **1963**.
58. Linstead, R.P., Lowe, A.R. *J. Chem. Soc.* **1934**, 1022–1027.
59. Leznoff, C.C.; D'Ascanio, A.M.; Yildiz, S.Z. *J. Porphyrins Phthalocyanines* **2000**, *4*, 103–111.
60. Chen, Y.; Fang, W.; Wang, K.; Liu W., Jiang, J. *Inorg. Chem.* **2016**, *55*, 9289–9296.
61. Hu, Y.; Lai, G.; Shen, Y.; Li, Y *J. Porphyrins Phthalocyanines* **2004**, *8*, 1042–1046.

62. Mikhalenko, S. A.; Luk'yanets, E. A. *Zh. Obsch. Khim.* **1969**, *39*, 2129–2136.
63. Mikhalenko, S. A.; Barkanova, S. V.; Lebedev, O. L.; Luk'yanets, E. A. *Zh. Obsch. Khim.* **1971**, *41*, 2735–2739.
64. Snow, A.W.; Jarvis, N.L. *J. Am. Chem. Soc.* **1984**, *106*, 4706–4611.
65. Ağirtas, M.S.; Yildiko, Ü.; Yilan, A. *J. Porphyrins Phthalocyanines* **2006**, *10*, 117–121.
66. Ağirtas, M.S. *Inorg. Chim. Acta* **2007**, *360*, 2499–2502.
67. Kasuga, K.; Yashiki, K.; Sugimori, T.; Handa, M. *J. Porphyrins Phthalocyanines* **2005**, *9*, 646–650.
68. Dubinina, T.V.; Paramonova, K.V.; Trashin, S.A.; Borisova, N.E.; Tomilova, L.G.; Zefirov, N.S. *Dalton Trans.* **2014**, *43*, 2799–2809.
69. Vagin, S.; Hanack, M. *Eur. J. Org. Chem.* **2004**, 600–606.
70. Adachi, K.; Watarai, H. *Bull. Chem. Soc. Jpn.* **2004**, *77*, 2011–2020.
71. Kostka, M.; Zimcik, P.; Miletin, M.; Klemera, P.; Kopecky, K.; Musil, Z. *J. Photochem. Photobiol. A. Chem.* **2006**, *178*, 16–25.
72. Adachi, K.; Chayama, K.; Watarai, H. *Langmiur* **2006**, *22*, 1630–1639.
73. Leznoff, C.C.; Black, L.S.; Hiebert, A.; Causey, P.W.; Christendat, D.; Lever, A.B.P. *Inorg. Chim. Acta*, **2006**, *359*, 2690–2699.
74. Causey, P.W.; Dubovyk, I.; Leznoff, C.C. *Can. J. Chem.* **2006**, *84*, 1380–1387.
75. Kuzmina, E.A.; Dubinina, T.V.; Zasedatelev, A.V.; Baranikov, A.V.; Makedonskaya, M.I.; Egorova, T.B.; Tomilova, L.G. *Polyhedron*, **2017**, *135*, 41–48.
76. Kasuga, K.; Asano, K.; Lin, L.; Sugimori, T.; Handa, M.; Abe, K.; Kikkawa, T.; Fujiwara, T. *Bull Chem. Soc. Jpn.* **1997**, *70*, 1859–1865.
77. Deng, X.; Porter III, W.W.; Vaid, T.P. *Polyhedron* **2005**, *24*, 3004–3011.
78. Hu, Y.; Lai, G.; Shen, Y.; Li, Y. *Monatsh. Chem.* **2004**, *135*, 1167–1172.
79. Varmuza, K.; Maresch, G.; Meller, A. *Monatsh. Chem.* **1974**, *105*, 327–333.
80. Platel, R.H.; Tasso, T.T.; Zhou, W.; Furuyama, T.; Kobayashi, N.; Leznoff, D.B. *Chem. Commun.* **2015**, *51*, 5986–5989.
81. Lapkina, L.A.; Sakharov, S.G.; Konstantinov, N.Yu.; Larchenko V.E.; Gorbunova, Yu.G.; Tsivadze, A.Yu. *Russ. J. Inorg. Chem.* **2007**, *52*, 1758–1768.
82. Kirin, I.S.; Moskalev, P.N. *Zh. Neorg. Khim.* **1971**, *16*, 3179–3181.
83. Pluschev, V.E.; Shklover, L.P. *Zh. Neorg. Khim.* **1964**, *9*, 2015–2016.
84. Kasuga, K.; Ando, M.; Morimoto, H. *Inorg. Chim. Acta* **1986**, *112*, 99–101.
85. Tutaβ, A.; Göldner, M.; Hückstädt, H.; Cornelissen, U.; Homborg, H. *Z. Anorg. Allg. Chem.* **2001**, *627*, 2323–2336.
86. Paillaud, J.L.; Drillon, M.; Decian, A.; Fischer, J.; Weiss, R.; Poinsot, R.; Herr, A. *Physica B*, **1991**, *175*, 337–348.
87. Jiang, J.; Xie, J.; Choi, M.T.M.; Yan, Y.; Sun, S.; Ng, D.K.P. *J. Porphyrins Phthalocyanines* **1999**, *3*, 322–328.
88. Liu, W.; Jiang, J.; Du, D.; Arnold, D.P. *Aust. J. Chem.* **2000**, *53*, 131–135.
89. Gorbunova, Yu.G.; Lapkina, L.A.; Golubeva, S.V.; Larchenko, V.E.; Tsivadze, A.Yu. *Mendeleev Commun.* **2001**, *11*, 214–215.
90. Kulinich, V.P.; Shaposhnikov, G.P.; Maizlish, V.E.; Smirnov, R.P. *Koordinatsionnaya Khim.* **1994**, *20*, 866–869.
91. Kasuga, K.; Ando, M.; Morimoto, H.; Isa, M. *Chem. Lett.* **1986**, 1095–1098.
92. Liu, W.; Jiang, J.; Pan, N.; Arnold, D.P. *Inorg. Chim. Acta* **2000**, *310*, 140–146.

93. Lan, Y.; Klyatskaya, S.; Ruben, M.; Fuhr, O.; Wernsdorfer, W.; Candini, A.; Corradini, V.; Rizzini, A.L.; Pennino del, U.; Troiani, F.; Joly, L.; Klar, D.; Wendeh, H.; Affronte, M. *J. Mater. Chem. C* **2015**, *3*, 9794–9801.
94. Tsivadze, A.Yu.; Martynov, A.G.; Polovkova, M.A.; Gorbunova, Yu. G. *Russ. Chem. Bull. Int. Ed.* **2011**, *60*, 2258–2262.
95. Pluyshchev, V.E.; Shklover, L.P. *Zh. Neorg. Khim.* **1964**, *9*, 335–339.
96. Shklover, L.P.; Pluyshchev, V.E. *Zh. Neorg. Khim.* **1964**, *9*, 340–346.
97. Kirin, I. S.; Moskalev, P. N.; Makashev, Yu. A. *Zh. Neorg. Khim.* **1965**, *10*, 1951–1953.
98. Nemykin, V. N.; Volkov, S. V. *Russ. J. Coord. Chem.* **2000**, *26*, 436–450.
99. Jiang, J.; Kasuga, K.; Arnold, D. P. *Supramolecular Photo-Sensitive and Electroactive Materials*; Nalwa, H. S., Ed.; Academic Press: New York, **2001**, pp. 113–210.
100. Ng, D.K.P., Jiang, J.; Kasuga, K.; Machida, K. *Handbook on the Physics and Chemistry of Rare Earths*; Gschneidner, Jr., K.A.; Eyring, L.; Lander, G.H., Eds.; Elsevier Science B.V.: Amsterdam, **2001**, Vol. 32, pp. 611–653.
101. Ishikawa, N.; Kaizu, Y. *Coord. Chem. Rev*. **2002**, *226*, 93–101.
102. Weiss, R.; Fischer, J. in *The Porphyrin Handbook*; Kadish, K. M.; Smith, K. M.; Guilard, R., Eds.; Academic Press: San Diego, **2003**; Vol. 16, pp. 171–246.
103. Ishii, K.; Kobayashi, N. in *The Porphyrin Handbook*; Kadish, K. M.; Smith, K. M.; Guilard, R., Eds.; Academic Press: San Diego, **2003**; Vol. 16, pp. 1–42.
104. L'Her, M.; Pondaven, A. in *The Porphyrin Handbook*; Kadish, K. M.; Smith, K. M.; Guilard, R., Eds.; Academic Press: San Diego, **2003**; Vol. 16, pp. 117–169.
105. Jiang, J.; Bao, M.; Rintoul, L.; Arnold, D. P. *Coord. Chem. Rev.* **2006**, *250*, 424–448.
106. Ishikawa, N. *Polyhedron* **2007**, *26*, 2147–2153.
107. Pushkarev, V. E.; Tomilova, L. G.; Tomilov, Yu V. *Russ. Chem. Rev.* **2008**, *77*, 875–907.
108. Jiang, J.; Ng, D. K. P. *Acc. Chem. Res.* **2009**, *42*, 79–88.
109. Bian, Y.; Zhang, Y.; Ou, Z.; Jiang, J. in *Handbook of Porphyrin Science*; Kadish, K. M.; Smith, K. M.; Guilard, R., Eds.; World Scientific: Singapore, **2011**; Vol. 14, pp. 249–460.
110. Gorbunova, Yu.G.; Martynov, A.G.; Tsivadze, A.Yu. in *Handbook of Porphyrin Science*; Kadish, K. M.; Smith, K. M.; Guilard, R., Eds.; World Scientific: Singapore, **2012**; Vol. 24, pp. 271–388.
111. Wang, H.; Wang, B.-W.; Bian, Y.; Gao, S.; Jiang, J. *Coord. Chem. Rev.* **2016**, *306*, 195–216.
112. Pushkarev, V. E.; Tomilova, L.G.; Nemykin, V.N. *Coord. Chem. Rev.* **2016**, *319*, 110–179.
113. Kolesnikova, E.E.; Shaposhnikov, G.P.; Kulinich, V.P.; Smirnov, R.P. *Izvestiya Vysshikh Uchebnykh Zavedenii, Khimiya i Khimicheskaya Tekhnologiya* **1990**, *33*, 22–26.
114. Köksoy, B.; Orman, E.B.; Kuruca, H.; Bulut, M.; Durmuş, M.; Özkaya, A.R. *J. Electrochem. Soc.* **2016**, *163*, H927-H936.
115. Bouvet, M.; Bassoul, P.; Simon, J. *Mol. Cryst. Liq. Cryst.* **1994**, *252*, 31–38.
116. Komeda, T.; Isshiki, H.; Liu, J.; Katoh, K.; Shirakata, M.; Breedlove, B.K.; Yamashita, M. *ACS Nano* **2013**, *7*, 1092–1099.
117. Sekhosana, K.E.; Amuhaya, E.; Nyokong, T. *Polyhedron* **2016**, *105*, 159–169.
118. Mantareva, V.; Durmuş, M.; Aliosman, M.; Stoineva, I.; Angelov, I. *Photodiagn. Photodyn. Ther.* **2016**, *14*, 98–103.
119. Bozoğlu, C.; Arici, M.; Uğur, A.L.; Erdoğmuş, A.; Koca, A. *Synth. Met.* **2014**, *190*, 56–65.
120. Zugle, R.; Nyokong, T. *J. Mol. Catal. A. Chem.* **2012**, *358*, 49–57.
121. Zugle, R.; Litwinski, C.; Torto, N.; Nyokong, T. *New J. Chem.* **2011**, *35*, 1588–1595.
122. Jiazan, N.; Feng, S.; Zhenxiang, L.; Shaoming, Y. *Inorg. Chim. Acta* **1987**, *139*, 165–168.

123. Subbotin, N.B.; Tomilova, L.G.; Kostromina, N.A.; Luk'yanets, E. A. *Zh. Obsch. Khim.* **1985**, *56*, 397–400.
124. Lomova, T.N. *Russ. J. Inorg. Chem.* **2015**, *60*, 1123–1128.
125. Korolev, V.V.; Lomova, T.N.; Ramazanova, A.G.; Mozhzhukhina, E.G. *Mendeleev Commun.* **2016**, *26*, 301–303.
126. Pushkarev, V.E.; Breusova, M.O.; Shulishova, E.V.; Tomilova, L.G. *Russ. Chem. Bull. Int. Ed.* **2005**, *54,* 2087–2093.
127. Nefedova, I.V.; Gorbunova, Yu.G.; Sakharov, S.G.; Tsivadze, A.Yu. *Russ. J. Inorg. Chem.* **2005**, *50*, 165–173.
128. Kirin, I.S.; Moskalev, P.N.; Makashev, Y.A. *Zh. Neorg. Khim.* **1967**, *12*, 707–712.
129. Tomilova, L.G.; Chernykh, E.V.; Gavrilov, V.I.; Shelepin, I.V.; Derkacheva, V.M.; Luk'yanets, E. A. *Zh. Obsch. Khim.* **1982**, *52*, 2606–2611.
130. Tomilova, L.G.; Chernykh, E.V.; Ioffe, N.T.; Luk'yanets, E. A. *Zh. Obsch. Khim.* **1983**, *53*, 2594–2601.
131. Trojan, K.L.; Kendall, J.L.; Kepler, K.D.; Hatfield, W.E. *Inorg. Chim. Acta* **1992**, *198–200*, 795–803.
132. Clarisse, C.; Riou, M.T. *Inorg. Chim. Acta* **1987**, *130*, 139–144.
133. Ovseevich, M.A.; Tomilova, L.G.; Kogan, E.G.; Zefirov, N.S. *Mendeleev Commun.* **1998**, 186–187.
134. Pushkarev, V.E.; Ivanov, A.V.; Zhukov, I.V.; Shulishov, E.V.; Tomilov, Yu.V. *Russ. Chem. Bull. Int. Ed.* **2004**, *53*, 554–560.
135. Tolkacheva, E.O.; Tsivadze, A.Y.; Bitiev, S.G.; Gorbunova, Y.G.; Zhilov, V.I.; Minin, V.V. *Zh. Neorg. Khim.* **1995**, *40*, 984–989.
136. Jiang, J.; Liu, R.C.W.; Mak, T.C.W.; Chan, T.W.D.; Ng, D.K.P. *Polyhedron* **1997**, *16*, 515–520.
137. De Cian, A.; Moussavi, M.; Fischer, J.; Weiss, R. *Inorg. Chem.* **1985**, *24*, 3162–3167.
138. Kalashnikova, I.P.; Zhukov, I.V.; Tomilova, L.G.; Zefirov, N.S. *Russ. Chem. Bull. Int. Ed.* **2003**, *52*, 1709–1714.
139. Sleven, J.; Görller-Walrand, C.; Binnemans, K. *Mater. Sci. Eng. C* **2001**, *18*, 229–238.
140. Maeda, F.; Hatsusaka, K.; Ohta, K.; Kimura, M. *J. Mater. Chem.* **2003**, *13*, 243–251.
141. Belarbi, Z.; Sirlin, C.; Simon, J.; Andre, J.-J. *J. Phys. Chem.* **1989**, *93*, 8105–8110.
142. Bian, Y.; Jiang, J.; Tao, Y.; Choi, M.T.M.; Li, R.; Ng, A.C.H. Zhu, P.; Pan, N.; Sun, X.; Arnold, D.P.; Zhou, Z.-Y.; Li, H.-W.; Mak, T.C.W.; Ng, D.K.P. *J. Am. Chem. Soc.* **2003**, *125*, 12257–12267.
143. Binnemans, K.; Sleven, J.; De Feytewr, S.; De Schryver, F.C.; Donnio, B.; Guillon, D. *Chem. Mater.* **2003**, *15*, 3930 3998
144. Lu, G.; Bai, M.; Li, R.; Zhang, X.; Ma, C.; Lo, P.-C.; Ng, D.K.P.; Jiang, J. *Eur. J. Inorg. Chem.* **2006**, 3703–3709.
145. Aroca, R.; Clavijo, R.E.; Jennings, C.A.; Kovacs, G.J.; Duff, J.M.; Loutfy, R.O. *Spectrochim. Acta A*, **1989**, *45*, 957–962.
146. Iwase, A.; Harnoode, C.; Kameda, Y. *J. Alloys Compd.* **1993**, *192*, 280–283.
147. Lapkina, L.A.; Niskanen, E.; Rönkkömäki, H.; Larchenko, V.E.; Popov, K.I.; Tsivadze, A.Yu. *J. Porphyrins Phthalocyanines* **2000**, *4*, 587–590.
148. Kirin, I.S.; Moskalev, P.N.; Ivannikova, N.V. *Zh. Neorg. Khim.* **1967**, *12*, 944–946.
149. Ishikawa, N.; Okubo, T.; Kaizu, Y. *Inorg. Chem.* **1999**, *38*, 3173–3181.
150. Bloor, J.E.; Schlabitz, J.; Walden, C.C.; Demerdache, A. *Can. J. Chem.* **1964**, *42*, 2201–2208.
151. Day, V.W.; Marks, T.J.; Wachter, W.A. *J. Am. Chem. Soc.* **1975**, *97*, 4519–4527.
152. Marks, T.J.; Stojakovic, D.R. *J. Am. Chem. Soc.* **1978**, *100*, 1695–1705.

153. Cuellar, E.A.; Marks, T.J. *Inorg. Chem.* **1981**, *20*, 3766–3770.
154. Furuyama, T.; Ogura, Y.; Yoza, K.; Kobayashi, N. *Angew. Chem. Int. Ed.* **2012**, *51*, 11110–11114.
155. Guilard, R.; Dormond, A.; Belkalem, M.; Anderson, J.E.; Liu, Y.H.; Kadish, K.M. *Inorg. Chem.* **1987**, *26*, 1410–1414.
156. Lux, F.; Dempf, D.; Graw, D. *Angew. Chem. Int. Ed.* **1968**, *7*, 819–820.
157. Kirin, I.S.; Kolyadin, A.B.; Moskalev, P.N. *Zh. Neorgan. Khim.* **1971**, *16*, 2731–2735.
158. Kadish, K.M.; Moninot, G.; Hu, Y.; Dubois, D.; Ibnlfassif, A.; Barbe, J.M.; Guilard, R. *J. Am. Chem. Soc.* **1993**, *115*, 8153–8166.
159. Shklover, L.P.; Pluschev, V.E.; Rozdin, I.A.; Novikova, N.A. *Zh. Neorg. Khim* **1964**, *9*, 478–479.
160. Block, B.P.; Meloni, E.G. *Inorg. Chem.* **1965**, *4*, 111–112.
161. Goedken, V.L.; Dessy, I.G.; Ercolani, C.; Fares, V.; Gastaldi, L. *Inorg. Chem.* **1985**, *24*, 991–995.
162. Ghosez, P.; Cote, R.; Gastonguay, L.; Veilleux, G.; Denes, G.; Dodelet, J.P. *Chem. Mater.* **1993**, *5*, 1581–1590.
163. Tomachinskaya L.A.; Chernii, V.Ya.; Volkov, S.V. *Zh. Neorg. Khim.* **2002**, *47*, 254–257.
164. Faraonov, M.A.; Konarev, D.V.; Fatatov, A.M.; Khasanov, S.S.; Troyanov, S.I.; Lyubovskaya, R.N. *Dalton Trans.* **2017**, *46*, 3547–3555.
165. Yao, J.; Yonehara, H.; Pac, C. *Bull. Chem. Soc. Jpn.* **1995**, *68*, 1001–1005.
166. Tau, P.; Nyokong, T. *Dalton Trans.* **2006**, 4482–4490.
167. Chauke, V.P.; Arslanoglu, Y.; Nyokong, T. *Polyhedron* **2011**, *30*, 2132–2139.
168. Yaşa, G.; Erdoğmuş, A.; Uğur, A.L.; Şener, M.K.; Avciata, U.; Nyokong, T. *J. Porphyrins Phthalocyanines* **2012**, *16*, 845–854.
169. Nombona, N.; Tau, P.; Sehlotho, N.; Nyokong, T. *Electrochim. Acta* **2007**, *53*, 3139–3148.
170. Barthel, M.; Dini, D.; Vagin, S.; Hanack, M. *Eur. J. Org. Chem.* **2002**, 3756–3762.
171. Winter, G.; Heckmann, H.; Haisch, P.; Eberhardt, W.; Hanack, M.; Lüer, L.; Egelhaaf, H.-J.; Oelkrug, D. *J. Am. Chem. Soc.* **1998**, *120*, 11663–11673.
172. Erdoğmuş, A.; Nyokong, T. *Inorg. Chim. Acta* **2009**, *362*, 4875–4883.
173. Altin, S.; Altindal, A.; Bulut, M. *Polyhedron* **2013**, *49*, 41–49.
174. Arslanoğlu, Y.; Sevim, A.M.; Hamuryudan, E. Gül, A. *Dyes Pigm.* **2006**, *68*, 129–132.
175. Arslanoğlu, Y.; Hamuryudan, E. *Dyes Pigm.* **2007**, *75*, 150–155.
176. Mbambisa, G.; Tau, P.; Antunes, E.; Nyokong, T. *Polyhedron* **2007**, *26*, 5355–5364.
177. Haisch, P.; Winter, G.; Hanack, M.; Lüler, L.; Egelhaaf, H.-J.; Oelkrug, D. *Adv. Mater.* **1997**, *9*, 316–321.
178. Ercolani, C.; Paoletti, A.M.; Pennesi, G.; Rossi, G.; Chiesi-Villa, A.; Rizzoli, C. *J. Chem. Soc., Dalton. Trans.* **1990**, 1971–1977.
179. Pluschev, V.E.; Shklover, L.P.; Rozdin, I.A. *Zh. Neorg. Khim* **1964**, *9*, 125–127.
180. Mühl, P. *Z. Chem.* **1967**, *7*, 352–353.
181. Tomachynski, L.A.; Tretyakova, I.N.; Chernii, V.Ya.; Volkov, S.V.; Kowalska, M.; Legendziewicz, J.; Gerasymchuk, Y.S.; Radzki, S. *Inorg. Chim. Acta.* **2008**, *361*, 2569–2581.
182. Gorsch, M.; Franken, A.; Sieversen, S.; Homborg, H. *Z. Anorg. Allg. Chem.* **1995**, *621*, 607–616.
183. Zhou, W.; Platel, R.H.; Tasso,T.T.; Furuyama, T.; Kobayashi, N.; Leznoff, D.B. *Dalton Trans.* **2015**, *44*, 13955–13961.
184. Villemin, D.; Hammadi, M.; Hachemi, M.; Bar, N. *Molecules* **2001**, *6*, 831–844.
185. Tomilova, L.G.; Ovchinnikova, N.A.; Luk'yanets, E.A. *Zh. Obsch. Khim.* **1987**, *57*, 2100–2104.
186. Silver, J.; Lukes, P.J.; O'Connor, J.M. *Polyhedron* **1989**, *8*, 1631–1635.
187. Ovchinnikova, N.A.; Tomilova, L.G.; Seregina, N.B.; Minin, V.V.; Larin, G.M.; Luk'yanets, E.A. *Zh. Obsch. Khim.* **1992**, *62*, 1631–1638.

188. Mikhalenko, S. A.; Luk'yanets, E. A. *Anilino-Kras. Prom.* (in Russian) **1977**, *3*, 1–4.
189. Mikhalenko, S. A.; Derkacheva, V. M.; Luk'yanets, E. A. *Zh. Obshch. Khim.* **1981**, *51*, 1650–1657.
190. Mikhalenko, S. A.; Korobkova, E. V.; Luk'yanets, E. A. *Zh. Obshch. Khim.* **1970**, *40*, 400–403.
191. Jiang, Z.; Ou, Z.; Chen, N.; Wang, J.; Huang, J.; Shao, J.; Kadish, K.M. *J. Porphyrins Phthalocyanines* **2005**, *9*, 352–360.
192. Law, K.-Y. *Inorg. Chem.* **1985**, *24*, 1778–1781.
193. Mbambisa, G.; Nyokong, T. *Polyhedron* **2008**, *27*, 2799–2804.
194. Kovshev, E.I.; Luk'yanets, E. A. *Anilino-Kras. Prom.* (in Russian) **1972**, *4*, 9–12.
195. Mikhalenko, S. A.; Yagodina, L.A.; Luk'yanets, E. A. *Zh. Obshch. Khim.* **1976**, *46*, 1598–1602.
196. Muto, T.; Sassa, T.; Wada, T.; Kimura, M.; Shirai, H. *Chem. Lett.* **2004**, *33*, 132–133.
197. Sleven, J.; Cardinaels, T.; Binnemans, K.; Guillon, D.; Donnio, B. *Liquid Cryst.* **2002**, *29*, 1425–1433.
198. Gal'pern, M.G.; Shalaev, V.K.; Shatskaya, T.A.; Shishkanova, L.S.; Skvarchenko, V.R.; Luk'yanets, E.A. *Zh. Obshch. Khim.* **1983**, *53*, 2601–2606.
199. Handa, M.; Suzuki, A.; Shoji, S.; Kasufga, K.; Sogabe, K. *Inorg. Chim. Acta* **1995**, *230*, 41–44.
200. Basova, T.; Plyashkevich, V.; Hassan, A. *Surface Sci.* **2008**, *602*, 2368–2372.
201. Tian, M.; Yanagi, S.; Sasaki, K.; Wada, T.; Sasabe, H. *J. Opt. Soc. Am. B.* **1998**, *15*, 846–853.
202. Keil, C.; Tsaryova, O.; Lapok, L.; Himcinschi, C.; Wöhrle, D.; Hild, O.R.; Zahn, D.R.T.; Gorun, S.M.; Schlettwein, D. *Thin Solid Films* **2009**, *517*, 4379–4384.
203. Reddy, D.; Chandrashekar, T.K. *Polyhedron* **1993**, *12*, 627–633.
204. Kobayashi, N.; Narita, F.; Ishii, K.; Muranaka, A. *Chem. Eur. J.* **2009**, *15*, 10173–10181.
205. Cellucci, L.; Ercolani, C.; Lukes, P.J.; Chiesi-Villa, A.; Rizzoli, C. *J. Porphyrins Phthalocyanines* **1998**, *2*, 9–19.
206. Gingl, F.; Strähle, J. *Z. Naturforsch. B* **1989**, *44*, 110–116.
207. Donzello, M.P.; Ercolani, C.; Lukes, P.J. *Inorg. Chim. Acta* **1997**, *256*, 171–172.
208. Donzello, M.P.; Ercolani, C.; Chiesi-Villa, A.; Rizzoli, C. *Inorg. Chem.* **1998**, *37*, 1347–1351.
209. Bauer, E.M.; Donzello, M.P.; Ercolani, C.; Masetti, E.; Panero, S.; Riccardi, G.; Rosa, A.; Chiesi-Villa, A.; Rizzoli, C. *Inorg. Chem.* **2003**, *42*, 283–293.
210. Janczak, J.; Kubiak, R. *Polyhedron* **2003**, *22*, 313–322.
211. Gingl, F.; Strähle, J. *Z. Naturforsch. B* **1988**, *43*, 445–448.
212. Tau, P.; Nyokong, T. *J. Porphyrins Phthalocyanines* **2006**, *10*, 69–75.
213. Chauke, V.P.; Antunes, E.; Nyokong, T. *J. Electroanal. Chem.* **2011**, *661*, 1–7.
214. Chauke, V.; Nyokong, T. *Inorg. Chim. Acta* **2010**, *363*, 3662–3669.
215. Meloni, E.G.; Ocone, L.R.; Block, B.P. *Inorg. Chem.* **1967**, *6*, 424–425.
216. Zhou, W.; Thompson, J.R.; Leznoff, C.C.; Leznoff, D.B. *Chem. Eur. J.* **2017**, *23*, 2323–2331.
217. Hanack, M.; Datz, A. *Chem. Ber.* **1986**, *119*, 1281–1290.
218. Wöhrle, D.; Suvorova, O.N.; Trombach, N.; Schupak, E.A.; Gerdes, R.; Semenov, N.M.; Bartels, O.; Zakurazhnov, A.A.; Schnurpfeil, G.; Hild, O.; Wendt, A. *J. Porphyrins Phthalocyanines* **2001**, *5*, 381–389.
219. Elvidge, J.A.; Lever, A.B.P. *J. Chem. Soc.* **1961**, 1257–1265.
220. Grunewald, H.; Homborg, H. *Z. Anorg. Allg. Chem.* **1992**, *608*, 81–87.
221. Suda, K.; Nakajima, S.; Satoh, Y.; Takanami, T. *Chem. Commun.* **2009**, 1255–1257.
222. Hill, H.A.O.; Norgett, M.M. *J. Chem. Soc. A* **1966**, 1476–1478.
223. Collins, G.C.S.; Schiffrin, D.J. *J. Electroanal. Chem.* **1982**, *139*, 335–369.
224. Börschel, V.; Strähle, J. *Z. Naturforsch. B* **1984**, *39*, 1664–1667.
225. Edmondson, S.J.; Mitchell, P.C.H. *Polyhedron* **1986**, *5*, 315–317.

226. Sheldon, R.A. *Recl. Trav. Chim. Pays-Bas* **1973**, *92*, 253–266.
227. Frick, K.; Verma, S.; Sundermeyer, J.; Hanack, H. *Eur. J. Inorg. Chem.* **2000**, 1025–1030.
228. Verma, S.; Hanack, H. *Z. Anorg. Allg. Chem.* **2003**, *629*, 880–892.
229. Darwish, W.; Seikel, E.; Harms, K.; Burghaus, O.; Sundermeyer, J. *Dalton Trans.* **2011**, *40*, 1183–1188.
230. Matsushita, O.; Derkacheva, V.M.; Muranaka, A.; Shimizu, S.; Uchiyama, M.; Luk'yanets, E.A.; Kobayashi, N. *J. Am. Chem. Soc.* **2012**, *134*, 3411–3418.
231. Rutter, H.A.; McQueen, J.D. *J. Inorg. Nucl. Chem.* **1960**, *12*, 361–363.
232. Grunewald, H.; Homborg, H. *Z. Naturforsch. B.* **1990**, *45*, 483–489.
233. Dolotova, O.V.; Bundina, N.I.; Derkacheva, V.M.; Negrimovsky, V.M.; Minin, V.V.; Larin, G.M.; Kaliya, O.L.; Lukyanets, E.A. *Zh. Obshch. Khim.* **1992**, *62*, 2064–2075.
234. Dolotova, O.V.; Bundina, N.I.; Derkacheva, V.M.; Kaliya, O.L.; Lukyanets, E.A. *Zh. Obshch. Khim.* **1988**, *58*, 2173.
235. Yoshihaba, K.; Wolf, G.K.; Baumgartner, F. *Radiochim. Acta* **1974**, *21*, 96–100.
236. Rummel, S.; Hermann, M.; Schmidt, K. *Z. Chem.* **1985**, *25*, 152–153.
237. Pfrepper, G. *Z. Chem.* **1970**, *10*, 76.
238. Göldner, M.; Kienast, A.; Homborg, H. *Z. Anorg. Allg. Chem.* **1998**, *624*, 141–146.
239. Ziener, U.; Hanack, M. *Chem. Ber.* **1994**, *127*, 1681–1685.
240. Frick, K.; Ziener, U.; Hanack, M. *Eur. J. Inorg. Chem.* **1999**, 1309–1313.
241. Batanova, E.A.; Borisenkova, S.A.; Dolotova, O.V.; Kaliya, O.L. *Vestnik Moskovskogo Universiteta, Seriya 2: Khimiya* (in Russian) **2000**, *41*, 123–126.
242. Zeiner, U.; Dürr, K.; Hanack, M. *Synth. Metal* **1995**, *71*, 2285–2286.
243. Göldner, M.; Hückstädt, H.; Murray, K.S.; Moubaraki, B.; Homborg, H. *Z. Anorg. Allg. Chem.* **1998**, *624*, 288–294.
244. Silver, J.; Jassim, Q.A.A. *Inorg. Chim. Acta* **1988**, *144*, 281–288.
245. Beck, A.; Mangold, K.-M.; Hanack, M. *Chem. Ber.* **1991**, *124*, 2315–2321.
246. Ohya, T.; Takeda, J.; Kobayashi, N.; Sato, M. *Inorg. Chem.* **1990**, *29*, 3734–3737.
247. Aktaş, A.; Acar, I.; Biyiklioğlu, Z.; Saka, E.T.; Kantekin, H. *Synth. Metal.* **2014**, *198*, 212–220.
248. Derkacheva, V.M.; Bundina, N.I.; Mekhryakova, N.G.; Gulina, T.Yu.; Kaliya, O.L., Luk'yanets, E.A. *Zh. Neorgan. Khim.* **1981**, *26*, 1687.
249. Matemadombo, F.; Maree, M.D.; Ozoemena, K.I.; Westbroek, P.; Nyokong, T. *J. Porphyrin Phthalocyanines* **2005**, *9*, 484–490.
250. Gal'pern, M.G.; Donyagina, V.F.; Shalaev, Skvarchenko, V.R.; Bundina, N.I.; Mekhryakova, N.G.; Kaliya, O.L.; Luk'yanets, E. A. *Zh. Obsch. Khim.* **1982**, *52*, 715–716.
251. Ahsen, V.; Yilmazer, E.; Ertas, M.; Bekâroĝlu, Ö. *J. Chem. Soc. Dalton Trans.* **1988**, 401–406.
252. Ogata, H.; Higashi, R.; Kabayashi, N. *J. Porphyrins Phthalocyanines* **2003**, *7*, 551–557.
253. Jones, J.G.; Twigg, M.V. *Inorg. Chem.* **1969**, *8*, 2018–2019.
254. Ercolani, C.; Floris, B. *Phthalocyanines — Properties and Applications*; Lever. A.B.P.; Leznoff, C.C., Eds.; VCH: New York, **1992**; Vol. 2, pp. 5–41
255. Floris, B.; Donzello, M.P.; Ercolani, C. in *The Porphyrin Handbook*; Kadish, K. M.; Smith, K. M.; Guilard, R., Eds.; Academic Press: San Diego, **2003**; Vol. 18, pp. 1–62.
256. Hanack, M.; Gül, A.; Hirsch, A.; Mandal, B.K.; Subramanian, L.R.; Witke, E. *Mol. Cryst. Liq. Cryst.* **1990**, *187*, 625–642.
257. Dieing, R.; Schmid, G.; Witke, E.; Feucht, C.; Dre*β*en, M.; Pohmer, J.; Hanack, M. *Chem. Ber.* **1995**, *128*, 589–598.
258. Hanack, M.; Knecht, S.; Polley, R.; Subramanian, L.R. *Synth. Metal.* **1996**, *80*, 183–189.
259. Köksoy, B.; Soyer, O.; Orman, E.B.; Özkaya, A.R.; Bulut, M. *Dyes. Pigm.* **2015**, *118*, 166–175.

260. Bottomley, L.A.; Gorce, J.N.; Goedken, V.L.; Ercolani, C. *Inorg. Chem.* **1985**, *24*, 3733–3737.
261. Sorokin, A.B.; Kudrik, E.V.; Bouchu, D. *Chem. Commun.* **2008**, 2562–2564.
262. Rossi, G.; Goedken, V.L.; Ercolani, C. *J. Chem. Soc. Chem. Commun.* **1988**, 46–47.
263. Ercolani, C.; Gardini, M.; Goedken, V.L.; Pennesi, G.; Rossi, G.; Russo, U.; Zanonato, P. *Inorg. Chem.* **1989**, *28*, 3097–3099.
264. Farrell, N.P.; Murray, A.J.; Thornback, J.R.; Dolphin, D.H.; James, B.R. *Inorg. Chim. Acta* **1978**, *28*, L144–L146.
265. Hanack, M.; Osio-Barcina, J.; Witke, E.; Pohmer, J. *Synthesis* **1992**, 211–214.
266. Omiya, S.; Tsutsui, M.; Meyer Jr, E.F.; Bernal, I, Cullen, D. *Inorg. Chem.* **1980**, *19*, 134–142.
267. Boucher, L.J.; Rivera, P. *Inorg. Chem.* **1980**, *19*, 1816–1818.
268. Polley, R.; Hanack, M. *Synthesis* **1997**, 295–296.
269. Safari, N.; Jamaat, P.R.; Shirvan, S.A.; Shoghpour, S.; Ebadi, A.; Darvishi, M.; Shaabani, A. *J. Porphyrins Phthalocyanines* **2005**, *9*, 256–261.
270. Shaabani, A.; Rezayan, A.H. *J. Porphyrins Phthalocyanines* **2005**, *9*, 617–620.
271. Hanack, M.; Vermehren, P. *Chem. Ber.* **1991**, *124*, 1733–1738.
272. Hanack, M.; Knecht, S.; Polley. R. *Chem. Ber.* **1995**, *128*, 929–933.
273. Dudnik, A.S.; Ivanov, A.V.; Tomilova, L.G.; Zefirov, N.S. *Russ. J. Coord. Chem.* **2004**, *30*, 110–114.
274. Dürr, K.; Hanack, M. *J. Porphyrins Phthalocyanines* **1999**, *3*, 224–229.
275. Knecht, S.; Polley, R.; Hanack, M. *Appl. Organomet. Chem.* **1996**, *10*, 649–660.
276. Rodriguez-Morgade, M.S.; Planells, M.; Torres, T.; Ballester, P.; Palomares, E. *J. Mater. Chem.* **2008**, *18*, 176–181.
277. Rodriguez-Morgade, M.S.; Torres, T.; Atienza-Castellanos, C.; Guldi, D. *J. Am. Chem. Soc.* **2006**, *128*, 15145–15154.
278. Weber, A.; Ertel, T.S.; Reinöhl, U.; Bertagnolli, H.; Leuze, M.; Hees, M.; Hanack, M. *Eur. J. Inorg. Chem.* **2000**, 2289–2294.
279. Rihter, B.D.; Kenney, M.E.; Ford, W.F.; Rodgers, M.A.J. *J. Am. Chem. Soc.* **1990**, *112*, 8064–8070.
280. Cammidge, A.N.; Berber, G.; Chambrier, I.; Hough, P.V.; Cook, M.J. *Tetrahedron* **2005**, *61*, 40676–4074.
281. Hanack, M.; Hees, M.; Witke, E. *New. J. Chem.* **1998**, 169–172.
282. Enakieva, Yu.Yu.; Gorbunova, Yu.G.; Sakharov, S.G.; Tsivadze, A.Yu. *Zh. Neorgan. Khim.* **2002**, *47*, 1973–1979.
283. Gorbunova, Yu.G.; Enakieva, Yu.Yu.; Sakharov, S.G.; Tsivadze, A.Yu. *J. Porphyrins Phthalocyanines* **2003**, *7*, 795–800.
284. Hanack, M.; Vermehren, P. *Inorg. Chem.* **1990**, *29*, 134–136.
285. Hanack, M.; Gül, A.; Subramanian, L.R. *Inorg. Chem.* **1992**, *31*, 1542–1544.
286. Orihashi, Y.; Nishikawa, M.; Ohno, H.; Tsuchida, E.; Matsuda, H.; Nakanishi, H.; Kato, M. *Bull. Chem. Soc. Jpn.* **1987**, *60*, 3731–3738.
287. Uchida, H.; Reddy, P.Y.; Nakamura, S.; Toru, T. *J. Org. Chem.* **2003**, 68, 8736–8738.
288. Somashekarappa, M.P.; Keshavayya, J. *Synth. React. Inorg. Met.-Org. Chem.* **1999**, *29*, 767–783.
289. Derkacheva, V.M.; Luk'yanets, E.A. *Zh. Obshch. Khim.* **1980**, *50*, 2313–2318.
290. Liu, L.-C.; Lee, C.-C.; Hu, A.T. *J. Porphyrins Phthalocyanines* **2001**, *5*, 806–807.
291. Ağirtaş, M.S. *Dyes Pigm.* **2007**, *74*, 490–493.
292. Wei, S.; Huanga, D.; Li, L.; Meng, M. *Dyes Pigm.* **2003**, *56*, 1–6.
293. Akçay, H.T.; Bayrak, R.; Demirbas¸ Ü.; Koca, A.; Kantekin, H.; Ismail Değirmencioğlu, I. *Dyes Pigm.* **2013**, *96*, 483–494.
294. Agboola, B.; Ozoemena, K.I.; Nyokong, T. *Electrochim. Acta* **2006**, *51*, 4379–4387.

295. Obirai, J.; Pereira Rodrigues, N.; Bedioui, F.; Nyokong, T. *J. Porphyrins Phthalocyanines* **2003**, *7*, 508–520.
296. Achar, B.N.; Bhandari, J.M. *Transition Met. Chem.* **1993**, *18*, 423–426.
297. Hanack, M.; Polley, R.; Knecht, S.; Schlick, U. *Inorg. Chem.* **1995**, *34*, 3621–3624.
298. Lin, M.-J.; Wang, J.-D.; Chen, N.-S. Huang J.-L. *Z. Anorg. Allg. Chem.* **2006**, *632*, 2315–2320.
299. Shinohara, H.; Tsaryova, O.; Schnurpfeil, G.; Wöhrle, D. *J. Photochem. Photobiol. A Chem.* **2006**, *184*, 50–57.
300. Kim, S.-J.; Matsumoto, M.; Shigehara, K. *J. Porphyrins Phthalocyanines* **2000**, *4*, 136–144.
301. Diaz-Garcia, M.A.; Ledoux, I.; Duro, J.A.; Torres, T.; Agullo-Lopez, F.; Zyss, J. *J. Phys. Chem.* **1994**, *98*, 8761–8764.
302. Gürek, A.G.; Bekaroğlu, Ö. *J. Chem. Soc. Dalton Trans.* **1994**, 1419–1423.
303. Golovin, M.N.; Seymour, P.; Jayaraj, K.; Fu. Y.; Lever, A.B.P. *Inorg. Chem.* **1990**, *29*, 1719–1727.
304. Tse, Y.-H.; Seymour, P.; Kobayashi, N.; Lam, H.; Leznoff, C.C.; Lever, A.B.P. *Inorg. Chem.* **1991**, *30*, 4453–4459.
305. Keen, M.; Malerbi, B.W. *J. Inorg. Nucl. Chem.* **1965**, *27*, 1311–1319.
306. Nyokong, T. *J. Chem. Soc. Dalton Trans.* **1994**, 1359–1365.
307. Tomilova, L.G.; Luk'yanets, E.A. *Zh. Neorg. Khim.* **1977**, *22*, 2586–2587.
308. Youssef, T.E.; Mohamed, H.H. *Polyhedron* **2011**, *30*, 2045–2050.
309. Münz, X.; Hanack, M. *Chem. Ber.* **1988**, *121*, 235–238.
310. Bouwkamp-Wijnoltz, A.L.; Visscher, W.; van Veen, J.A.R. *Electrochim. Acta* **1994**, *39*, 1641–1645.
311. Murata, K.; Ishii, K. *Eur. J. Inorg. Chem.* **2017**, 5103–5107.
312. Haisch, P.; Knecht, S.; Schlick, U.; Subramanian, L.R.; Hanack, M. *Mol. Cryst. Liq. Cryst.* **1995**, *270*, 7–16.
313. Kobayashi, N.; Higashi, R.; Tomura, T. *Bull. Chem. Soc. Jpn.* **1997**, *70*, 2693–2698.
314. Ogunbayo, T.B.; Ogunsipe, A.; Nyokong, T. *Dyes Pigm.* **2009**, *82*, 422–426.
315. Hanack, M.; Beck, A.; Lehmann, H. *Synthesis* **1987**, 703–705.
316. Rodriguez-Morgade, S.; Hanack, M. *Chem.-Eur. J.* **1997**, *3*, 1042–1051.
317. Eberhardt, W.; Hanack, M. *Synthesis* **1997**, 95–100.
318. Hao, R.; Ye, C.; Wang, X.; Zhu, L.; Chen, S.; Yang, J.; Tao, X. *J. Phys. Chem. C* **2017**, *121*, 13524–13531.
319. Xu, H.-T.; Xie, W.-W.; Guo, Y.-C.; Ci, Y.; Chen, L.-J.; Peng, B.-X. *J. Chin. Chem. Soc.* **2007**, *54*, 211–217.
320. Ogunbayo, T.B.; Nyokong, T. *Polyhedron* **2009**, *28*, 2710–2718.
321. Ogunbayo, T.B.; Nyokong, T. *J. Mol. Catal. A. Chem.* **2011**, *337*, 68–76.
322. van Leeuwen, M.; Beeby, A.; Fernandes, I.; Ashworth, S.H. *Photochem. Photobiol. Sci.* **2014**, *13*, 62–69.
323. Zheng, X.; Wang, Y.; Hu, J.; Yang, G.; Guo, Z.; Xia, J.; Xu, Z.; Fang, G. *J. Mater. Chem. A* **2017**, *5*, 24416–24424.
324. Hanack, M.; Haisch, P.; Lehmann, H.; Subramanian, L.R. *Synthesis* **1993**, 387–390.
325. Zorlu, Y.; Dumoulin, F.; Durmuş, M.; Ahsen, V. *Tetrahedron*, **2010**, *55*, 3248–3258.
326. Milaeva, E.R.; Chechulina, I.N.; Prokof'ev, A.I.; Bundel, Yu.G. *Zh. Obshch. Khim.* **1989**, *59*, 2794–2795.
327. Vacus, J.; Doppelt, P.; Simon, J.; Memetzidis, G. *J. Mater. Chem.* **1992**, *2*, 1065–1068.
328. Dent, C.E.; Linstead, R.P. *J. Chem. Soc.* **1934**, 1027–1031.
329. Allen, R.L.M. *Colour Chemistry*. Appleton-Century-Crofts: New York, **1971**. pp. 231–240.
330. Gregory, P. *J. Porphyrins Phthalocyanines*, **2000**, *4*, 432–437.

331. Xia, D.; Yu, S.; Shen, R.; Ma, C.; Cheng, C.; Ji, D.; Fan, Z.; Wang, X.; Du, G. *Dyes Pigm.* **2008**, *78*, 84–88.
332. Negrimovskii, V.M.; Derkacheva, V.M.; Kaliya, O.L.; Luk'yanets, E.A. *Zh. Obshch. Khim.* **1991**, *61*, 460–470.
333. Oksengendler, I.G.; Kondratenko, N.V.; Luk'yanets, E.A.; Yagupolskii, L.M. *Zh. Organich. Khim.* **1977**, *13*, 1554–1558.
334. Oksengendler, I.G.; Kondratenko, N.V.; Luk'yanets, E.A.; Yagupolskii, L.M. *Zh. Organich. Khim.* **1978**, *14*, 1046–1051.
335. Achar, B.N.; Fohlen, G.M.; Parker, J.A.; Keshavayya, J. *Polyhedron* **1987**, *6*, 1463–1467.
336. Derkacheva, V.M.; Iodko, S.S.; Kaliya, O.L.; Luk'yanets, E.A. *Zh. Obshch. Khim.* **1981**, *51*, 2319–2324.
337. McKeown, N.B.; Chambrier, I.; Cook, M.J.; *J. Chem. Soc. Perkin Trans. 1* **1990**, 1169–1177.
338. Cook, M.J.; Dunn, A.J.; Howe, S.D.; Thomson, A.J.; Harrison, K.J. *J. Chem. Soc. Perkin Trans. 1* **1988**, 2453–2458.
339. Birchall, J.M.; Haszeldine, R.N.; Morley, J.O. *J. Chem. Soc. C.* **1970**, 2667–2672.
340. Fu, G.; Fu, Y.; Jayaraj, K.; Lever, A.B.P. *Inorg. Chem.* **1990**, *29*, 4090–4095.
341. MacCragh, A.; Koski, W. S. *J. Am. Chem. Soc.* **1965**, *87*, 2496–2497.
342. Ouemega, B.; Bayo-Bangoura, M.; Bayo, K. *J. Soc. Ouest-Afr. Chim.* **2011**, *031*, 42–48.
343. Wong, E.W.Y.; Miura, A.; Wright, M.D.; He, Q.; Walsby, C.J.; Shimizu, S.; Kobayashi, N.; Leznoff, D.B. *Chem.-Eur. J.* **2012**, *18*, 12404–12410.
344. Basova, T.V.; Parkhomenko, R.G.; Krasnov, P.O.; Igumenov, I.K.; Kadem, B.; Hassan, A.K. *Dyes Pigm.* **2015**, *122*, 280–285.
345. Bilal, C.; Sevim, A.M.; Gül, A. *Inorg. Chem. Commun.* **2016**, *69*, 10–12.
346. Shaabani, A.; Bahadoran, F.; Bazgir, A.; Safari, N. *Iran. J. Chem. & Chem. Eng.*, **1999**, *18*, 104–107.
347. Clare, P.; Glocking, F. *Inorg. Chim. Acta,* **1975**, 14, L12.
348. Meller, A.; Ossko, A. *Monatsh. Chem.*, **1972**, 103, 150–155.
349. Claessens, C.G.; González-Rodríguez, D.; del Rey, B.; Torres, T.; Mark, G.; Schuchmann, H.-P.; von Sonntag, C.; MacDonald, J.G.; Nohr, R.S. *Eur. J. Org. Chem.* **2003**, 2547–2551.
350. Morse, G.E.; Helander, M.G.; Maka, J.F.; Lu, Z.-H.; Bender, T.P. *ACS Appl. Mater. Interfaces*, **2010**, *2*, 1934–1944.
351. Chauhan, S. M. S. Kumari, P. *Tetrahedron*, **2009**, *65*, 2518–2524
352. Giribabu, L.; Kumar, C.V.; Surendar, A.; Reddy, V.G.; Chandrasekharam, M; Reddy, P.Y. *Synth. Commun.* **2007**, *37*, 4141–4147.
353. Geyer, M.; Plenzig, F.; Rauschnabel, J.; Hanack, M.; del Rey, B.; Sastre, A.; Torres, T. *Synthesis*, **1996**, 1139–1151.
354. Gommans, H.; Aernouts, T.; Verreet, B.; Heremans, P.; Medina, A.; Claessens, C.G.; Torres, T. *Adv. Funct. Mater.* **2009**, *19*, 3435–3439.
355. Rodríguez-Morgade, M.S.; Claessens, C.G.; Medina, A.; González-Rodríguez, D.; Gutiérrez-Puebla, E.; Monge, A.; Alkorta, I.; Elguero, J.; Torres, T. *Chem.-Eur. J.* **2008**, *14*, 1342–1350.
356. Rodríguez-Morgade, M.S.; Esperanza, S.; Torres, T.; Barbera, J. *Chem. Eur. J.* **2005**, *11*, 354–360.
357. González-Rodríguez, D.; Torres, T.; Denardin, E.L.G.; Samios, D.; Stefani, V.; Corrêa, D.S. *J. Organomet. Chem.* **2009**, *694*, 1617–1622.
358. González-Rodríguez, D.; Carbonell, E.; Guldi, D.M.; Torres, T. *Angew. Chem. Int. Ed.* **2009**, 48, 8032–8036.

359. Liu, J.-Y.; Yeung, H.-S.; Xu, W.; Li, X.; Ng, D.K.P. *Org. Lett.* **2008**, 10, 5421–5424.
360. El-khouly, M.E.; Kim, J.-H.; Kim, J.-H.; Kay, K.-Y.; Fukuzumi, S. *J. Phys. Chem. C* **2012**, *116*, 19709–19717.
361. Guilleme, J.; González-Rodríguez, D.; Torres, T. *Angew. Chem. Int. Ed.* **2011**, *50*, 3506–3509.
362. Linsky, J.P.; Paul, T.R.; Nohr, R.S.; Kenney, M.E. *Inorg Chem.* **1980**, *19*, 3131–3135.
363. Yanagi, H.; Imamura, M.; Ashida, M. *J. Appl. Phys,* **1994**, *75*, 2061–2068.
364. Yanagi, H.; Duko, S.; Ueda, Y.; Ashida, M.; Wöhrle, D. *J. Phys. Chem.* **1992**, *96*, 1366–1372.
365. Napier A.; Collins, R.A. *Thin Solid Films*, **1994**, *248*, 166–177.
366. Janczak J.; Kubiak, R. *Inorg. Chim. Acta*, **1999**, *288*, 174–180.
367. Kubiak R.; Janczak, J. *J. Alloy. Compd.* **1993**, *200*, L7-L8.
368. Janczak J.; Kubiak, R. *J. Chem. Soc., Dalton Trans.* **1994**, 2539–2543.
369. Janczak J.; Kubiak, R. *Acta Chim. Scand.* **1995**, *49*, 871–877.
370. Janczak J.; Kubiak, R. *J. Alloy. Compd.* **1993**, *202*, 69–72.
371. Owen, J.E.; Kenney, M.E. *Inorg. Chem.* **1962**, *1*, 334–336.
372. Kasuga, K.; Moriwaki, N.; Honda, M. *Inorg. Chim. Acta.*, **1996**, *244*, 137–139.
373. Joyner, R.D.; Kenney, M.E. *Inorg. Chem.* **1962**, *1*, 236–238.
374. Lowery, M.K; Starshak, A.J.; Esposito, J.N.; Krueger, P.C.; Kenney, M.E. *Inorg. Chem.* **1965**, *4*, 128.
375. Ejsmont, K.; Kubiak, R. *Acta Crystallogr. C* **1997**, *53*, 1051–1054.
376. Jancazk, J.; Razik, M.; Kubiak, R. *Acta Crystallogr. C* **1999**, *55*, 359–361.
377. Joyner, R.D.; Kenney, M.E. *J. Am. Chem. Soc.*, **1960**, *82*, 5790–5791.
378. Linck, R.G.; Esposito, J.N.; Kenney, M.E. *J. Inorg. Nucl. Chem.,* **1962**, *24*, 299–302.
379. Kroenke, W. J.; Kenney, M. E. *Inorg. Chem.* **1964**, *3*, 251–254.
380. Kluson, P.; Drobek, M.; Kalaji, A.; Karaskova, M.; Rakusan, J. *Res. Chem. Intermed*, **2009**, *35*, 103–116.
381. Hofman, J.-W.; van Zeeland, F.; Turker, S.; Talsma, H.; Lambrechts, S.A.G.; Sakharov, D.V.; Hennink W.E.; van Nostrum, C.F. *J. Med. Chem.* **2007**, *50*, 1485–1494.
382. Bello K.A.; Bello, I.A. *Dyes Pigm.* **1997**, *35*, 261–267.
383. Hanack, M.; Metz, J.; Pawlowski, G. *Chem. Ber.* **1982**, *115*, 2836–2853.
384. Kasuga, K.; Lin, L.; Hand, M.; Sugimori, T.; Isa, K.; Matsuura, K.; Takinami, Y. *Inorg. Chim.*, **1999**, *38*, 4174–4176.
385. Kubiak, R.; Janczak, J.; Razik, M. *Inorg. Chim. Acta* **1999**, *293*, 155–159.
386. Kubiak, R.; Razik, M. *Acta Crystallogr. C* **1998**, *54*, 483–485.
387. Kagaya, Y. Isago, H. *Bull. Chem. Soc. Jpn.* **1997**, *70*, 2179–2185.
388. Kagaya Y., Isago, H. *Chem. Lett.* **1994**, *23*, 1957–1960.
389. Knör, G. *Inorg. Chem.* **1996**, *35*, 7916–7918.
390. Isago, H. Kagaya, Y. *Bull. Chem. Soc. Jpn.* **1994**, *67*, 383–389.
391. Janczak, J. Kubiak, R. Hahn, F. *Inorg. Chim. Acta* **1998**, *281*, 195–200.
392. Janczak, J. Kubiak, R. Richter, J. Fuess, H. *Polyhedron*, **1999**, *18*, 2775–2780.
393. Benihya, K.; Mossoyan-Deneux, M.; Hahn, F.; Boucharat, N.; Terzian, G. *Eur. J. Inorg. Chem.* **2000**, 1771–1779.
394. Li, J.; Subramanian, L.R.; Hanack, M. *Eur. J. Org. Chem.*, **1998**, 2759–2767.
395. Isago, H.; Kagaya, Y. *J. Porphyrins and Phthalocyanines*, **2009**, *13*, 382–389.
396. Gouterman, M.; Sayer, P.; Shankland, E.; Smith, J.P. *Inorg. Chem.*, **1981**, *20*, 87–92.
397. Sayer, P.; Gouterman, M.; Connell, C.R. *Acc. Chem. Res.* **1982**, *15*, 73–79.
398. Fujiki, M.; Tabei, H.; Isa, K. *J. Am. Chem. Soc.* **1986**, *108*, 1532–1536.

399. Janczak, J.; Kubiak, R.; Jezierski, A. *Inorg. Chem.* **1999**, *38*, 2043–2049.
400. Isago, H.; Kagaya, Y. *Inorg. Chem.* **2012**, *51*, 8447–8454.
401. Kobayshi, N.; Furuyama, T.; Satoh, K. *J. Am. Chem. Soc.* **2011**, *133*, 19642–19645.
402. Furuyama, T.; Satoh, K.; Kushiya, T.; Kobayashi, N. *J. Am. Chem. Soc.* **2014**, *136*, 765–776.
403. Baumann, F. *U.S. Patent* 2,613,128 (**1952**).
404. Fukada, N. *Nippon Kagaku Zasshi* **1954**, *76*, 1141–1143.
405. Bernauer, K.; Fallab, S. *Helv. Chim. Acta* **1961**, *44*, 1287–1292.
406. Fukada, N. *Bull. Chem. Soc. Jpn.* **1961**, *34*, 884–887.
407. Weber, J.H.; Busch, D.H. *Inorg. Chem.* **1965**, *4*, 469–471.
408. Kim, Y.D.; Kim, J.P.; Kwon, O.S.; Cho, I.H. *Dyes Pigm.* **2009**, *81*, 45–52.
409. Ali, H. Langlois, R.; Wagner, J.R.; Brasseur, N.; Paquette, B.; J.E. van Lier, *Photochem. Photobiol.* **1988**, *47*, 713–717.
410. Brasseur, N;. Ali, H.; Langlois, R.; Wagner, R.J.; Rousseau, J.; van Lier, J.E. *Photochem. Photobiol.* **1987**, *45*, 581–586.
411. Paquette, B.; Ali, H.; Langlois, R.; van Lier, J.E. *Photochem. Photobiol.* **1988**, *47*, 215–220.
412. Ozoemena, K.; Kuznetsova, N.; Nyokong, T. *J. Mol. Catal. A: Chem.* **2001**, *176*, 29–40.
413. Ogunsipe, A.; Nyokong, T. *J. Photochem. Photobiol. A: Chem.* **2005**, *173*, 211–220.
414. Ambroz, M.; Beeby, A.; McRobert, A.J.; Simpson, M.S.C.; Svensen, R.K.; Phillips, D. *J. Photochem. Photobiol. B: Biol.* **1991**, *9*, 87–95.
415. Achar, B.N.; Fohlen, G.M.; Parker, J.A. *J. Polym. Sci.* **1983**, *21*, 589–597.
416. Shirai, H.; Maruyama, A.; Kobayashi, K.; Hojo, N.; Urushido, K. *Makromol. Chem.* **1980**, *181*, 575–584.
417. Sun, X.; Wang, L.; Tan, Z. *Catal. Lett.* **2015**, *145*, 1094–1102.
418. Bai, M.; Song, R.; Zhang, Y.; Han, S.; Song, X.; Meng, F. *Inorg. Chem. Commun.* **2013**, *28*, 99–103.
419. Song, X.; She, Y.; Ji, H.; Zhang, Y. *Org. Process Res. Dev.* **2005**, *9*, 297–301.
420. Bahadoran F.; Dialameh, S. *J. Porphyrins Phthalocyanines* **2005**, *9*, 163–169.
421. Sakamoto, K.; Ohno, E. *Prog. Org. Coat.* **1997**, *31*, 139–145.
422. Masilela, N.; Nyokong, T. *Dyes Pigm.* **2010**, *84*, 242–248.
423. Shaposhnikov, G.P.; Maizlish, V.E.; Kulinich, V.P. *Russ. J. Gen. Chem.* **2005**, *75*, 1480–1488.
424. Opris, D.M.; Nüesch, F.; Löwe, C.; Molberg, M.; Nagel, M. *Chem. Mater.* **2008**, *20*, 6889–6896.
425. Leznoff, C.C.; Vigh, S.; Svirskaya, P.I.; Greenberg, S.; Drew, D.M.; Ben-Hur, E.; Rosenthal, I. *Photochem. Photobiol.* **1989**, *49*, 279–284.
426. Kobayashi, N.; Ojima, F., Osa, T.; Vigh, S.; Leznoff, C.C. *Bull. Chem. Soc. Jpn.* **1989**, *62*, 3469–3474.
427. Sugimori, T.; Nojima, J.; Ozawa, T.; Handa, M.; Kasuga, K. *Chem. Lett.* **2004**, *33*, 1014–1015.
428. Sesalan, S.B.; Koca, A.; Gül, A. *Dyes Pigm.* **2008**, *79*, 259–264.
429. Banfi, S.; Caruso, E.; Buccafurni, L.; Ravizza, R.; Garibaldi, M.; Monti, E. *J. Organomet. Chem.* **2007**, *692*, 1269–1276.
430. Scalise I.; Durantini, E.N. *Bioorg. Med. Chem.* **2005**, *13*, 3037–3045.
431. Gaspard, S.; Thi, T.T. *J. Chem. Soc., Perkin Trans. II*, **1989**, 383.
432. Durmuş, M.; Nyokong, T. *Inorg. Chem. Commun.* **2007**, *10*, 332–338.
433. Sehlotho, N.; Durmus, M.; Ahsen, V.; Nyokong, T. *Inorg. Chem. Commun.* **2008**, *11*, 479–483.
434. Tolbin, A.Yu.; Tomilova, L.G. *Russ. Chem. Bull. Int. Ed.* **2007**, *56*, 2433–2437.
435. Özçesmeci, I.; Güner, S.; Okur, A.I.; Gül, A. *J. Porphyrins Phthalocyanines* **2007**, *11*, 531–536.
436. George, L.; Müller, A.; Röder, B.; Santala, V.; Efimov, A. *Dyes and Pigments,* **2017**, 147, 334–342.

437. Pereira, J.B.; Carvalho, E.F.A.; Faustino, M.A.F.; Fernandes, R.; Neves, M.G.P.M.S.; Cavaleiro, J.A.S.; Gomes, N.C.M.; Cunha, *B.*; Almeida, A.; Tomé, J.P.C. *Photochem. Photobiol.* **2012**, *88*, 537–547.
438. Sharman, W.M.; van Lier, J.E. *J. Porphyrins Phthalocyanines* **2005**, *9*, 651–658.
439. Sharman, W.M.; van Lier, J.E. *J. Porphyrins Phthalocyanines* **2000**, *4*, 441–453.
440. McKeown, N.B.; Painter, J. *J. Mater. Chem.* **1994**, *4*, 1153–1156.
441. Kobayashi, N.; Higashi, R.; Ishii, K.; Hatsusaka, K.; Ohta, K. *Bull. Chem. Soc. Jpn.* **1999**, *72*, 1263–1271.
442. Kroon, J.M.; Koehorst, R.B.M.; van Dijk, M.; Sanders, G.M.; Sudhölter, E.J.R. *J. Mater. Chem.* **1997**, *7*, 615–624.
443. Clarkson, G.J.; McKeown, N.B.; Treacher, K.E. *J. Chem. Soc. Perkin Trans. 1* **1995**, 1817–1823.
444. Vacus, J. Simon, J. *Adv. Mater.* **1995**, *7*, 797–800.
445. Gürek, A.G.; Durmus, M.; Ahsen, V. *New J. Chem.* **2004**, *28*, 693–699.
446. Atilla, D.; Durmuş, M.; Gürek, A.G.; Ahsen, V.; Nyokong, T. *Dalton Trans.* **2007**, 1235–1243.
447. Atilla, D.; Durmuş, M.; Yılmaz, Ö.; Gürek, A.G.; Ahsen, V.; Nyokong, T. *Eur. J. Inorg. Chem.* **2007**, 3573–3581.
448. Huang, J.D.; Wang, S.Q.; Lo, P.C.; Fong, W.P.; Ko, W.H.; Ng, D.K.P. *New J. Chem.* **2004**, *28*, 348–354.
449. Brasseur, N.; Ouellet, R.; Madeleine, C.L.; van Lier, J.E. *Brit. J. Cancer*, **1999**, *80*, 1533–1541.
450. Doan, P.; Karjalainen, A.; Chandrasselan, J.G.; Sandberg, O.; Yli-Harja, O.; Rosholm, T.; Franzen, R.; Candeias, N.R.; Kandhavelu, M. *Eur. J. Med. Chem.* **2016**, *120*, 296–303.
451. Dhahagani, K.; Kumar, S.M.; Chakkaravarthi, G.; Anitha, K.; Rajesh, J.; Ramu, A.; Rajagapal, G. *Spectrochim. Acta A*, **2014**, *117*, 87–94.
452. Ladopoulou, E.M.; Matralis, A.N.; Nikitakis, A.; Kourounakis, A.P.; *Bioorg. Med. Chem.* **2015**, *23*, 7015–7023.
453. Hamidian, H.; Azizi, S. *Bioorg. Med. Chem.* **2015**, *23*, 7089–7094.
454. Zheng, B.; Ke, M.; Lan, W.; Hou, L.; Guo, J.; Wan, D.; Cheong, L.; Huang, J. *Eur. J. Med. Chem.* **2016**, *114*, 380–389.
455. Bart, B. Demirbag, U. Ozel, A. Kantekin, H. *Int. J. Biol. Macromol.* **2017**, *105*, 499–508.
456. Zhu, Y.-J.; Huang, J.-D.; Jiang, X.-J.; Sun, J.-C. *Inorg. Chem. Commun.* **2006**, *9*, 473–477.
457. Ercolani, C.; Monacelli, F.; Dzugan, S.; Goedken, V. L.; Pennesi, G.; Rossi, G. *J. Chem. Soc., Dalton Trans.* **1991**, 1309–1315.
458. Moubaraki, B.; Benlian, D.; Baldy, A.; Pierrot, M. *Acta Crystallogr. C* **1989**, *45*, 393–394.
459. Lever, A.B.P.; Licoccia, S.; Ramaswamy, B.S. *Inorg. Chim. Acta* **1982**, *64*, L87–L90.
460. Kasha, M.; Rawls, H.R.; Et-Bayoumi, M.A. *Pure Appl. Chem.* **1965**, *11*, 371–392.
461. Kobayashi, N.; Shirai, H.; Hojo, N. *J. Chem. Soc., Dalton Trans.* **1984**, 2107–2110.
462. Kimura, M.; Shirai, M. in *The Porphyrin Handbook*; Kadish, K. M.; Smith, K. M.; Guilard, R., Eds.; Academic Press: San Diego, **2003**, Vol. 19, pp. 151–177.

216 Computational Guidance in the Design of Functional Tetrapyrroles

Angel Zhang and Martin J. Stillman

Department of Chemistry, The University of Western Ontario,
1151 Richmond Street, London, Ontario, Canada N6A 5B7

List of Abbreviations

BM	bohr magneton
Chl	chlorophyll
DCM	dichloromethane

DSSCs	dye-sensitized solar cells
DFT	density functional theory
ESOP	excited state oxidation potential
GSOP	ground state oxidation potential
HOMO	highest occupied molecular orbital
ΔHOMO	energy difference between HOMO and HOMO-1
LUMO	lowest unoccupied molecular orbital
ΔLUMO	energy difference between LUMO and LUMO+1
MCD	magnetic circular dichroism
Nc	naphthalocyanine
Pc	phthalocyanine
PCE	power conversion efficiency
TD-DFT	time-dependent density functional theory
THF	tetrahydrofuran
ZnTPP	zinc tetraphenylporphyrin

I. Introduction

Tetrapyrroles have many functions in Nature, including oxygen transport and storage (hemoglobin and myoglobin), photosynthesis (chlorophyll), electron transport (cytochromes), and catalysis (vitamin B12).[1] Their diverse roles are due to their highly versatile ring structure and ability to bind a variety of metals.[2] Tetrapyrroles are known in many subclasses, for example, porphyrins, and chlorins in Nature and a range of other rings, including phthalocyanines in the synthetic laboratory (Figure 1). Their optical spectra vary considerably in complexity and relative absorbance across the blue and red regions of the visible spectrum into the near IR up to almost 900 nm for bacteriochlorophylls. The remarkable chemical and photophysical properties of tetrapyrroles make them promising candidates for many applications including the dyes in

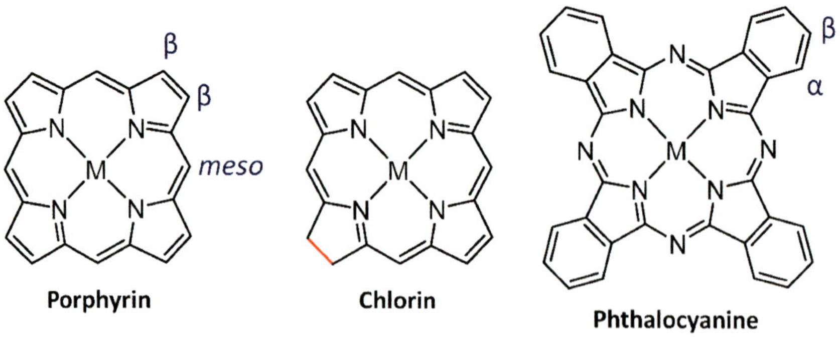

Figure 1. Structures of a metalated porphyrin, chlorin, and phthalocyanine. The α-, β-, and *meso*-positions are labeled.

solar cells and photodynamic therapy.[3–5] Strong absorption in the red-to-near infrared region for both of these applications, is necessary for high efficiency. In this chapter, we will focus on the design of tetrapyrroles for dye-sensitized solar cells (DSSCs).

Due to the increasing demand for renewable energy, methods for solar energy to electricity conversion are of increasing importance. Traditional silicon-based, solid-state solar cells are expensive and energy-intensive (high temperature and pressure processes) so efforts have been made in developing nanocrystalline-based DSSCs.[6] DSSCs are cheaper and easier to fabricate than silicon-based solar cells but the power conversion efficiency (PCE) is still not as high.[7–8] In a process unlike traditional solar cells, light absorption and electron transport are handled by different materials in the cell.[6, 9–10] The stability and optical properties of the sensitizing dye can, therefore, be tuned for maximum electron transfer to a wide-band-gap semiconductor, for example TiO_2 nanoparticles. As mentioned above, an ideal dye should absorb strongly in the red region. For high PCEs, the dye should also exhibit panchromatic response in the visible region, have anchoring groups so it can bind strongly to the surface of the semiconductor, and be photostable so it can sustain many years of exposure to sunlight.[11]

The purpose of this chapter is to describe three cases in which tetrapyrroles were designed to exhibit high intensity, red located Q bands in order to provide efficient dyes in DSSCs. Our objectives are to illustrate how the combination of computational methods with experimental data allows prediction of functionally relevant properties. Computational methods coupled with a detailed understanding of the underlying electronic structures in tetrapyrroles means that computationally guided molecular design for specific functions is now possible. Our procedure to design efficient electron transfer in DSSCs involved first analyzing the experimental optical spectra of core compounds. These compounds are based on tetrapyrroles that will be used in the second stage. Tetrapyrroles are chosen as the core aromatic system for the novel compounds envisaged and their absorption, magnetic circular dichroism (MCD), and emission spectra are used as an indicator of the underlying electronic structure. With good parameters for the electronic structures on hand, obtained by computational analysis, we can add these data to the established trend data to connect observable properties with computational results. This allows for the prediction of experimental properties of the novel fictive compounds and quantitative assessment of the precision of the described electronic structure. We next briefly describe the key features of the optical properties of tetrapyrroles that relate to their use as dyes in DSSCs.

II. Absorption and Magnetic Circular Dichroism Spectra of Tetrapyrroles

Due to the highly conjugated π-system in tetrapyrroles, both natural and synthetic tetrapyrroles are deeply colored. The slight changes to the ring structure between porphyrins, chlorins, and phthalocyanines (Pc) result in different spectral properties (Figure 2).[12] Although they all exhibit absorption in the blue (B band) and red (Q band) region, the relative intensities of these bands differ significantly. MCD spectroscopy can be used to measure the change in electronic structure and underlying angular momentum of the tetrapyrroles that results in the different absorption property.[13]

A. Origin of the Optical Spectra

In 1959, Gouterman proposed the four-orbital model to explain the absorption spectra of tetrapyrroles.[14] His model states that the absorption bands of porphyrins arise from transitions between two highest occupied molecular orbitals (HOMO, HOMO-1) and two lowest unoccupied molecular orbitals (LUMO, LUMO+1), Figure 3.[15] These four orbitals can be referred to as Gouterman's MOs.[16] For metalloporphyrins of D_{4h} symmetry, the HOMO pair are of a_{1u} or a_{2u} symmetry character and can be accidentally degenerate while the LUMO pair have e_g character (e_{gx} or e_{gy}) and are usually doubly degenerate. In addition, the HOMO pair and LUMO pair inherit angular momenta (M_L) of ±4 BM and ±5 BM, respectively, from the $C_{16}H_{16}^{2-}$ cyclic polyene parent. The angular momentum of each orbital also corresponds to the number of nodes; the two HOMOs have four nodes and the two LUMOs have five nodes each. The optical and redox properties of both natural and synthetic tetrapyrroles are dependent on the orientation of the nodal planes. The occupied MO of a_{1u} character is identified with nodes that pass through the pyrrole nitrogen atoms and the occupied MO of a_{2u} character is identified with four nodes at 22.5° from the a_{1u} nodes. With respect to the five nodes of the LUMO shown in Figure 3, the orientation of the nodes no longer results in major differential change in the energies of LUMO and LUMO+1 as a function of peripheral substitutions. However, the introduction of major changes to the tetrapyrroles ring, for example the reduction in chlorins, splits the typically degenerate LUMO/LUMO+1.

Specifically in porphyrins, electronic transitions between the four Gouterman MOs give rise to two excited states, S_1 and S_2. A transition from the ground state (S_0) to the first (S_1) and second (S_2) excited states give rise to the observed Q and B bands, respectively, We should note that the contributions from the elements in the π ring will change this straightforward picture. A specific case is that of the orbital origins of the Q and B_1 and B_2 states in, where as just noted, there are two B states (comprised of two x/y pairs). The S_1 and S_2 excited states are of E_u symmetry and they consist of a degenerate pair of states named Q_x, Q_y and B_x, B_y, respectively.[15] The x and y subscripts denote the direction of polarization of the transition within the tetrapyrrole ring (the direction is reversed for chlorophyll with respect to porphyrins).[17,18] Normally when there is not much splitting of the four MOs, the x and y components are degenerate and the absorption spectrum will show one Q band peak, with Q_x and Q_y superimposed on one another. The Q_x and Q_y bands are observed separately when the four MOs are split significantly. However, although the B_x and B_y bands are split as well, the extent is not so great.

The B and Q bands originate from allowed and forbidden transitions between the Gouterman MOs, respectively. In the case of porphyrins where the HOMO pair and LUMO pair are each close to degeneracy, the forbidden Q band is weak while the B band is intense. The splitting of the usually degenerate MOs breaks the selection rule that leads to the Q band being forbidden and allows borrowing of intensity from the allowed B band. The intensity of the Q and B bands are coupled so an

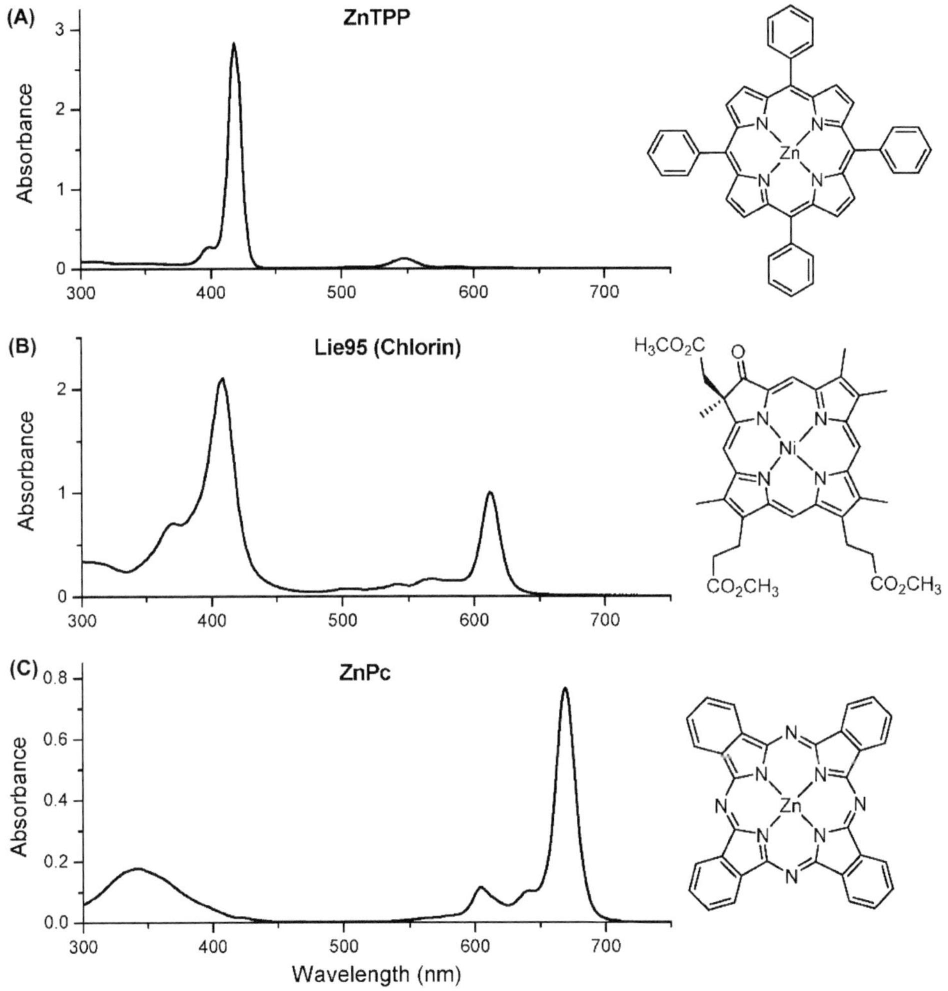

Figure 2. The absorption spectra of **(A)** ZnTPP, **(B)** a synthetic chlorin, **Lie95**, and **(C)** ZnPc. The absorption spectrum of **Lie95** resembles that of a simple methyl-substituted chlorin so it is used here as an example of a chlorin spectrum.

increase in intensity of the Q band will decrease the intensity of the B band. The reduction of the pyrrole in the chlorin macrocycle splits the HOMO and LUMO pairs which results in a stronger Q band, but slightly weaker B band (Figure 2). Of the three tetrapyrroles mentioned above, phthalocyanines exhibit the most intense Q band with two weak overlapping B bands.[19] The aza-substitution of the four methine bridges in the ring selectively affects the a_{2u} MO of the HOMO pair more than the a_{1u} MO and introduces a significant energy split of over 2 eV.[20]

A complication in the assessment of the optical spectra of many tetrapyrroles is the appearance of excited state vibronic bands lying to the blue of the Q and B bands. This is not a particular issue for metallophthalocyanines because the Q_{00} band intensity is essentially allowed and, therefore, already intense with molar absorptivities in the 10^5 range. For the most symmetric and least substituted porphyrins, in which there is accidental degeneracy of the two highest occupied MOs, the Q_{00} oscillator strength will approach zero, leaving the Q_{vib} bands (01, 02, 03, *etc.*)

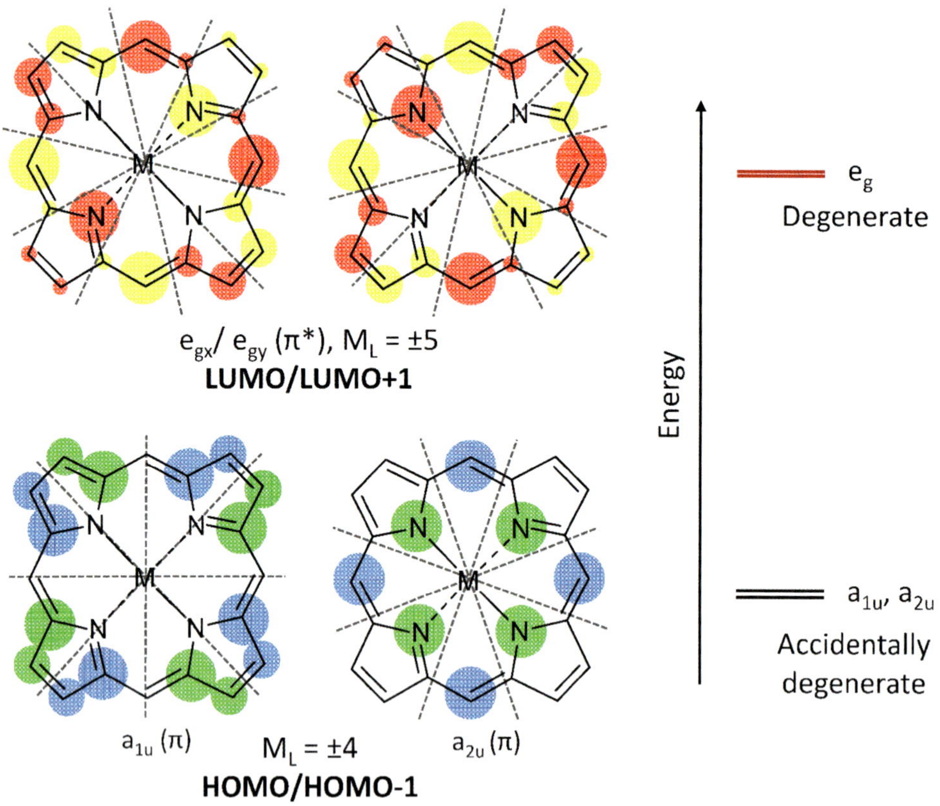

Figure 3. Simplified electron density surfaces for the four Gouterman MOs. The blue and green shading represent the occupied π MOs while the red and yellow shading represent the unoccupied π^* MOs. The nodes for the HOMOs (four nodes) and LUMOs (five nodes) are shown with grey dotted lines. Reproduced from Ref. 24 with permission of The Royal Society of Chemistry.

as the most intense bands. This effect is most readily seen in the trends in this chapter that are based on the trends first reported by Mack and coworkers in 2005.[20] When the two highest occupied MOs are split, then Q_{00} will exhibit measureable intensity and now two or more bands will be measured to the blue of the 1st excited state energy. No matter the intensity of the Q_{00} band, its angular momentum retains the value of the 1st excited state of the tetrapyrrole, so is reflected experimentally in the magnitude of A term where this can be estimated from −A/D, with A given by the peak-to-trough magnitude in the MCD spectrum and where D, the dipole strength, can be estimated from the peak intensity of the absorption. As described above, a ΔHOMO > ΔLUMO (the usual situation) would represent a positive A term with the MCD derivative shaped band signs alternating negative to positive with increasing energy. So, even with relatively extremely weak Q_{00} the MCD spectrum allows identification of the exact energy of the 1st excited state.

B. MCD Spectra Band Morphologies

The temperature independent MCD spectral band morphologies of the Faraday A and B terms[21,22] have been used extensively to characterize and subsequently identify the origins of electronic transition for symmetric tetrapyrroles, most notably for porphyrins and phthalocyanines.[12,13] For tetrapyrroles with peripheral substitution the MCD spectrum will comprise just B terms from transitions to nondegenerate excited states. These excited states do, however, owe their polarization and inherent angular momentum to the degenerate excited states of the symmetric parent. For the 18 π tetrapyrroles, the lowest π–π* energy states will typically exhibit angular momentum approaching the ±9 BM of the $C_{16}H_{16}^{2-}$ cyclic polyene. The series of degenerate e_u and e_g MOs of the cyclic polyene are replaced to differing extents by nondegenerate MOs in tetrapyrroles following ring substitution of the carbons with heteroatoms.[23] More common examples include substitution with the nitrogens of the pyrroles and the aza bridges in Pcs. The loss of the degeneracy of the top-filled e_u and lowest unfilled e_g MOs can be readily explained by reference to the nodal patterns of these orbitals, with 18 π electrons, the top filled orbitals exhibit four nodes and the lowest unfilled exhibit five nodes, Figure 3.

Peripheral substitution with electronically active groups and the alignment of the nodes in the occupied and unoccupied MOs introduce sensitivity to the differing geometric distributions of electron densities. This sensitivity changes the relative energies of all MOs, but from a spectroscopic viewpoint, the effect is most apparent with the HOMOs and LUMOs. The effect of peripheral substitution with groups that introduce electronic "push" or "pull" on the aromatic π ring electrons are seen in the relative electron density of each occupied MO. Electron donor ("push") groups destabilize or raise the MO energies by adding density to the core 18 π ring, while electron withdrawing ("pull") groups stabilize the MOs by depleting the core electron density. The end result of all this is that for most tetrapyrroles, the highest occupied MOs (nominally with e_u symmetry) split up to several eV (for example about 2 eV in ZnPc) reducing their inherent angular momenta of the parent degenerate MO. The unoccupied π^* orbitals (nominally with e_g symmetry), however, split only slightly, leaving the angular momentum magnitude intact.

There are three ways in which the HOMO pair and LUMO pair can split and each type of splitting results in a distinct MCD pattern.[24] Figure 4 illustrates the origins of B terms in low symmetry molecules starting from the typical positive A

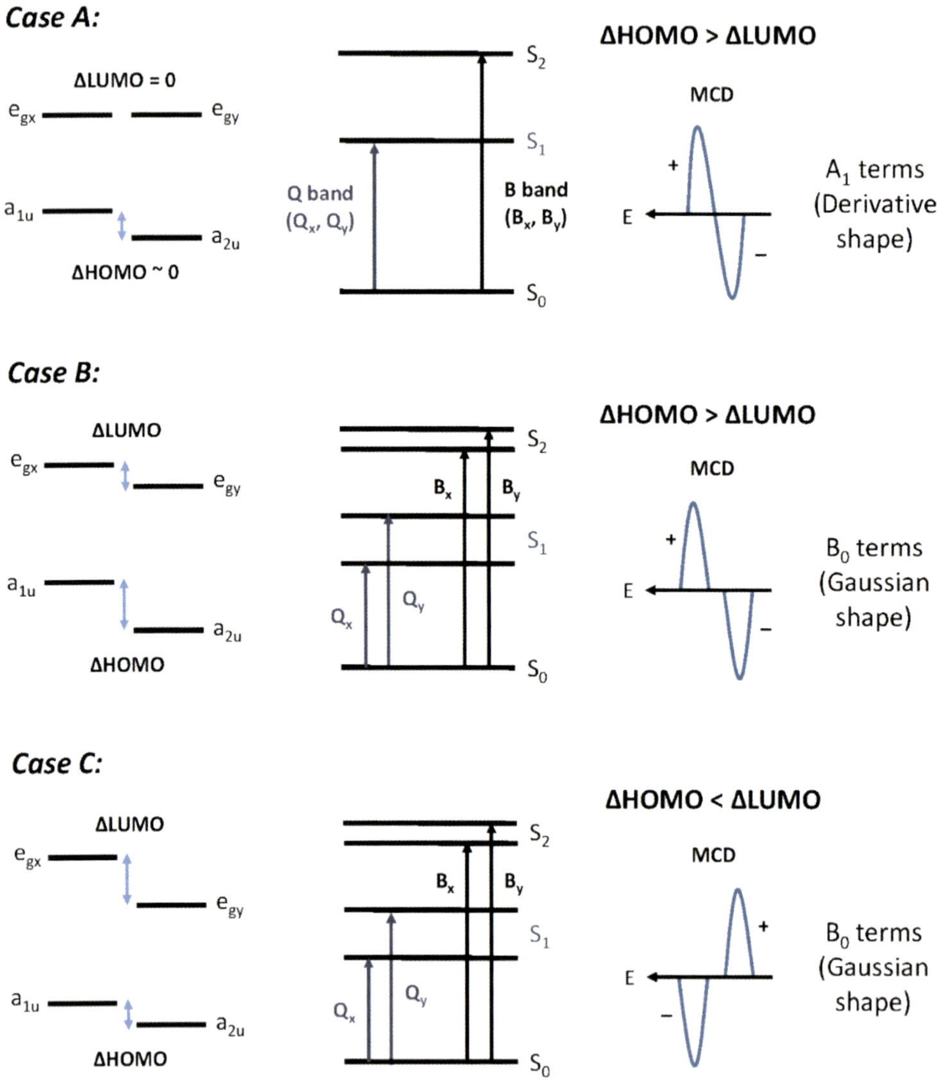

Figure 4. The three splitting patterns of the four Gouterman MOs and the resulting excited state degeneracies and the resultant MCD spectral morphologies. Reproduced from Ref. 24 with permission of The Royal Society of Chemistry.

term of the symmetric tetrapyrrole. This splitting pattern, *Case A*, will result in strong positive A term with a negative-to-positive sign pattern sequence with increasing energy. For this ΔHOMO > ΔLUMO case, the MCD spectral bands then follow a normal sign pattern reflecting the situation that the unoccupied π* orbitals have a greater angular momentum than the occupied π MOs. The resulting B term sign sequence derives from a set of "positive A terms."[25,26] *Case B* represents the situation where both the LUMO and HOMO degeneracies are broken but the splitting between the HOMO pair is greater than the splitting between the LUMO pair. This is typical for very low symmetry porphyrins, and a pair of equal and oppositely signed B terms will be observed. The B term magnitudes will be weaker as the energy separation of the x,y components increases, but the sign pattern will remain unchanged. The sign sequence will again depend on the sign pattern of the A term parent. Under rare geometric substitution patterns or specific electronic interactions between the peripheral substituents (*Case C*), the LUMO e_g pair of orbitals will split more than the HOMO e_u pair, resulting in an inversion in the angular momentum between the ground and excited states. The geometrical and electronic effects split the LUMO pair significantly, so that ΔLUMO > ΔHOMO. Under these conditions, a positive-to-negatively signed pseudo A term, or pair of B terms, will be measured with increasing energy.[24,25]

C. Relative Intensity of MCD Bands

The measured intensity of Q_x *vs.* Q_y in an MCD spectrum of the tetrapyrroles can be explained using the matrix elements that describe each MCD band (Figure 5). Note that the matrix elements shown here are only one part of the whole equation, which has been described by Piepho and Schatz.[21] S_1 and S_2 contain x and y components, labeled $S_{1(X)}$, $S_{1(y)}$, $S_{2(x)}$, and $S_{2(y)}$. In the presence of a magnetic field, splitting of the x and y states is observed. Usually for tetrapyrroles, the splitting of $S_{2(x)}$ and $S_{2(y)}$ is not very extensive so B_x and B_y absorption bands are observed as a single band. When the extent of splitting between $S_{1(X)}$ and $S_{1(y)}$ is small (Figure 5A), the resulting Q_x and Q_y absorption bands are close to each other and well-separated energetically from the B_x and B_y bands. This gives rise to A or pseudo A terms, comprising equal and opposite B terms, in the corresponding MCD spectrum. The matrix element that describes the Q_x MCD band indicates that $S_{1(X)}$ couples only with $S_{1(y)}$, and for the Q_y MCD band, $S_{1(y)}$ couples only with $S_{1(x)}$. In other words, the Q_x and Q_y bands couple with each other and their resulting intensities are the same. Coupling to B_x and B_y is still present, but is weak. However, when the splitting between $S_{1(X)}$ and $S_{1(y)}$ is large (Figure 5B), the Q_y band lies much closer to the B_x band. This allows the Q_y MCD band to couple with Q_x, as well as with B_x. The matrix element for this Q_y band shows the additional coupling of $S_{1(y)}$ with $S_{2(x)}$, together with the original coupling with $S_{1(x)}$. The additional

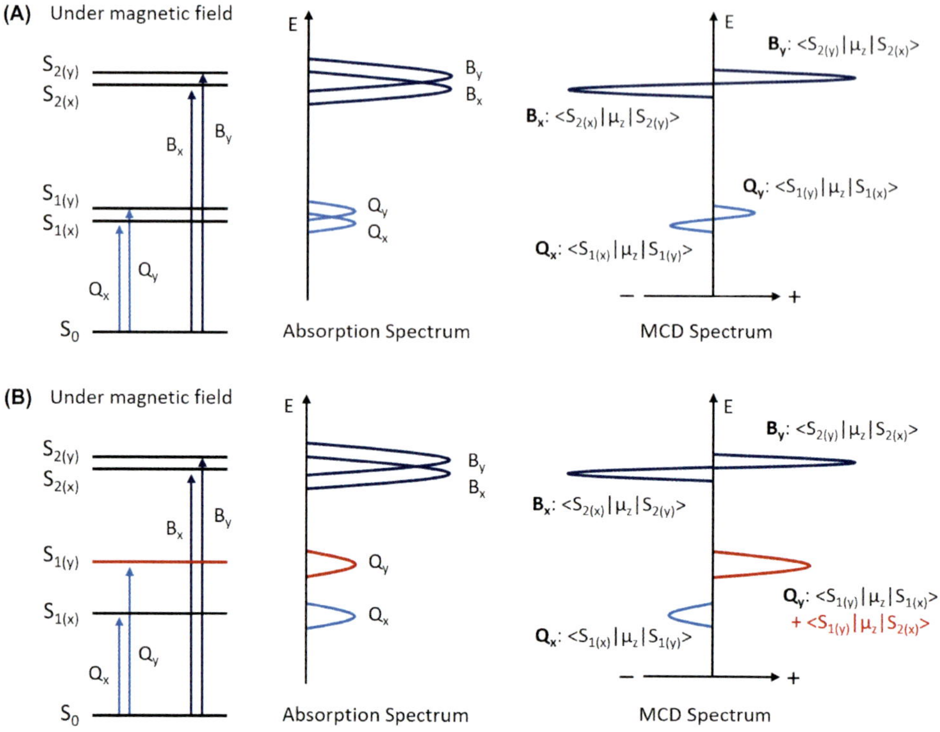

Figure 5. Simplified diagrams of the first (S_1) and second (S_2) excited states under a magnetic field and the corresponding absorption and MCD spectrum for **(A)** small splitting of S_1 and **(B)** large splitting of S_1. Amount of splitting is not to scale. The main matrix elements that give rise to the Q and B MCD bands are labeled beside the respective bands.

coupling of Q_y with the strong B_x band increases the intensity of Q_y MCD B term when compared to the intensity of the Q_x MCD B term. Because this is a magnetically induced effect, the absorption spectrum is unaffected, meaning that the higher energy Q component has a weaker relative absorption.

III. Thiophene Substituted Phthalocyanine

The first case we describe concerns the application of thiophene substituted ZnPc compounds, ***α*PcS1** and ***α*PcS2**, reported by Yamamoto *et al.*[27] (Figure 6) as dyes in DSSCs. In these compounds, the α substituted thiophene was designed to provide a connection to the TiO_2 semiconductor and also shift the Q band to a longer wavelength. Thiophene substitution on the α-position of the Pc was found to destabilize the HOMO more than substitution at the β-position, leading to a lower energy Q band. The additional substitution of diisopropylphenoxyl on each of the benz groups of the Pc was designed to reduce aggregation. The extra extended π conjugation of ***α*PcS2** was expected to further

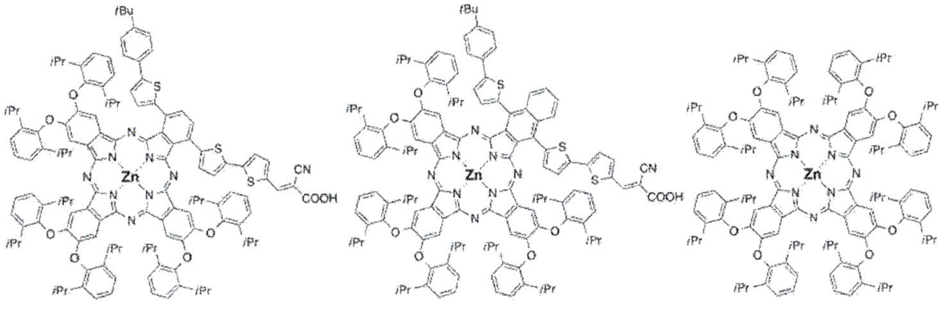

αPcS1 αPcS2 PcRef

Figure 6. Structures of ***α*PcS1** and ***α*PcS2** and **PcRef** (originally labeled number **5** in the cited paper). Reproduced from Ref. 27 with permission of Wiley.

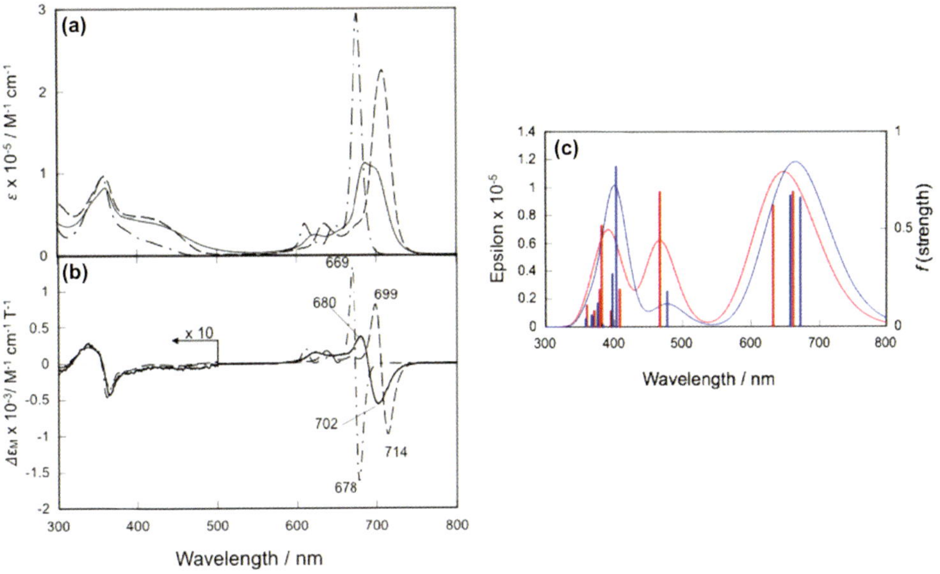

Figure 7. **(a)** Absorption and **(b)** MCD spectra of ***α*PcS1** (solid line), ***α*PcS2** (dashed line), and **5** (dash-dotted line) in THF. **(c)** Theoretical spectra of ***α*PcS1** (red) and ***α*PcS2** (blue). Reproduced from Ref. 27 with permission of Wiley.

expand the light-harvesting region for DSSC application. In the cells, the PCEs for ***α*PcS1** and ***α*PcS2** were recorded to be 5.5% and 3.8%, respectively. In this section, we will examine the relationship between the experimental data and the computational properties.

The absorption, MCD, and theoretical spectra of ***α*PcS1** and ***α*PcS2** are shown in Figure 7. Experimental data for the symmetrical ZnPc, **5** (Figure 6), is also included as reference. Attachment of the thiophene units red-shifted the Q band.

The bandwidths of the MCD Q bands of **αPcS1** and **αPcS2** are larger than that of **5**, suggesting that they are pseudo A terms, comprising equal and opposite B terms. **αPcS1**, **αPcS2**, and **5** exhibit similar MCD spectra in the B region but **αPcS1** and **αPcS2** show a new broad absorption band in the 380–480 nm region. The lack of MCD signal under this broad absorption band implies that negligible angular momentum is involved in these electronic transitions. TD-DFT results support this conclusion in that the HOMO → LUMO+2 transition (MO 546→549 for **αPcS1** and MO 571→574 for **αPcS2**) dominates the formation of this band (Figure 7C, Figure 8, and Table 1). LUMO+2 is a thiophene-based MO so this is an intramolecular charge transfer transition from the electron-rich Pc ring to the thiophene side chain.

Consistent with the optical data, the TD-DFT results show that **αPcS2** has a smaller HOMO–LUMO gap than **αPcS1** which lead to the slightly red-shifted

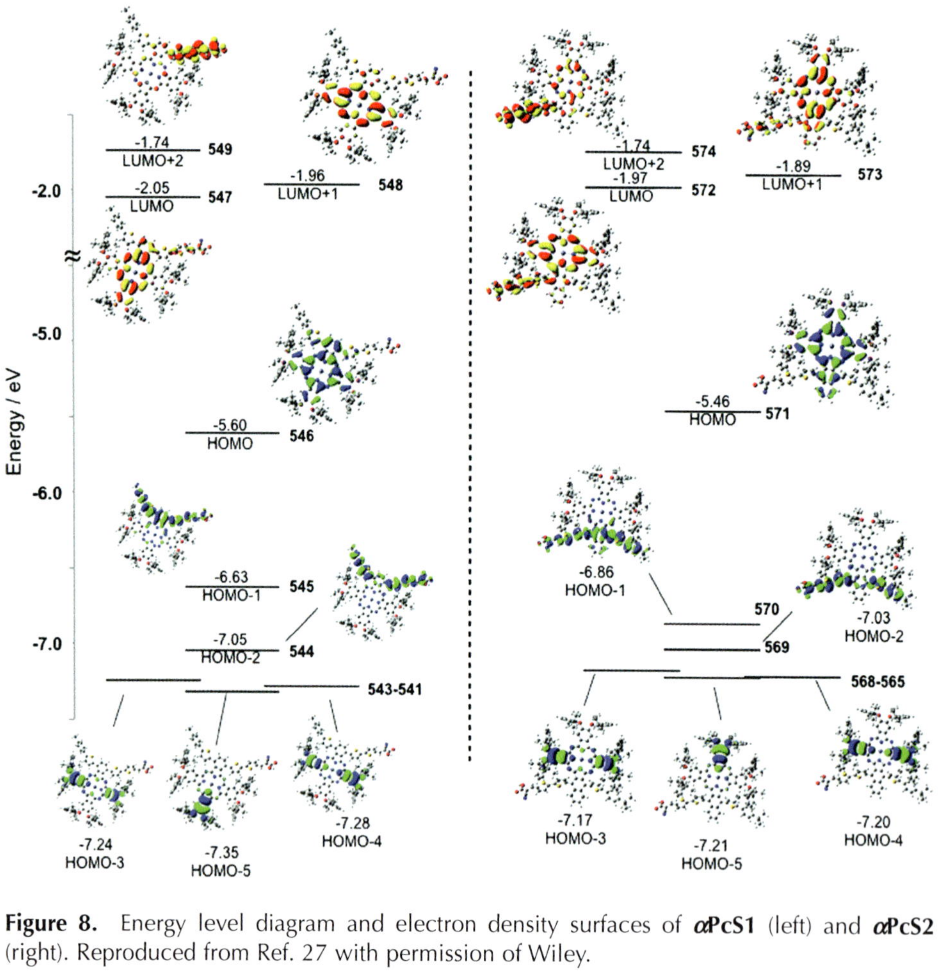

Figure 8. Energy level diagram and electron density surfaces of **αPcS1** (left) and **αPcS2** (right). Reproduced from Ref. 27 with permission of Wiley.

Table 1. Calculated absorption band energies, oscillator strengths (f), and transition contributions. Refer to Figure 8 for the electron distributions in each of these MOs. Gouterman MOs are highlighted in bold. Reproduced from Ref. 27 with permission of Wiley.

Dye	Energy [eV (nm)]	f	Configurations (main contributions)
***α*PcS1**	1.87 (661.50)	0.69	**546** (H)→**547** (L) (90%)
	1.96 (632.22)	0.62	**546**→**547** (3%), **546**→**548** (89%)
	2.65 (467.27)	0.69	545→549 (9%), **546**→549 (62%), **546**→550 (8%)
	3.04 (408.40)	0.19	545→**547** (73%)
	3.14 (395.40)	0.08	545→**548** (37%), 545→549 (20%)
	3.25 (382.10)	0.52	542→**548** (21%), 545→**548** (26%), 545→549 (17%)
	3.27 (379.10)	0.19	541→**547** (73%)
	3.29 (376.47)	0.04	542→**548** (50%), 545→**548** (19%)
	3.34 (371.17)	0.08	543→**548** (66%)
	3.44 (360.45)	0.11	**546**→550 (65%)
***α*PcS2**	1.84 (672.50)	0.66	**571** (H)→**572** (L) (82%), **571**→**573** (4%), **571**→574 (8%)
	1.89 (657.54)	0.67	**571**→**573** (84%), **571**→574 (9%)
	2.59 (477.96)	0.18	**571**→**572** (9%), **571**→574 (66%), **571**→575 (12%)
	3.07 (403.38)	0.82	565→574 (2%), 569→572 (11%), 569→574 (15%),
	3.12 (397.76)	0.27	570→**572** (19%), 570→574 (9%), **571**→574 (11%),
	3.23 (383.90)	0.01	571→575 (13%)
	3.30 (376.28)	0.12	569→574 (7%), **571**→575 (65%)
	3.36 (368.73)	0.05	557→**572** (62%)
	3.37 (367.42)	0.06	568→**572** (60%)
	3.46 (358.47)	0.04	566→**572** (11%), 566→**573** (47%), 566→574 (14%)
			566→573 (14%), 570→573 (40%)
			570→572 (41%), 570→574 (15%)

Q band (Figure 8). The annulation with the benzene unit in ***α*PcS2** destabilized the HOMO and resulted in a smaller HOMO–LUMO gap. The typical Gouterman HOMO-1, identified by four nodes, could not be found despite searching down to HOMO-30. This is likely because of the interactions between the peripheral groups and the ring.

IV. Electronically Active Substituents on Porphyrins

The case described above involved the addition of polythiophene to a phthalocyanine which together red-shifted and intensified the Q band and resulted in dyes with 5.5% PCE when used in DSSCs. In the case described now, we show the effect of electronically active β-substituents on the electronic structure of Zn tetratolylporphyrin.[24] The objective of the design of **1–4** (Figure 9) was to introduce push–pull substituents (electron donating and electron withdrawing, respectively) to extend the π-conjugation, red-shift the Q band, and introduce a dipole moment.

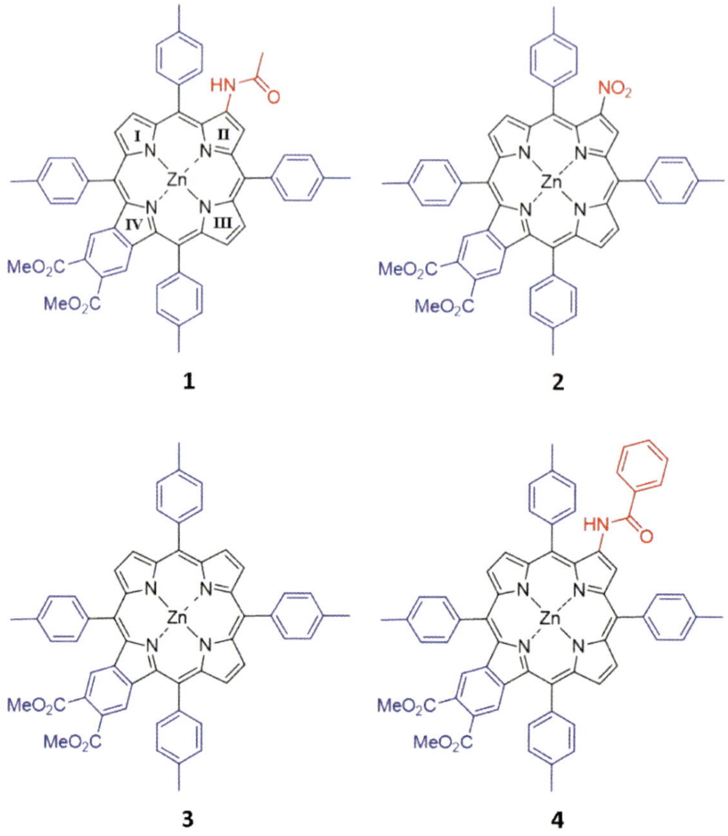

Figure 9. Structures of compounds **1–4**. The Zn-porphyrins were substituted with dimethoxy-benzo and a methylamido (**1**), a nitro (**2**), a proton (**3**), or a phenylamido (**4**) group at the β position on the opposite pyrrole. The four pyrroles are labeled I, II, III and IV. Reproduced from Ref. 24 with permission of The Royal Society of Chemistry.

A. The Optical Data of Compounds 1–4

Figure 10 shows the experimental and computational optical data for the four β-substituted porphyrins. For each of the compounds (**1–4**), the experimental MCD (top panel), absorption (2nd panel), and emission (3rd panel) spectra are compared with the computational absorption data (4th panel) obtained from TD-DFT calculations. Because of a systematic overestimation of the energy of the 1st excited state (the Q band) by TD-DFT calculations, each of the computational spectra were red-shifted manually so that the Q bands lined up. The extent of the red-shift is slightly different (ranging from 41 for **1** to 77 nm for **2**) and is noted for each of the four compounds. The energy states are shown as lines and the line broadened spectra are shown with a band width of 500 cm^{-1} that best matches the Q band width.

The absorption spectra of the four compounds in Figure 10, are typical of a metalated tetraarylporphyrin with a B_{00}:Q_{00} ratio > 10 and well-resolved Q_{vib} bands.[20] The Q band of **2** is red-shifted by over 20 nm compared with **1**, **3** and **4**, with Q_{00} at 614 nm and is the most intense of the four compounds, although still very weak. The MCD spectra of **1**, **3** and **4** (top row Figure 10) also follow the pattern typical for zinc tetraarylporphyrins, exhibiting the expected sequence of two positive A terms aligned with the Q band (Q_{00}) and the higher energy vibronic band (Q_{vib}). Both the absorption and MCD Q bands for compounds **1** and **4** show evidence for splitting into the Q_x and Q_y components, where the MCD band morphology is referred to as a "pseudo A term" where it is composed of equal and opposite B terms (Figure 4). The MCD spectrum of compound **2** is quite different as it is dominated by a negative A term (or pseudo negative A term) in the Q band region that arises because ΔLUMO > ΔHOMO.

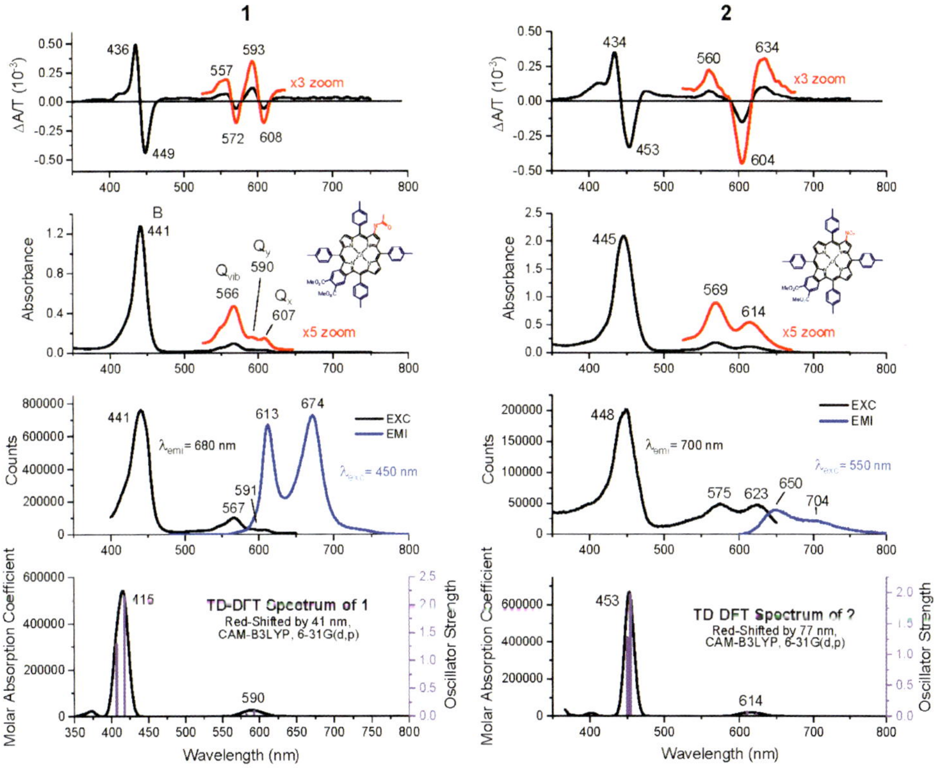

Figure 10. MCD, absorption, emission, excitation and calculated absorption spectra for the four peripherally substituted Zn-tetraaryl porphyrins, **1**, **2**, **3** and **4**. The Q region in the MCD and absorption spectra are enlarged by 3x and 5x, respectively (red lines). In **1**, the splitting of $Q_{(0,0)}$ is indicated by Q_y at 590 nm and Q_x at 607 nm. In the theoretical spectra shown in row 4, the very weak Q band and the B band are each composed of two states, indicated by purple vertical lines. Reproduced from Ref. 24 with permission of The Royal Society of Chemistry.

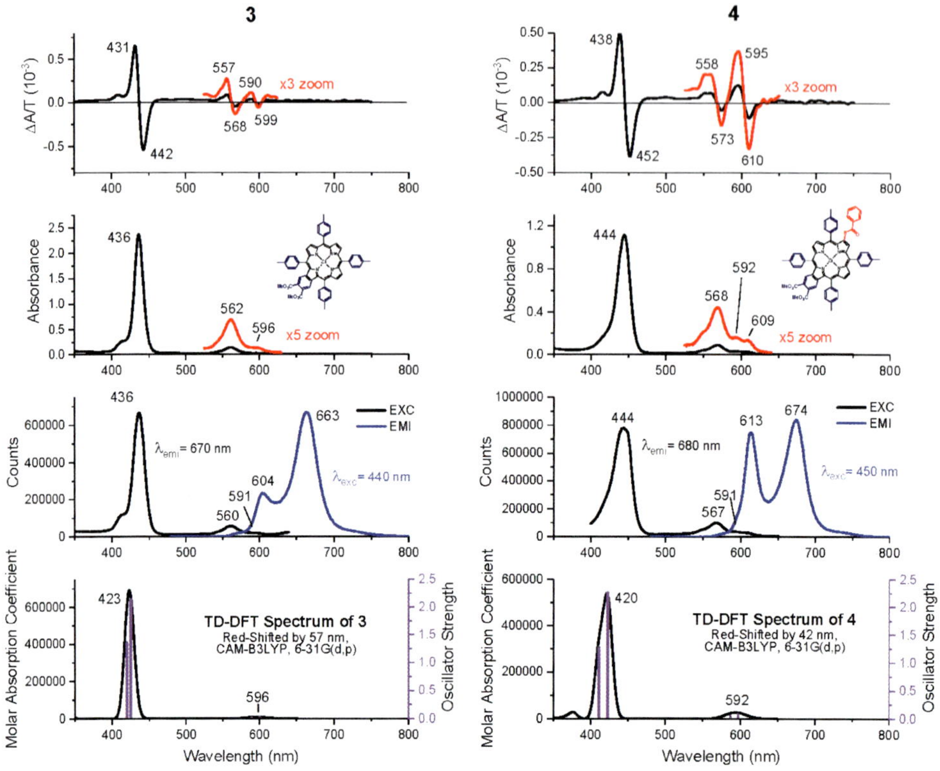

Figure 10. (*Continued*)

Row three in Figure 10 shows the emission and excitation spectra of all four compounds. The emission spectral data of compounds **1**, **3** and **4** are typical of Zn-tetraarylporphyrins, with a pair of strong bands that mirror the Q_x excited state and ground state vibrations (for **1** at 613 and 674 nm, for **3** at 604 and 663 nm, and for **4** at 613 and 674 nm). However, the peripheral substituents of compound **3** enhanced the ground state vibronic structure, making emission to the ground state vibration (663 nm) more intense. The emission of **2** is unlike the emission of the other species in that the 650 nm band is stronger than the 704 nm band and both are very weak. Also, there is a significant Stokes shift between the absorption and emission (~25 nm) whereas the Stokes shift of the other compounds is *ca*. 5–6 nm. In addition, the excitation spectrum of **2** does not match the absorption spectrum closely, which suggests that the excited state may be relaxed. This is important because the MCD spectra also indicated that the excited state is different when compared with **1**, **3** and **4**. The energy of the Q band emission provides the relaxed energy of the 1st excited state. Measurement of the Q band emission confirms that the lowest energy state has been located. Of course, emission spectra can only be obtained at room temperature for closed shell, diamagnetic metals such as Zn(II).

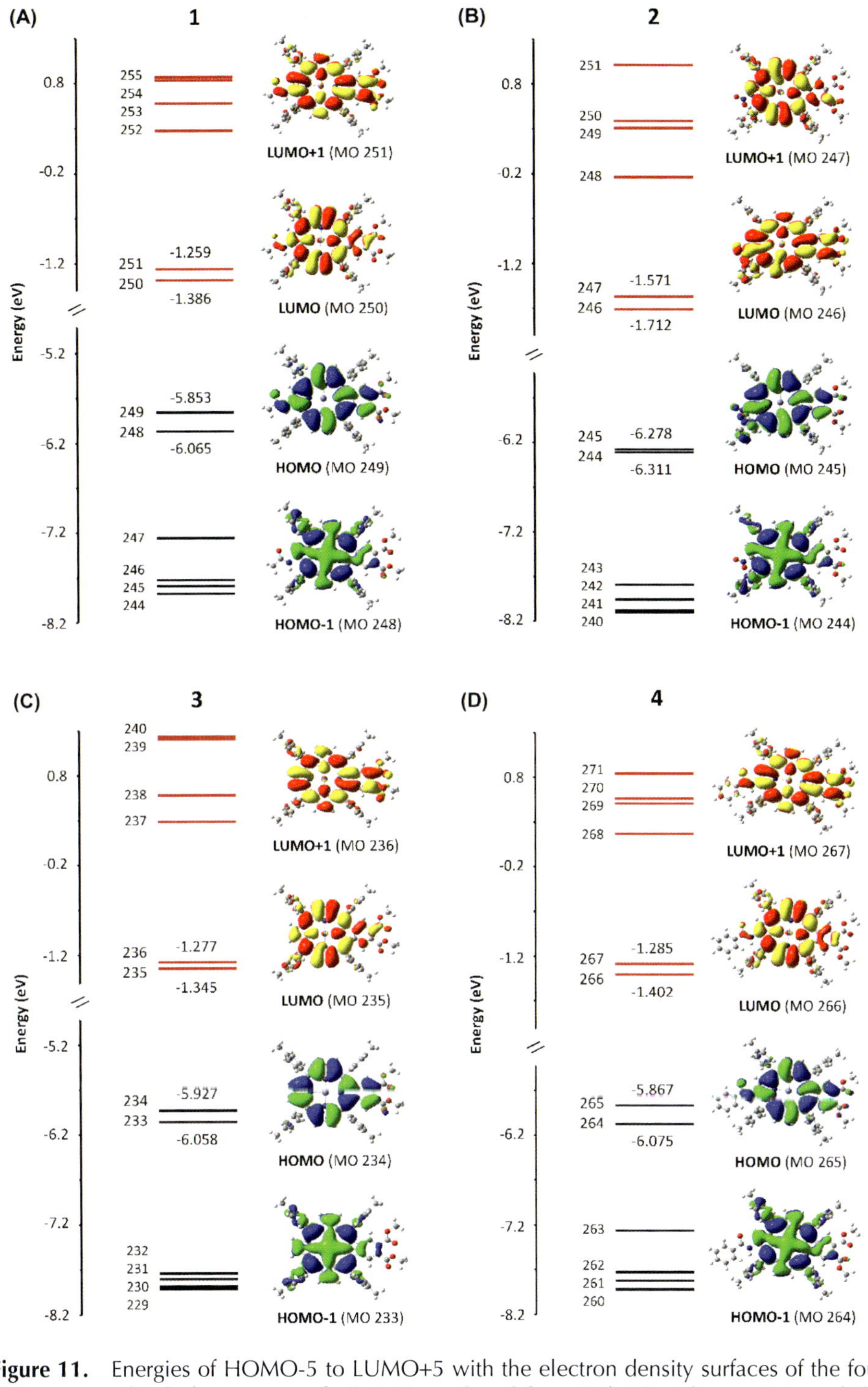

Figure 11. Energies of HOMO-5 to LUMO+5 with the electron density surfaces of the four Gouterman orbitals for compounds **1–4**. Reproduced from Ref. 24 with permission of The Royal Society of Chemistry.

B. The Electronic Structures of Compounds 1–4

The first set of computational data to be examined is the energy level dependence of the occupied and unoccupied π MOs and their electron density distributions, the surfaces. Figure 11 shows these data for **1–4** (A to D). As is found frequently (but not in the case of the **αPcS1** and **αPcS2** molecules described earlier), the four Gouterman MOs are recognized by the four and five nodal patterns. For each of these compounds the nominally-degenerate pairs of HOMO/HOMO–1 and LUMO/LUMO+1 maintain their relatively isolated energies. The TDDFT calculations are summarized in Table 2 for the Q and B states and plotted in Figure 12.

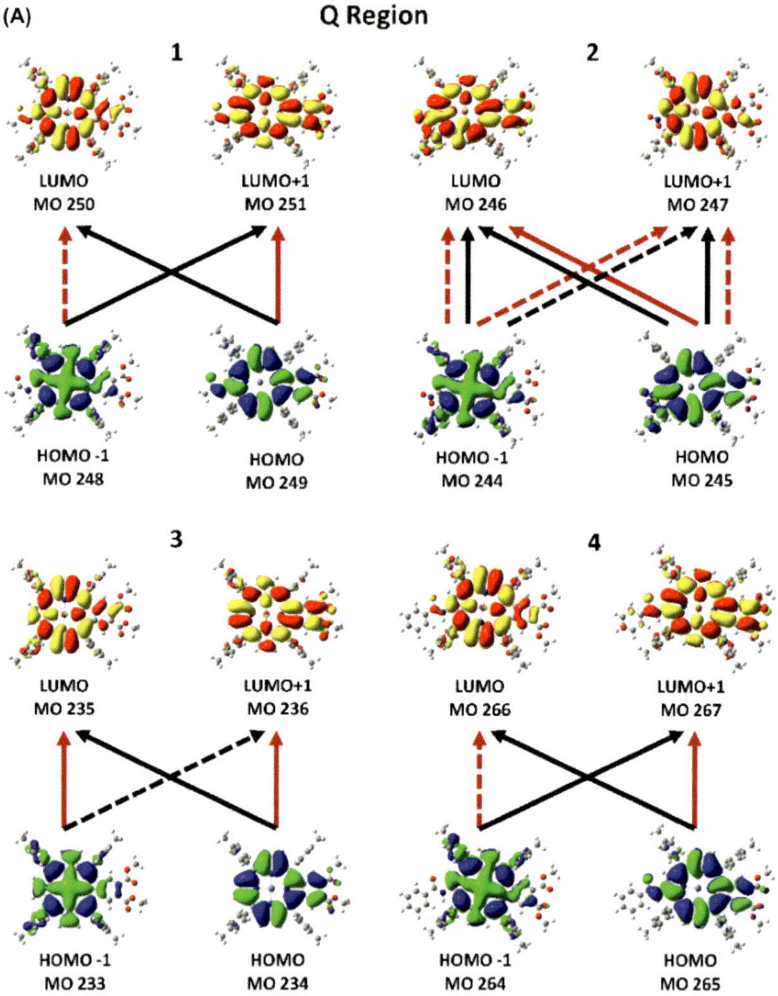

Figure 12. Major transitions between the four Gouterman MOs that give rise to the two Q and two B band states obtained from TD-DFT calculations. Black lines are transitions that contribute to the lower excited state, Q_{-1} and B_{-1}, and red lines show transitions that contribute to the higher excited state, Q_{+1} and B_{+1}. Dotted lines show transitions that contribute with a negative phase of the MO. Reproduced from Ref. 24 with permission of The Royal Society of Chemistry.

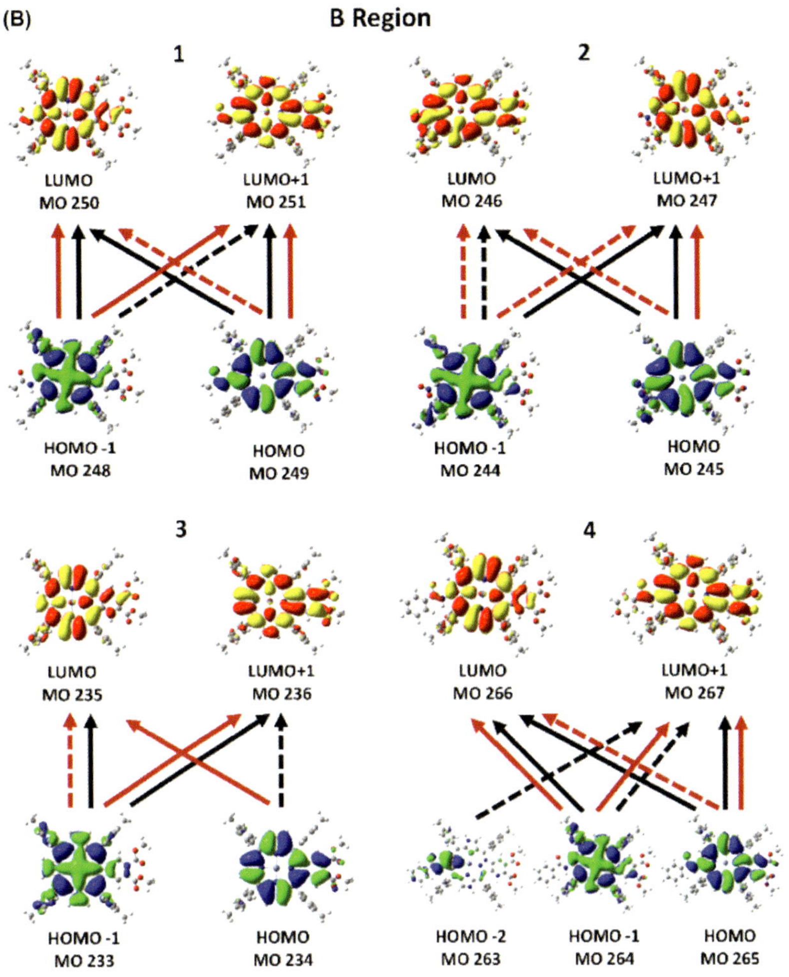

Figure 12. (*Continued*)

Table 2. Molecular orbital contributions to Q_{-1}, Q_{+1}, B_{-1} and B_{+1}. Reproduced from Ref. 24 with permission of The Royal Society of Chemistry.

Compound	State	State Energy /nm	f	Transition (Composition)
1	Q_{-1}	551.93	0.0188	248 → 251 (0.421)
				249 → 250 (0.558)
	Q_{+1}	541.40	0.0075	248 → 250 (–0.482)
				249 → 251 (0.506)
	B_{-1}	376.44	2.1560	248 → 250 (0.430)
				248 → 251 (–0.265)
				249 → 250 (0.218)
				249 → 251 (0.416)

(*Continued*)

Table 2. *(Continued)*

Compound	State	State Energy /nm	f	Transition (Composition)
	B_{+1}	365.98	1.2955	248 → 250 (0.260) 248 → 251 (0.476) 249 → 250 (–0.364) 249 → 251 (0.251)
2	Q_{-1}	546.06	0.0082	244 → 246 (0.395) 244 → 247 (–0.288) 245 → 246 (0.325) 245 → 247 (0.382)
	Q_{+1}	532.94	0.0138	244 → 246 (–0.325) 244 → 247 (–0.364) 245 → 246 (0.410) 245 → 247 (–0.288)
	B_{-1}	376.98	1.9642	244 → 246 (–0.164) 244 → 247 (0.485) 245 → 246 (0.431) 245 → 247 (0.176)
	B_{+1}	373.03	1.2718	244 → 246 (–0.448) 244 → 247 (–0.174) 245 → 246 (–0.173) 245 → 247 (0.479)
3	Q_{-1}	543.55	0.0050	233 → 236 (–0.446) 234 → 235 (0.533)
	Q_{+1}	532.69	0.0032	233 → 235 (0.484) 234 → 236 (0.498)
	B_{-1}	368.12	2.1424	233 → 235 (0.492) 233 → 236 (0.101) 234 → 236 (–0.483)
	B_{+1}	362.93	1.3659	233 → 235 (–0.102) 233 → 236 (0.528) 234 → 235 (0.446)
4	Q_{-1}	553.08	0.0172	264 → 267 (0.419) 265 → 266 (0.551)
	Q_{+1}	542.73	0.0099	264 → 266 (–0.475) 265 → 267 (0.503)
	B_{-1}	379.05	2.2710	263 → 267 (–0.112) 264 → 266 (0.436) 264 → 267 (–0.251) 265 → 266 (0.206) 265 → 267 (0.418)
	B_{+1}	367.27	1.2806	264 → 266 (0.249) 264 → 267 (0.479) 265 → 266 (–0.371) 265 → 267 (0.242)

The Q states for **1**, **2** and **4** are formed from the same sets of transitions. Interestingly, Figure 12 shows that the Q state of **3** involves a different set of phases than for **1** and **4**, and the MCD and emission spectra echo this change, supporting the connection between state structure and observable spectral data. Compound **2** is a very different case, where the Q state is formed from twice as many transitions as the other Q states, indicating the complexity in **2**'s Q state. This complication presumably arises from the very near degeneracy of the HOMO and HOMO–1 (MO 245 and 244).

The computational TD-DFT data accurately reproduces the absorption spectra except for the blue-shift. For **1**, **2** and **4** a red-shift of *ca.* 45 nm aligns the Q band. The split Q_x and Q_y components observed in the absorption and MCD spectra are mirrored by the calculations, where two separate states are seen under the very weak Q band (row four in Figure 10).

The B region near 440 nm in these 4 compounds exhibits the typical positive, pseudo-A term, even for **2**. Once again, the transitions forming the B states are typical for tetrapyrroles, involving only the four Gouterman MOs except for **4**. The low energy and the energetic isolation of the Q state clearly influences the complexity of the state construction (Figure 12), when compared with the doubling of transitions for the B state. Compounds **1** and **4** are the same except for the introduction of a component from MO 263 to 267 in **4** that is absent in all the other states.

C. Examining Effects of Active Substituents on the Electronic Structure of Porphyrins

Analysis of the experimental data is important when assessing the contributions of electronically active substituents for functional activation, as we are in this work. The goal for dyes in DSSCs is that they exhibit highly absorbing, red-located absorption coupled with panchromatic absorption at wavelengths between 400 nm and 600 nm and that the excited state electron can be delivered to the conduction band of the semiconductor. Synthetically it has been a goal to introduce pushing (electron donating) and pulling (electron withdrawing) groups on the periphery of porphyrins to aid in this property. In the next part of this present case we examine how sensitive the four Gouterman MOs are to substituents on the periphery of the porphyrin. In the paper we are discussing, we focused on two aspects: (i) how did the NO_2 substituent in **2** interact with the ring π MOs and (ii) could the dipole magnitude be increased through placement of additional electron donating and withdrawing groups. Quantification of the effects of changes in the substituents was achieved using our previously published trends in experimental property as a function of computational parameters. Figure 13 connects the Q band energy, the

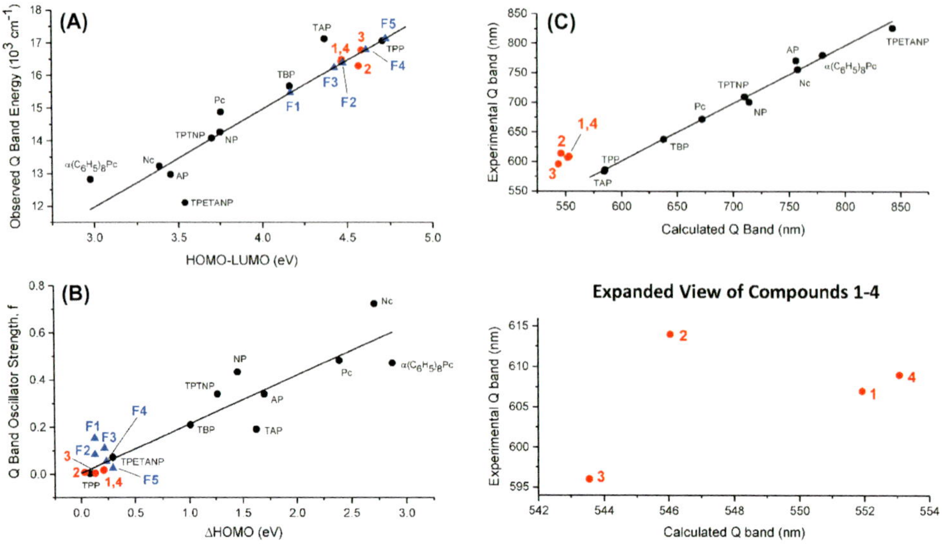

Figure 13. **(A)** Observed Q band energy *versus* the calculated HOMO–LUMO gap. **(B)** Calculated Q band oscillator strength, f, against calculated ΔHOMO. The 10 legacy compounds are plotted as black circles. Red circles represent compounds **1–4** and blue triangles represent the fictive molecules, **F1–F5**. **(C)** Experimental Q band wavelength *versus* calculated Q band wavelength. An expanded view of the plot for compounds **1–4** is included below, showing how the experimental Q band wavelength of **2** does not fit the trend line. Reproduced from Ref. 24 with permission of The Royal Society of Chemistry.

Q band oscillator strength, and a comparison between Q band experimental and TD-DFT values with computational values from the four Gouterman MOs. The use of a series of porphyrins and phthalocyanines provides a valuable calibration of the predicted properties for the fictive compounds.

The goals for a dye in a DSSC application are to have an intense, red Q band. Clearly, the rings reported in the paper do not approach those goals. The changes introduced in the fictive compounds (**F1–F5**), Figure 14, were designed to investigate the effect of electronically active β-substituents on the electronic structure. Figure 14 shows that the fictive β-substituents resulted in dramatic changes in the energies of the ring π MOs. However, even though the LUMO and LUMO+1 were split by nearly 1 eV (**F1**), as in chlorophyll, the Q state energy was hardly changed in comparison with the compound with two amine and two nitro groups (**F2**). The data reported showed that indeed the dipole was significant for **F2** but the Q band energy and more significantly, its oscillator strength did not match the goals for application in DSSCs.

In summary, using a set of calibrated trends provides a reliable means of assessing the potential of a specific porphyrin possessing properties in the electronic spectrum that are necessary for a specific function. We see that by using

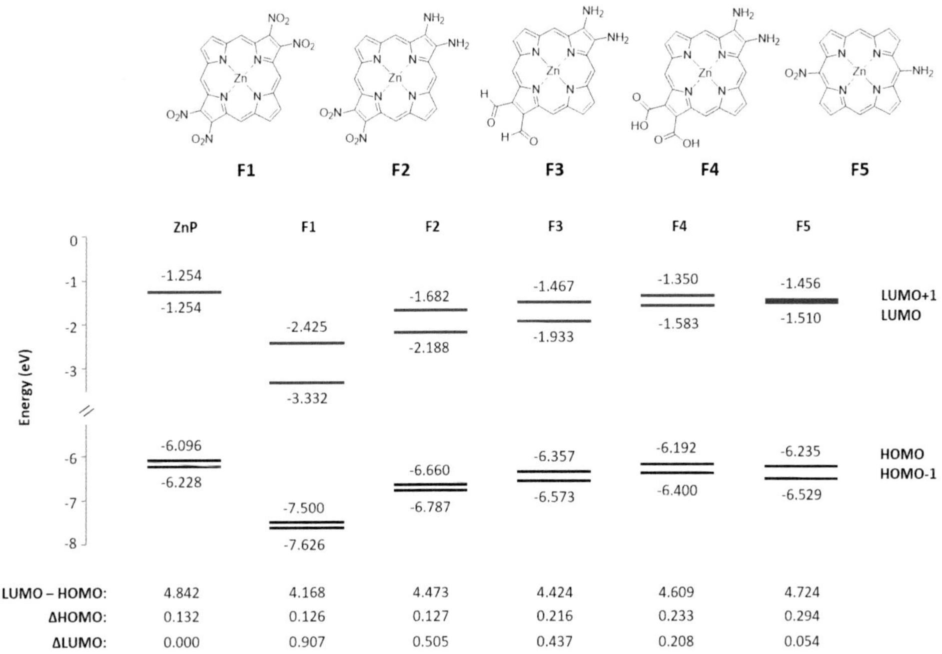

Figure 14. The structure, energy level diagram, and values of two HOMOs and two LUMOs in eV for the fictive molecules **F1–F5**. The frontier MO energies of ZnP is shown for comparison. Reproduced from Ref. 24 with permission of The Royal Society of Chemistry.

experimental data from the same core structure we can reliably predict optical properties from the computational data of fictive compounds.

In the final case we describe use of the chlorin ring as the underlying electronic structure from which to build properties that will approach the functional requirements of the dye in DSSCs. As above, we use the established trend data to assess the predicted properties.

V. Electronically Activated Chlorins

The first example of the substituted phthalocyanine showed that while a red located Q band with significant intensity was achieved, the polythiophene orbitals were located at LUMO+2, above the typical Gouterman 5-node LUMO. This meant that the first excited state for ***α*PcS1** at a calculated 661 nm did not involve the polythiophene orbital at MO 549. Indeed, MO 549 only becomes involved in the 3rd excited state at 467 nm (TD-DFT).

Changing the core ring to a chlorin kept the red-located, intense Q band of the phthalocyanine but removed the massively split HOMO/HOMO–1 introduced by the aza nitrogens of the phthalocyanine. This meant that the relative energies of the thiophene MOs are now very different.

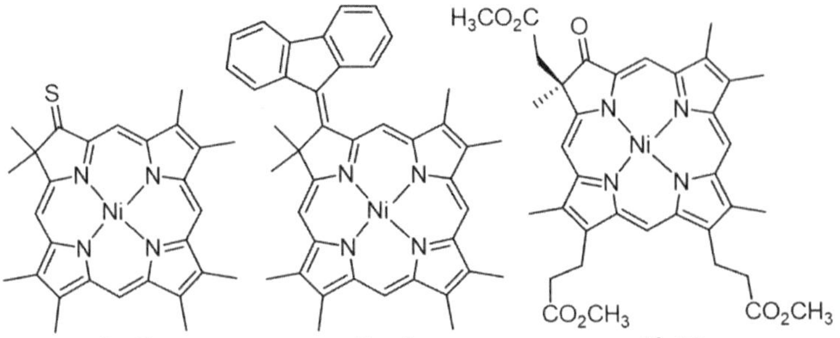

Syn3 Syn6 Lie95

Figure 15. Structures of the synthetic chlorins, **Syn3**,[28] **Syn6**[28] and **Lie95**.[29] Reproduced from Ref. 30 with permission of The Royal Society of Chemistry.

A. Optical Properties

Following the procedure with the β-substituted porphyrins described above, we analyzed the absorption and MCD spectral data of three substituted chlorins[28,29] coupled with TD-DFT calculations.[30] The absorption spectral properties of the three compounds (Figure 15) were identified as primarily being due to the chlorin ring and the specific substituents, and not noticeably to the Ni(II) central metal, Figure 16. Substitution with the thioketone (**Syn3**) and the fluorene group (**Syn6**) introduced broader absorption in the B band region. The B band is red-shifted to *ca.* 460 nm from the typical 400 nm for porphyrins and 350 nm for phthalocyanines. Coupled with the intense, red Q band near 680 nm, these two chlorins provide a wider range of panchromatic absorption than **αPcS1** and **αPcS2** (Figure 6, Figure 16). The overlapped Q_y and vibronic bands lie in the important 500–650 nm region of greatest solar intensity, increasing the panchromatic property of these chlorins. We calculated the optical spectral properties of the Ni(II) and respective Zn(II) coordinated **Syn3**, **Syn6** and **Lie95**. Figure 16 shows that the predicted spectral properties for the Zn(II) chlorins are almost identical to those calculated for the Ni(II) chlorins. However, the TD-DFT calculation did show the presence of charge transfer states with low oscillator strengths in the Ni(II) chlorins. Recognizing that the coordinated Ni(II) did not contribute extensive absorption, we used Zn(II) in the fictive compounds to be described below.

Trusting the computational data to be accurate is always a problem with DFT and especially TD-DFT calculations. Our approach described in Section IV, of using the trends of the archival compounds was used again to assess the predicted experimental spectral properties of the fictive compounds.

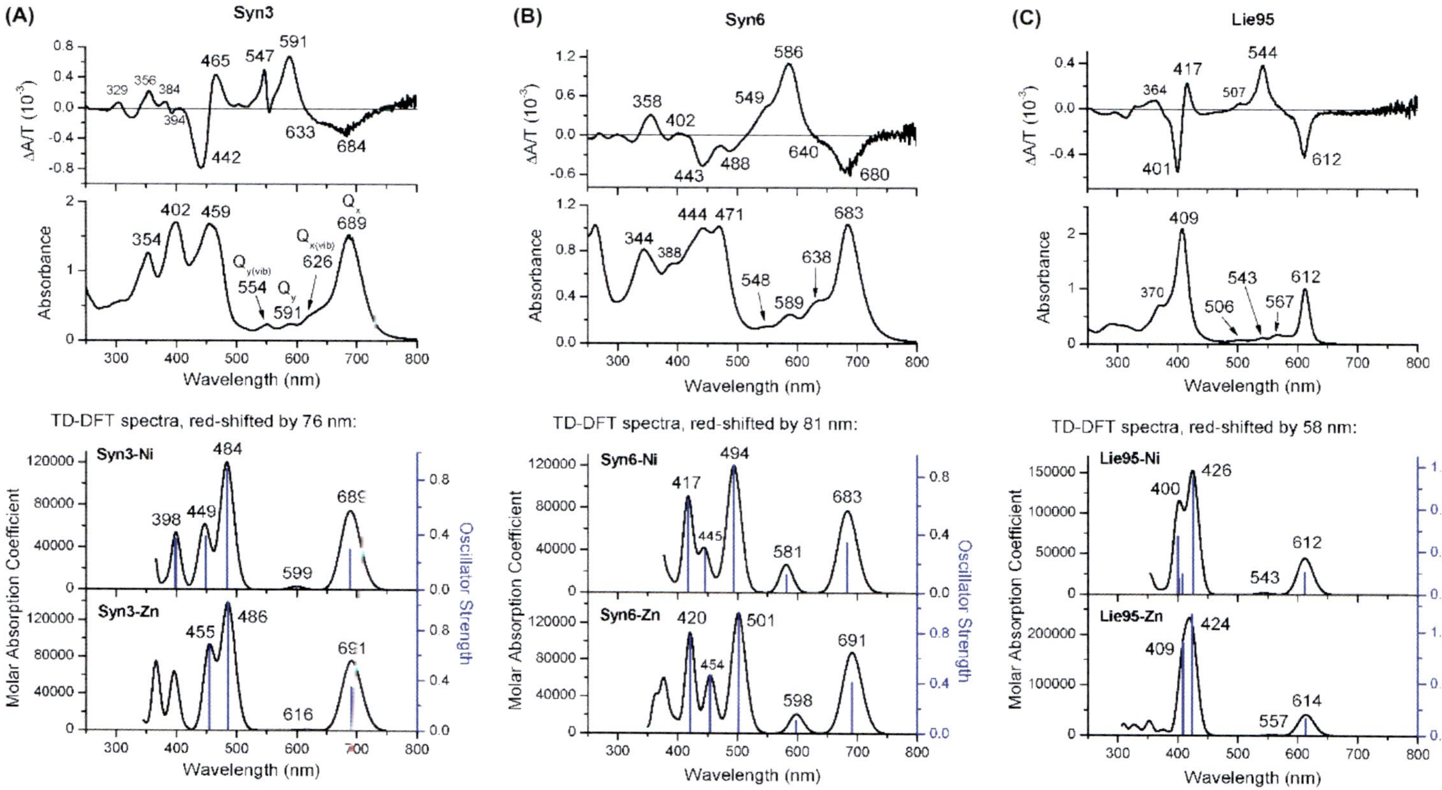

Figure 16. MCD, absorption and calculated absorption spectra for Ni(II) coordinated **Syn3 (A)**, **Syn6 (B)**, and **Lie95 (C)** in DCM. The theoretical spectra of the respective Zn(II) coordinated **Syn3**, **Syn6** and **Lie95** are shown in the last row. Reproduced from Ref. 30 with permission of The Royal Society of Chemistry.

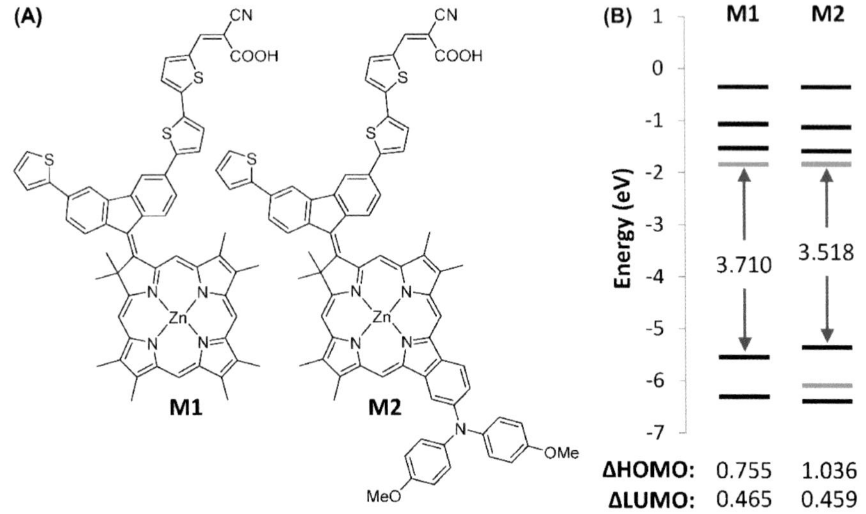

Figure 17. (**A**) Structures of M1 and M2. (**B**) The energies of the frontier MOs. Substituents-based MOs are highlighted in green. Reproduced from Ref. 30 with permission of The Royal Society of Chemistry.

B. Electronically Activated Chlorins with Thiophene Substituents

As described for the ***α*PcS1** and ***α*PcS2** compounds in Section III, the dye needs to conduct the excited electron to the conduction band of the semiconductor, here TiO_2 with a CB at −4.0 eV.[31,32] Thiophenes with carboxylic acid terminal groups serve this purpose well.[33,34] The fluorene rings on the chlorin (**Syn6**) provided an excellent attachment position for the thiophene units so for the next step of this study, and following the work of Yamamoto *et al.* with the phthalocyanines,[27] we used in this case the fluorene substituted chlorin (tagged **Syn6**) and added three thiophene units as shown in Figure 17A for two fictive designs (**M1** and **M2**, in the paper). To introduce the dipole moment of the NO_2 β-substituted porphyrins described above (Section IV), we additionally added to the design the strongly electron donating methoxy-substituted, triphenylamine on the opposite pyrrole (**M2**). The effect of these two modifications was to dramatically reduce the HOMO–LUMO energy and, as shown in Figure 17B, introduce the thiophene MOs as the LUMO (MO 258 in **M1**, MO 323 in **M2**). In both **M1** and **M2**, LUMO+1, +2 and +3 each shared the five node electron distribution typical of 18 π electron rings, which was interpreted to be a result of the further interactions with the thiophene delocalization (Figure 18). In both **M1** and **M2,** LUMO+4 was a purely thiophene delocalized excited MO (MO 262 in **M1**, MO 327 in **M2**). For **M1** the HOMO and HOMO−1 retained the typical 18 π "a_{1u}" / "a_{2u}" nodal pattern sequence, respectively. MOs localized on the thiophenes are found at HOMO−2 and −3 (MOs 255 and 254). The effect of the additional electron donor in **M2** was

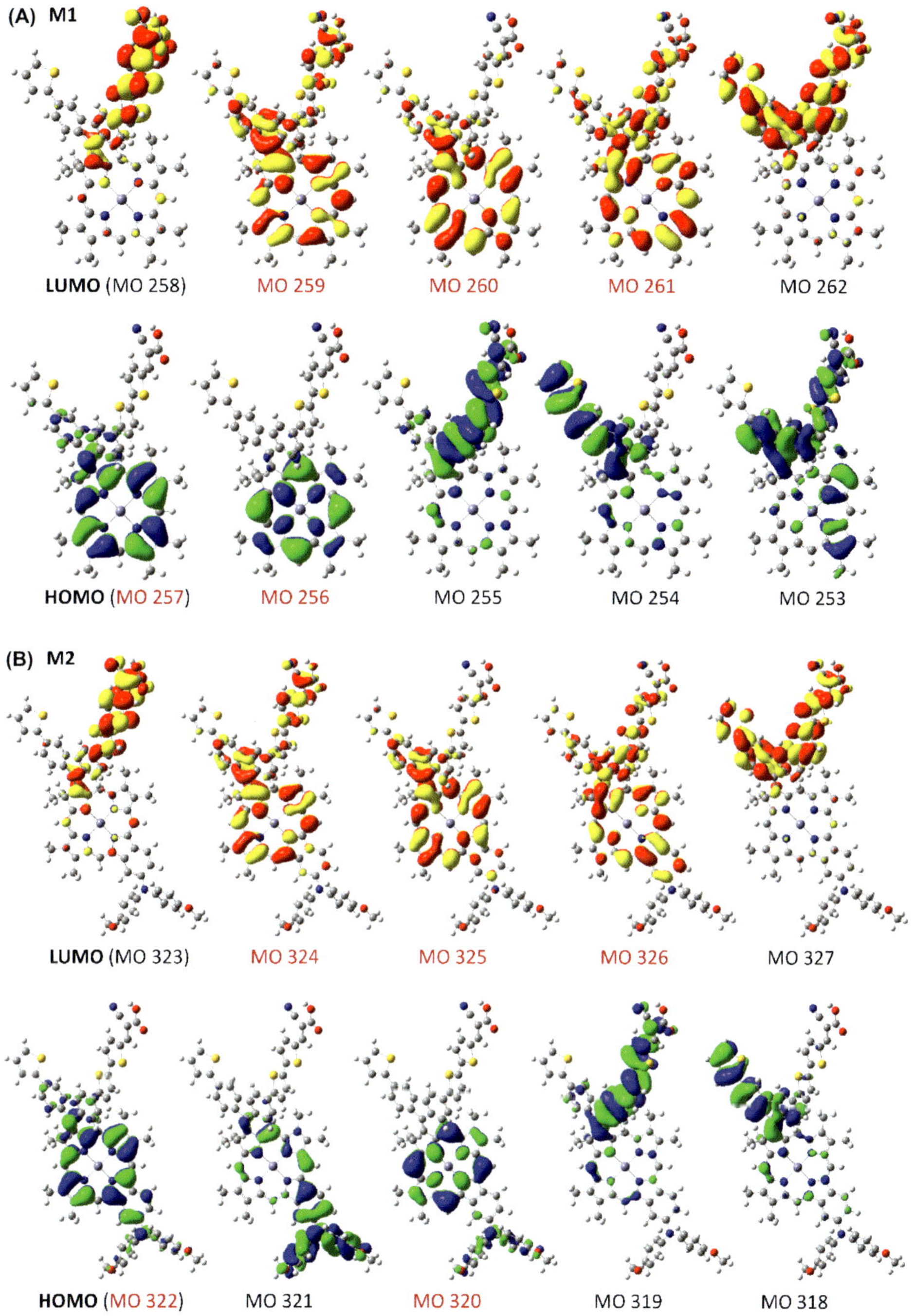

Figure 18. The electron density surfaces of the frontier MOs for **M1** and **M2**. Gouterman MOs, identified by four nodes for the occupied MOs and five nodes for the unoccupied MOs are highlighted in red. Surfaces are shown with an isovalue of 0.02. Reproduced from Ref. 30 with permission of The Royal Society of Chemistry.

Table 3. Calculated molecular orbital transitions that give rise to the Q bands of **M1** and **M2**. Substituent-based MOs are highlighted in bold. Refer to Figure 18 for the electron distributions in each of these **MOs**. Reproduced from Ref. 30 with permission of The Royal Society of Chemistry.

Compound	State	State Energy/nm	f	Transition (contribution)
M1	Q_{-1}	629.79	0.6978	256→260 (0.18852)
				257→**258** (–0.31380)
				257→259 (0.58077)
	Q_{+1}	525.86	0.1255	256→**258** (0.14046)
				256→259 (–0.33163)
				256→261 (0.12320)
				257→**258** (–0.14317)
				257→260 (0.55427)
				257→261 (–0.10127)
M2	Q_{-1}	676.72	0.8027	320→325 (–0.15197)
				322→**323** (0.34700)
				322→324 (0.56346)
	Q_{+1}	558.72	0.2730	320→**323** (0.12517)
				320→324 (0.24460)
				321→324 (–0.13141)
				322→**323** (–0.14240)
				322→325 (0.57942)
				322→326 (–0.11211)

to raise the Gouterman HOMO (MO 322) so that now triphenylamine MOs were located as HOMO–1 (321), followed by the "a_{2u}" as HOMO–2 (320). The thiophene MOs are found below this (MO 319 and 318). This new sequence means that the lowest energy HOMO–LUMO based transition in **M2** includes a significant fraction of charge transfer to the thiophene LUMO 322 with additional intensity from LUMO+1 and +2 (324 and 325), Table 3.

The computational absorption spectra of **M1** and **M2** (Figure 19) were drawn using the 500 cm^{-1} of the Q band bandwidth (so too narrow in reality for the B band region) and have not been manually red-shifted. The spectra of these fictive chlorins show similar panchromaticity to the synthetic chlorins (Figure 16) but with stronger absorption predicted between 400 nm to 500 nm and for the Q_x and Q_y bands. So this combination of thiophene, fluorene and triphenylamine associated with the chlorin ring provides a compound approaching the ideal properties of a dye for DSSC application.

The data shown for **M1** and **M2** are computational, so to overcome this weak link in the proposed design and to determine the probable experimental values, we again placed the computational parameters on the trend lines (Figure 20). This is key to the work in these papers, that is, relating the computational data for known compounds to the data of the fictive compounds greatly improves the reliability of the

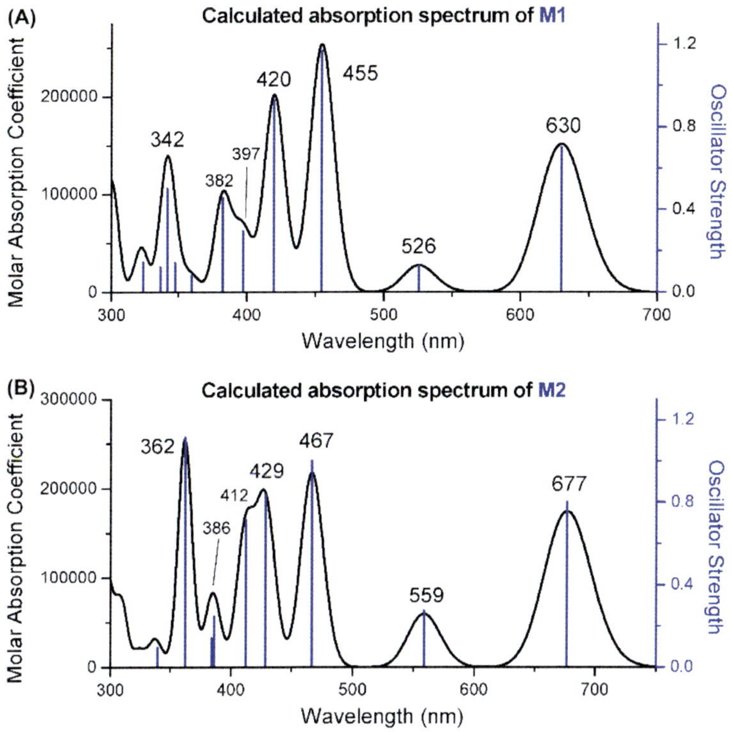

Figure 19. The theoretical absorption spectra of **(A) M1**, and **(B) M2** from CAM-B3LYP/6-31G(d,p) TD-DFT calculations. The wavelengths are directly from TD-DFT calculations. A bandwidth of 500 cm^{-1} was used for the spectral presentation. Reproduced from Ref. 30 with permission of The Royal Society of Chemistry.

prediction. Figure 20 shows clearly how the change in the HOMO–LUMO energies and the presence of the thiophene LUMO dramatically improves the properties of the fictive chlorins. Using the trend line of Figure 20C, the experimental Q bands of **M1** and **M2** were estimated to be 716 and 777 nm, respectively. Even more importantly, the calculated oscillator strengths for **M1** and **M2** (0.7 and 0.8, respectively) are about the same as reported for naphthalocyanine (Nc) Q band at 756 nm.[20]

Key differences, though, between Nc and **M2** are that there is almost continuous absorption from <400 nm to 780 nm because the red-shifted B bands extend to the Q_Y, Q_X and their vibronic components. In addition, and probably more significantly, the HOMO and HOMO–1 are split by less than 1 eV compared with the 2.75 eV of the Nc. We feel this allows for new designs of chlorin dyes in which the HOMO/HOMO–1 energy can be increased resulting in an even greater red-shift and intensification of the Q band.

A final assessment of the possible efficiencies of these compounds as dyes in DSSCs can be obtained by comparing the ground state and excited state oxidation potential (GSOP and ESOP, respectively) energies to the TiO_2 conduction band

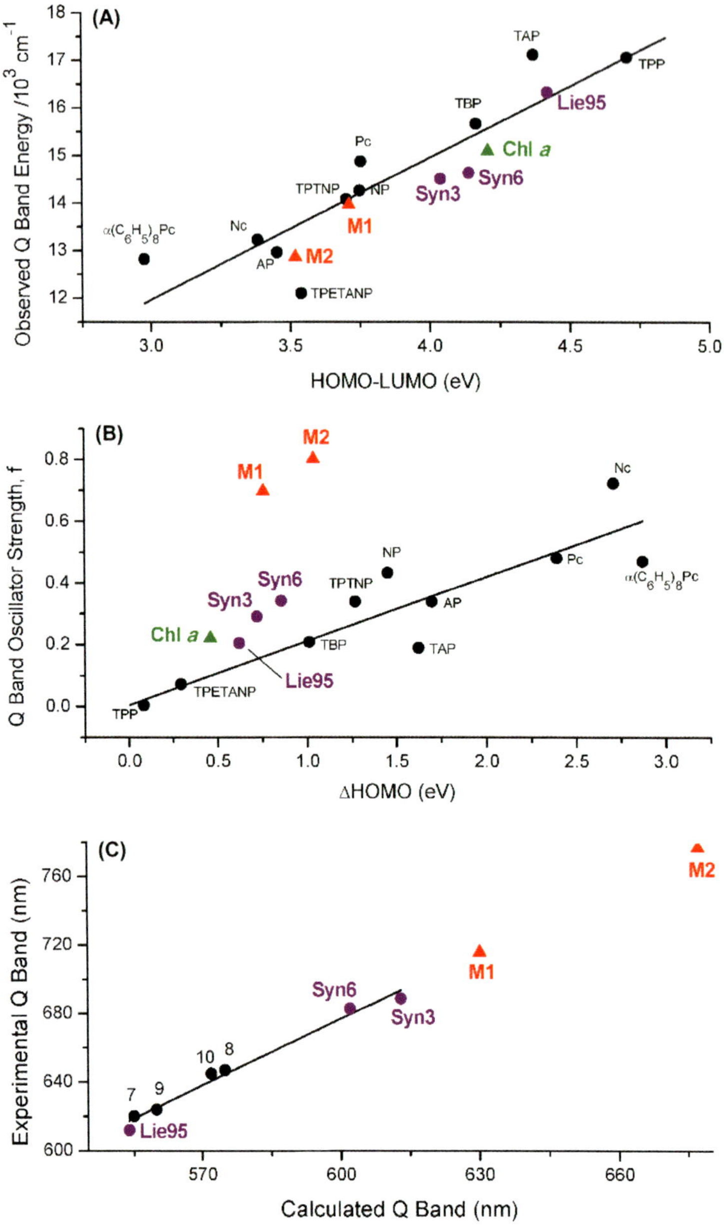

Figure 20. **(A)** Observed Q band energy *vs.* the calculated HOMO–LUMO gap. **(B)** The oscillator strength, f, of the Q band *vs.* the calculated ΔHOMO. **(C)** The experimental Q band wavelength *vs.* the theoretical Q band wavelength. Compounds **7–10** were previously synthesized chlorins by Montforts and coworkers (structures are shown in Figure 21).[28, 35] Reproduced from Ref. 30 with permission of The Royal Society of Chemistry.

energy (–4.0 eV) and the recharge potential of the I^-/I_3^- electron donor (–4.8 eV).[32] While it is commonplace to use the HOMO and LUMO energies directly from the DFT calculations, we feel this is very risky because these energies depend very much on the functionals used. Rather, the GSOP and ESOP energies should provide a more accurate estimate of the efficiency of the electron transfer and dye regeneration. The starting point however, leans on application of Koopmans' Theorem for DFT calculations that the 1st oxidation potential, or the 1st ionization energy, is the same as the negative of the energy of HOMO.[36,37] Therefore, the vertical transition GSOP can be estimated using the HOMO energy. The ESOP value can be estimated from the GSOP and the energy of the 1st excited state, $E_{0\text{–}0}$, where ESOP = GSOP-$E_{0\text{–}0}$ (in eV).[38] While $E_{0\text{–}0}$ can be an experimental value it may be approximated by the TD-DFT calculated value. We caution that for many tetrapyrroles, and certainly for the molecules described here, the TD-DFT value for $E_{0\text{–}0}$ is overestimated, meaning a red-shift needed to be applied to match the experimental Q band energy. The trends in Figure 20 clearly illustrate this because the experimental Q band wavelength of **Syn6** is 683 nm whereas the calculated TD-DFT value is 602 nm.

Figure 22 summarizes the GSOP (from the HOMO) and the ESOP values for **M1**, **M2**, **αPcS1**, and **αPcs2**. The ESOPs were calculated using both the

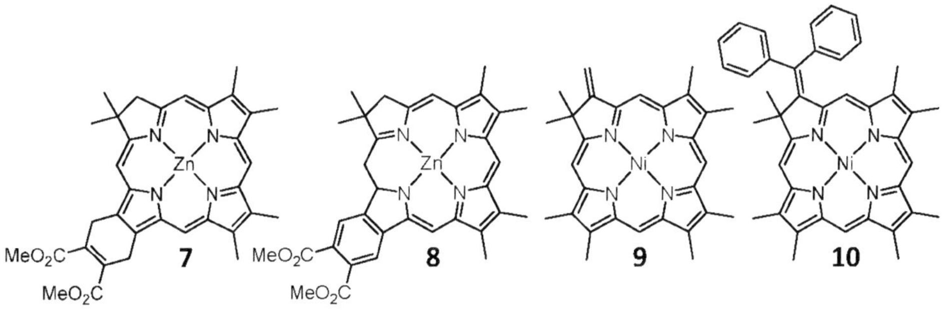

Figure 21. Structures of compounds **7–10**, from Montforts and coworkers.[28,35]

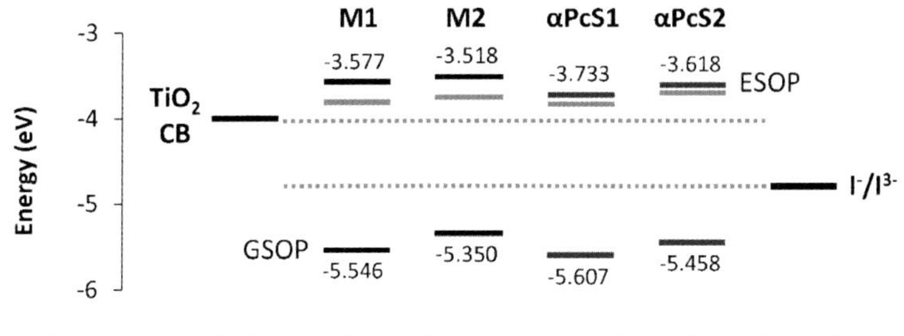

Figure 22. The GSOP and ESOP values of **M1**, **M2**, **αPcS1** and **αPcS2** with respect to the energies of the TiO_2 conduction band (–4.0 eV) and I^-/I^{3-} redox potential (–4.8 eV). Gray lines show the ESOP value calculated using the experimental Q band energy. Adapted from Ref 30.

theoretical E_{0-0} (black/purple lines) and experimental E_{0-0} (gray lines). The experimental E_{0-0} for **M1** and **M2** are extracted from the trend in Figure 20C. As expected, the ESOP value obtained using experimental E_{0-0} is lower than the theoretical ESOP as TD-DFT overestimates the energy of the lowest excited state. All four compounds straddle the CB energy and regeneration energy.

VI. Summary

Our objective in describing the three different tetrapyrroles DSSC dye candidate systems here was to illustrate a process of designing novel functional tetrapyrroles guided by computational analysis. Our approach uses experimental data from model compounds to calibrate the computational trend lines. These synthetic tetrapyrroles might not be good candidates as dyes in DSSCs initially, but the important part is that their experimental optical properties are well-known. The computational parameters of these compounds then establish reliable trend lines so we can use just the computational results from the novel fictive compounds with confidence to predict properties. We also have established a much larger range of compounds where the oscillator strength of the Q band is proportional to ΔHOMO, the splitting between the Gouterman HOMO and HOMO−1 pair. Typically this splitting is increased by withdrawing electrons from those atoms on the aromatic ring using electronegative nitrogen atoms, for example the four aza nitrogens in Pcs, or by adding electron withdrawing groups on the pyrroles. But in each case, the splitting required for the intense absorption wanted, required nearly 3 eV of splitting between the HOMO pair. This means that the options for chemistry using both the HOMO (a_{1u}) and HOMO−1 (a_{2u}) orbital properties is severely limited. The results in Section V show that intensely absorbant Q bands can be achieved with substituted chlorins. The substitution of the chlorin ring, therefore, opens up a new avenue to highly absorbent Q bands.

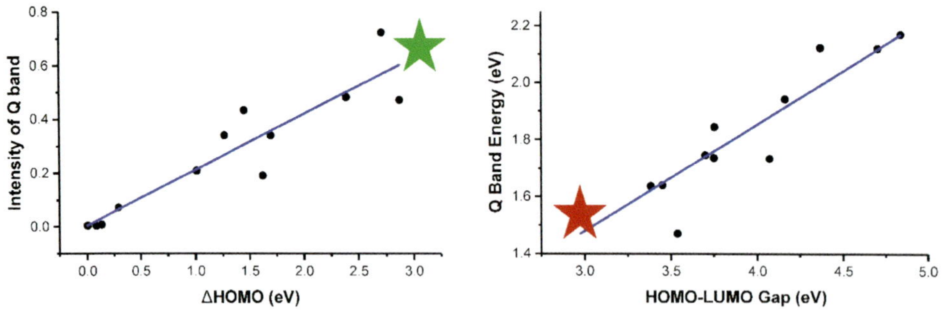

Figure 23. Trends connecting the Q band intensity and energy with the computed electronic structure. Green and red stars show where an ideal dye for DSSCs would lie on the trend.

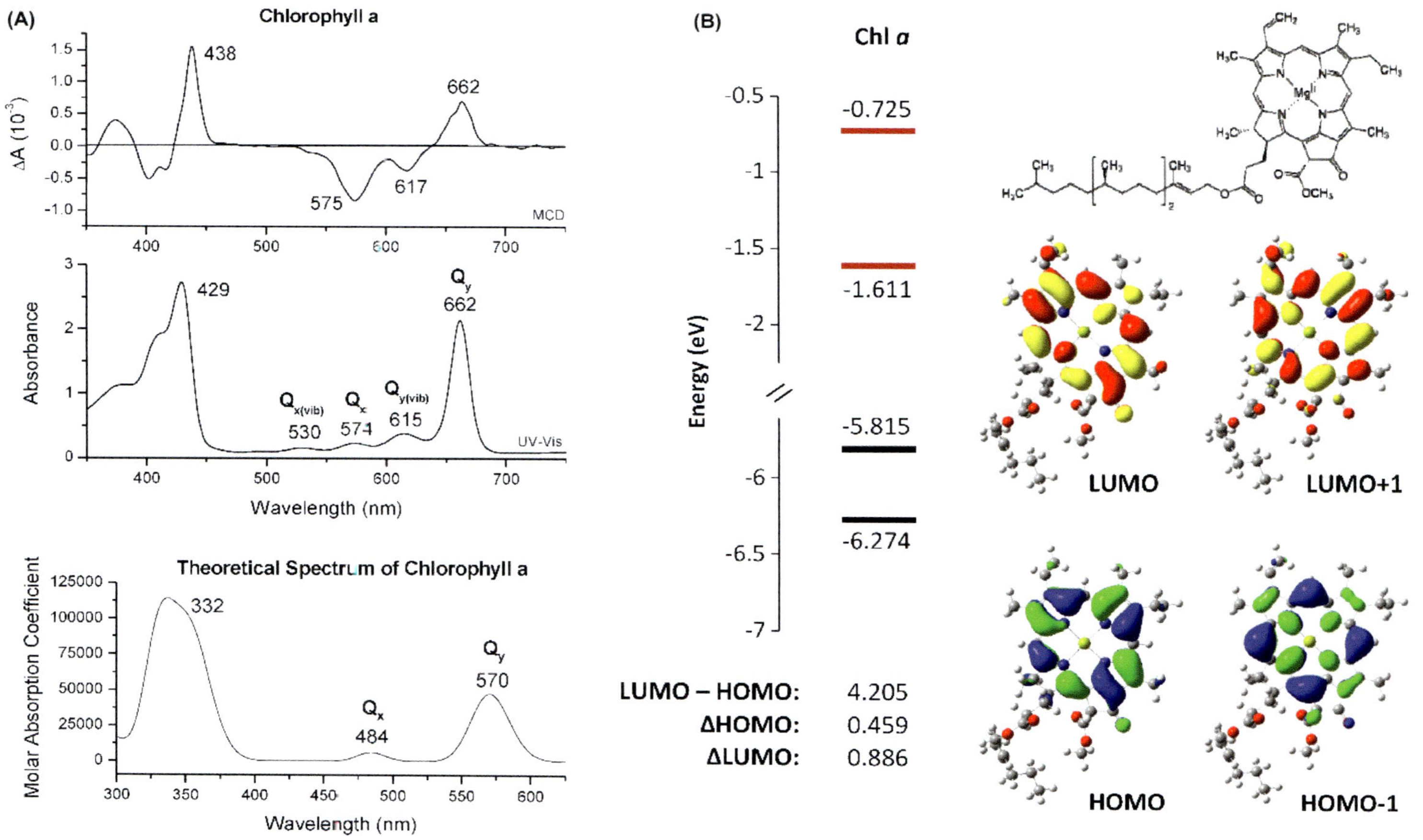

Figure 24. **(A)** The MCD, absorption, and calculated absorption spectrum of chlorophyll *a*. **(B)** The structure, energy level diagram and electron density surfaces of the Gouterman MOs for Chl *a*. Occupied MOs are colored blue and green and unoccupied MOs are colored red and yellow.

Figure 23 summarizes on those trend lines the goals of tetrapyrrole synthesis for DSSC application: namely, a red-located, intense 1st excited state — the Q band. We can take away from this figure that for the archival compounds used, a red shift in the Q band is linearly connected with the HOMO–LUMO energy even though we caution that the HOMO–LUMO gap is not the energy for the Q band. The HOMO–LUMO energy can be decreased by introducing splitting in both the 18 π "e_u" HOMO/HOMO–1 and the "e_g" LUMO/LUMO+1 pairs. A common approach has been to increase the delocalization of the ring π electrons by adding conjugated aromatic rings to the periphery. The data reported in our final case showed that this was also possible by building on the chlorin ring without the need for extensive molecular constructions. Finally, the panchromatic property is provided best by the substituted chlorins because the B bands are also red-shifted into the visible region.

In our final figure (Figure 24), we return to the data for chlorophyll itself. The spectral properties and calculations all indicate a reasonably successful dye for DSSC, except that chlorophyll is not stable enough outside the thylakoid membrane. This leaves the technologically challenged task of reproducing and improving the electronic structure while reducing degradation properties.

VII. Acknowledgments

We thank the Natural Science and Engineering Research Council (NSERC) of Canada for a Discovery Grant to M.J.S. and the Ontario Government for the Ontario Graduate Scholarship to A.Z. We thank Mr. John Vanstone and the Electronics and IT Team for their excellent maintenance of our computers and instruments. We acknowledge and thank Professor Peter Montforts and Professor Hong Wang for providing us with the chlorins and porphyrins, respectively. We thank Fujitsu Poland for providing the up to date version of Scigress. We thank Professor Nagao Kobayashi and Professor Mutsumi Kimura for long time collaborations.

VIII. References

1. Milgrom, L. R. *The Colours of Life: An Introduction to the Chemistry of Porphyrins and Related Compounds*. Oxford University Press: New York, 1997.
2. Tanaka, T.; Osuka, A. Conjugated porphyrin arrays: synthesis, properties and applications for functional materials. *Chem. Soc. Rev.* **2015**, *44*, 943–969.
3. Higashino, T.; Imahori, H. Porphyrins as excellent dyes for dye-sensitized solar cells: recent developments and insights. *Dalton Trans.* **2015**, *44*, 448–463.
4. Imahori, H.; Kurotobi, K.; Walter, M. G. Porphyrin- and Phthalocyanin-Based Solar Cells. In *Handbook of Porphyrin Science*, Kadish, K. M.; Smith, K. M.; Guilard, R., Eds. World Scientific Publishing: Singapore, 2012; Vol. 18, pp. 57–121.

5. Ethirajan, M.; Chen, Y.; Joshi, P.; Pandey, R. K. The role of porphyrin chemistry in tumor imaging and photodynamic therapy. *Chem. Soc. Rev.* **2011**, *40*, 340–362.
6. O'Regan, B.; Grätzel, M. A low-cost, high-efficiency solar cell based on dye-sensitized colloidal TiO_2 films. *Nature* **1991**, *353*, 737–740.
7. Mathew, S.; Yella, A.; Gao, P.; Humphry-Baker, R.; Curchod, B. F.; Ashari-Astani, N.; Tavernelli, I.; Rothlisberger, U.; Nazeeruddin, M. K.; Gratzel, M. Dye-sensitized solar cells with 13% efficiency achieved through the molecular engineering of porphyrin sensitizers. *Nat. Chem.* **2014**, *6*, 242–247.
8. Green, M. A.; Emery, K.; Hishikawa, Y.; Warta, W.; Dunlop, E. D. Solar cell efficiency tables (Version 45). *Prog. Photovolt: Res. Appl.* **2015**, *23* (1), 1–9.
9. Grätzel, M. Photoelectrochemical cells. *Nature* **2001**, *41*, 338–344.
10. Grätzel, M. Dye-sensitized solar cells. *J. Photochem. Photobiol. C.* **2003**, *4*, 145–153.
11. Hagfeldt, A.; Boschloo, G.; Sun, L.; Kloo, L.; Pettersson, H. Dye-sensitized solar cells. *Chem. Rev.* **2010**, *110*, 6595–6663.
12. Stillman, M. J. Theoretical Aspects of the Optical Spectroscopy of Porphyrinoids. In *Handbook of Porphyrin Science*, Kadish, K. M.; Smith, K. M.; Guilard, R., Eds. World Scientific Publishing: Singapore, 2011; Vol. 14, pp. 461–524.
13. Tiedemann, M. T.; Stillman, M. J. Application of magnetic circular dichroism spectroscopy to porphyrins, phthalocyanines and hemes. *J. Porphyrins Phthalocyanines* **2011**, *15*, 1134–1149.
14. Gouterman, M. Study of the effects of substitution on the absorption spectra of porphin. *J. Chem. Phys.* **1959**, *30*, 1139–1161.
15. Gouterman, M. Spectra of porphyrins. *J. Mol. Spectrosc.* **1961**, *6*, 138–163.
16. Gouterman, M. Optical spectra and electronic structure of porphyrins and related rings. In *The Porphyrins*, Dolphin, D., Ed. Academic Press: New York, 1978; Vol. 3, pp. 1–165.
17. Fragata, M.; Nordén, B.; Kurucsev, T. Linear dichroism (250–700 nm) of chlorophyll *a* and pheophytin *a* oriented in a lamellar phase of glycerylmonooctanoate/H_2O. characterization of electronic transitions. *Photochem. Photobiol.* **1988**, *47*, 133–143.
18. Weiss, C. The molecular orbital theory of chlorophyll. *Ann. N. Y. Acad. Sci.* **1975**, *244*, 204–213.
19. Nyokong, T.; Gasyna, Z.; Stillman, M. J. Analysis of the absorption and magnetic circular dichroism spectra of zinc phthalocyanine and the π-cation-radical species $[ZnPc(-1)]^+$ *Inorg. Chem.* **1987**, *26*, 1087–1095.
20. Mack, J.; Asano, Y.; Kobayashi, N.; Stillman, M. J. Application of MCD spectroscopy and TD-DFT to a highly non-planar porphyrinoid ring system. New insights on red-shifted porphyrinoid spectral bands. *J. Am. Chem. Soc.* **2005**, *127*, 17697–17711.
21. Piepho, S. B.; Schatz, P. N. *Group Theory in Spectroscopy with Applications to Magnetic Circular Dichroism.* Wiley: New York, 1983.
22. Kobayashi, N.; Muranaka, A.; Mack, J. *Circular Dichroism and Magnetic Circular Dichroism Spectroscopy for Organic Chemists.* Royal Society of Chemistry: Cambridge, 2012.
23. Mack, J., Bunya, M., Shimizu, Y., Uoyama, H., Komobuchi, N., Okujima, T., Uno, H., Ito, S., Stillman, M.J., Ono, N., Kobayashi, N. Application of MCD spectroscopy and TD-DFT to nonplanar core-modified tetrabenzoporphyrins: Effect of reduced symmetry on nonplanar porphyrinoids. *Chem. Eur. J.* **2008**, *14*, 5001–5020.
24. Zhang, A.; Kwan, L.; Stillman, M. J. The spectroscopic impact of interactions with the four Gouterman orbitals from peripheral decoration of porphyrins with simple electron withdrawing and donating groups. *Org. Biomol. Chem.* **2017**, *15*, 9081–9094.

25. Michl, J. Magnetic circular dichroism of cyclic π-electron systems. 2. Algebraic solution of the perimeter model for the B terms of systems with a (4N + 2)-electron [n]annulene perimeter. *J. Am. Chem. Soc.* **1978**, *100*, 6812–6818.
26. Saegusa, Y.; Ishizuka, T.; Komamura, K.; Shimizu, S.; Kotani, H.; Kobayashi, N.; Kojima, T. Ring-fused porphyrins: extension of π-conjugation significantly affects the aromaticity and optical properties of the porphyrin pi-systems and the Lewis acidity of the central metal ions. *Phys. Chem. Chem. Phys.* **2015**, *17*, 15001–15011.
27. Yamamoto, S.; Zhang, A.; Stillman, M. J.; Kobayashi, N.; Kimura, M. Low-symmetry omega-shaped zinc phthalocyanine sensitizers with panchromatic light-harvesting properties for dye-sensitized solar cells. *Chem. Eur. J.* **2016**, *22*, 18760–18768.
28. Dullweber, F.; Montforts, F.-P. Synthesis of chlorins extended by highly substituted double bonds. *Synlett* **2008**, *2008*, 3213–3215.
29. Bats, J. W.; Haake, G.; Meier, A.; Montforts, F.-P.; Scheurich, G. Synthesis of oxochlorins from the red blood pigment heme and their transformation into potentially biologically active chlorin derivatives. *Liebigs Ann.* **1995**, *1995*, 1617–1631.
30. Zhang, A.; Stillman, M. J. Exploring function activated chlorins using MCD spectroscopy and DFT methods: design of a chlorin with a remarkably intense, red Q band. *Phys. Chem. Chem. Phys.* 2018.
31. Kim, S.; Lee, J. K.; Kang, S. O.; Ko, J.; Yum, J.-H.; Fantacci, S.; Angelis, F. D.; Censo, D. D.; Nazeeruddin, M. K.; Gratzel, M. Molecular engineering of organic sensitizers for solar cell application. *J. Am. Chem. Soc.* **2006**, *128*, 16701–16707.
32. Meng, S.; Kaxiras, E.; Nazeeruddin, M. K.; Grätzel, M. Design of dye acceptors for photovoltaics from first-principles calculations. *J. Phys. Chem. C* **2011**, *115*, 9276–9282.
33. Mishra, A.; Pootrakulchote, N.; Fischer, M. K.; Klein, C.; Nazeeruddin, M. K.; Zakeeruddin, S. M.; Bauerle, P.; Gratzel, M. Design and synthesis of a novel anchoring ligand for highly efficient thin film dye-sensitized solar cells. *Chem. Commun.* **2009**, 7146–7148.
34. Balanay, M. P.; Kim, D. H. Computational study of absorption energies of organic sensitizers used in photovoltaic applications. *J. Phys. Chem. C* **2011**, *115*, 19424–19430.
35. Kutzki, O.; Montforts, F.-P. Synthesis of a Chlorin-quinone dyad and a benzochlorin derivative. *Synlett.* **2001**, *2001*, 53–56.
36. De Angelis, F.; Fantacci, S.; Selloni, A. Alignment of the dye's molecular levels with the TiO_2 band edges in dye-sensitized solar cells: a DFT-TDDFT study. *Nanotechnology* **2008**, *19*, 424002.
37. Tsuneda, T.; Song, J. W.; Suzuki, S.; Hirao, K. On Koopmans' theorem in density functional theory. *J. Chem. Phys.* **2010**, *133* (17), 174101.
38. Pastore, M.; Fantacci, S.; Angelis, F. D. *Ab Initio* Determination of ground and excited state oxidation potentials of organic chromophores for dye-sensitized solar cell. *J. Phys. Chem. C* **2010**, *114*, 22742–22750.

217 Structure and Electronic Circular Dichroism of Chiral Porphyrins and Chiral Porphyrin Dimers

Milan Balaz,* Yoonbin A. Joh,*,† Krisztina Varga,† Nina Berova,‡ Roberto Purrello§ and Alessandro D'Urso§

*Yonsei University, Underwood International College, Integrated Science and Engineering Division, Seoul, 03722, Republic of Korea

†University of New Hampshire, Department of Molecular, Cellular, and Biomedical Sciences, Durham, NH 03824, USA

‡Columbia University, Department of Chemistry, New York, NY 10027, USA

§University of Catania, Chemical Science Department, Catania, Italy

List of Abbreviations

BHP	basket-handle porphyrin
BINOL	1,1′-bi-2-naphthol
CD	circular dichroism
COSY-NMR	correlation spectroscopy — nuclear magnetic resonance spectroscopy
DFT	Density Functional Theory
DMAP	4- dimethylaminopyridine
DMF	dimethylformammide
ECD	electronic circular dichroism
EDC	3-(ethyliminomethyleneamino)-*N*,*N*-dimethylpropan-1-amine
HPLC	high performance liquid chromatography
ICD	induced circular dichroism
IR	infrared spectroscopy
L-SDP	sodium dodecyl L-proline
LB	Langmuir-Blodgett
LIF-CE	laser induced fluorescence — capillary electrophoresis

NOE	nuclear Overhauser effect
QND	quinuclidine
SDS	sodium dodecyl sulfate
SEM	scanning electron microscopy
SWNT	single-walled carbon nanotube
TEM	transmission electron microscopy
THF	tetrahydrofuran
UV-vis	ultraviolet-visible spectroscopy
VCD	vibrational circular dichroism

I. Introduction

Since Nature adopts mainly homochirality[1] to produce and to develop living organisms, chirality has become one of the most intriguing properties of matter which attracts many scientists from different research fields, ranging from medicine to material science. In particular, chirality plays a fundamental role in designing pharmaceuticals, because often one of the enantiomers possesses a higher therapeutic index.[2] Even the development of chiral materials provides the opportunity to achieve unusual properties such as a negative refractive index or nonlinear optical effects which prompts interesting applications in optoelectronics.[3] Consequently, technological developments in the synthetic approaches, separation techniques and detection methods of chiral molecules are required.

The word chirality derives from the Greek term $\chi\varepsilon\iota\rho$ (hand) and refers to objects that cannot be superimposed upon their mirror image ("enantiomers") by mere rotations and translations in space, just like our two hands. Molecular chirality originates from the non-symmetric arrangement of chemical groups covalently linked to an atomic center. Chiral molecules are able to rotate the electric and magnetic vectors of plane-polarized light in either a clockwise or counterclockwise manner (optical activity). In addition, when an alternating beam of left and right circularly polarized light interacts with a chiral molecule, the difference in absorption of the two beams (*i.e.*, $\Delta A = A_l - A_r$, where A_l and A_r denote the absorption of left and right circularly polarized light, respectively) generates circular dichroism (CD). Unlike the corresponding UV-vis bands, the observed CD bands are visible indicators of chirality and are known as positive or negative Cotton effects, depending on the relative values of A_l and A_r.[4] CD within a specific absorption region is defined as the presence of optical activity. Various CD selection rules have been developed and used to correlate the sign of the CD with chirality of the sample. Moreover, the two parameters associated with the electronic transition μ (electric dipole transition moment) and m (magnetic dipole transition moment), must not be orthogonal, because in an optically active molecule the electron rearrangement during the excitation process undergoes a helical pathway. Thus, the parameter determining the possibility to observe CD is the rotational strength R defined by the Rosenfeld Equation (1):

$$R = I_m\{\mu \,.\, m\} \tag{1}$$

where I_m means that one should use only the imaginary part of the complex number and μ and m are the electric and magnetic dipole transition moments, respectively.[4] Finally, it is useful to mention a general rule of CD spectroscopy: the integral of the CD over the whole electromagnetic spectrum is zero or, in algebraic terms, $\Sigma R = 0$(sum known also as the Kuhn rule). The "sum rule" is particularly significant in analysis of the CD intensity associated with exciton coupling.

Exciton coupling method is a simple and the most commonly used approach for analyzing the absorption and CD spectra of through space interacting chromophores (not necessarily chemically bonded) constituting a chiral system.[5] When two or more chromophores are located close in space and absorb strongly, their electric dipole transition moments can couple, which results in a split of the excited state energies and the appearance of a typical bisignate CD curve (see below). This means that there are two possible electronic transitions and hence two bands in the absorption spectrum (Figure 1). For chiral systems the transitions to the two excited state energy levels have CD Cotton effect with opposite signs. The signs and the shifts in the UV and CD spectra depend on the orientation of the chromophores in space. When multiple chromophores are arranged in a cofacial manner (H-aggregation), the observed spectral bands shift to higher energy. In contradistinction, if the chromophores are involved in a near parallel head-to-tail alignment (J-aggregation), a shift of the spectral bands to lower energy is observed compared to what is measured for the monomeric chromophore. A more general orientation of the chromophores occurs when they are arranged obliquely in space. Theoretically, two absorption bands should be observed at both shorter and longer wavelengths and the CD spectral band has a bisignate profile in comparison with the absorption band maximum; however the interchromophoric distance and the spatial arrangement of the chromophores determine whether both are clearly observed.

Porphyrins and their metalated derivatives have an essentially square planar coordination environment and tend to adopt D2h or D2 symmetry for the free base ligands and D4h or D4 symmetry for metalloporphyrins. These macrocycles seem to be the ideal compounds which have applications in many scientific fields owing to their particular spectroscopic characteristics and enormous synthetic versatility. These factors allow one to functionalize the planar achiral macrocycle with the desired functional groups and to detect, by spectroscopic techniques, the interactions with other molecules. In particular, the application of porphyrins is significantly broadened when chiral substituents are linked to the porphyrin ring. Upon covalent linking, the chiral substituents

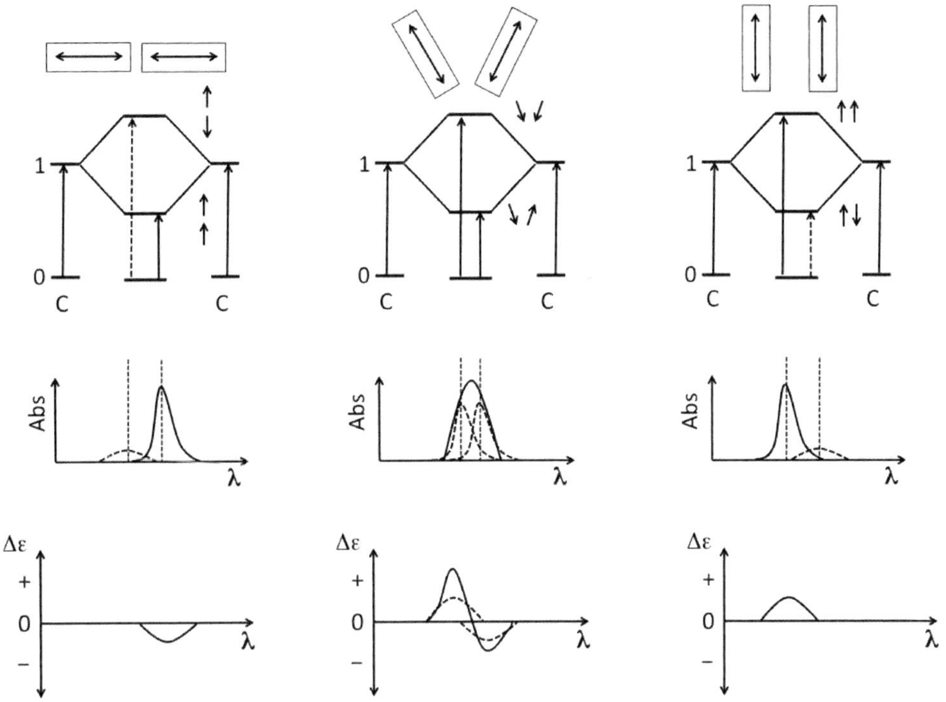

Figure 1. Orientation dependence in exciton coupling between two chromophores (rectangular boxes) and their induced electric dipoles (double-headed arrow). The limiting orientation leads to red-shifted (left), broadened (center) or blue-shifted (right) bands. The solid arrows connecting ground (0) and excited (1) states represent allowed transitions; the dashed arrows represent forbidden transitions. UV and CD spectra are reported for each situation. Adapted from Ref. 4.

convert the porphyrin ring into a chiral porphyrin system, which opens more opportunities for chiral recognition, optical resolution and stereoselective synthesis.

In this chapter we summarize numerous publications regarding chiral porphyrins, both monomers and dimers, describe their circular dichroism properties and introduce their applications in the area of sensing and molecular recognition.

II. Preparation of Chiral Porphyrins: Synthesis, Functionalization and Chiral Resolution

Three major approaches have been utilized to access chiral, enantiomerically pure porphyrins, these are: (i) synthesis of a porphyrin skeleton by condensation of pyrrole or pyrrole derivatives (*e.g.*, dipyrromethenes) with a chiral, enantiomerically pure aldehyde, (ii) attachment of a chiral molecule to a preformed porphyrin and (iii) chiral resolution, by chiral HPLC or capillary electrophoresis with chiral additives, of racemic porphyrins prepared by non-stereoselective synthesis as a mixture of two enantiomers. The following section briefly introduces representative examples of each approach.

A. Chiral Reactants

1. *Chiral Aldehydes*

In 1997, Hayashi, Ogoshi and coworkers prepared a chiral porphyrin dimer linked *via* a binaphthyl derivative using either the (*S*)- or (*R*)-enantiomer of 2,2′-dimethoxy-[1,1′-binaphthalene]-3,3′-dicarbaldehyde as a chiral starting material (Scheme 1).[6]

Commercially available (*R*)- or (*S*)-*N*-BOC-phenylalaninal was used to prepare chiral, enantiopure 5,15-bis-substituted porphyrins by condensation with achiral dipyrromethene.[7] The subsequent functionalization of the free-*meso*-positions also allowed synthesis of chiral porphyrin dimers (tweezers; see Scheme 2).

2. *Chiral Pyrroles*

Bürgi, Kräutler and coworkers explored a series of biconcave D4-symmetric and C4-symmetric metalloporphyrins with rigid frameworks. Chiral porphyrins have been prepared as single enantiomers by condensation of chiral pyrrole that was synthesized *via* a highly diastereo-discriminating Diels–Alder reaction with a chiral (–)methyl ester of fumaric acid (Scheme 3).[8]

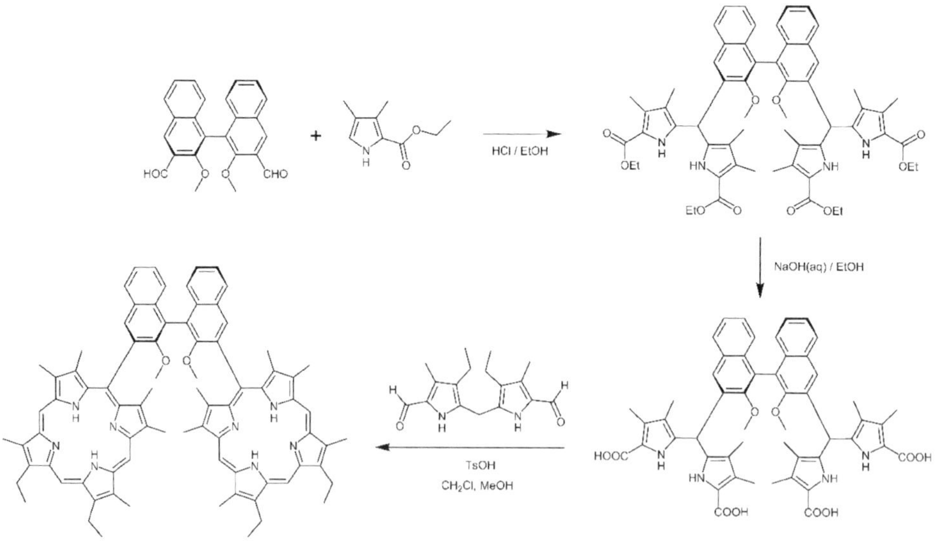

Scheme 1. Representative example for the synthesis of a chiral porphyrin dimer from a (*S*)-2,2′-dimethoxy-[1,1′-binaphthalene]-3,3′-dicarbaldehyde. Adapted from Ref. 6.

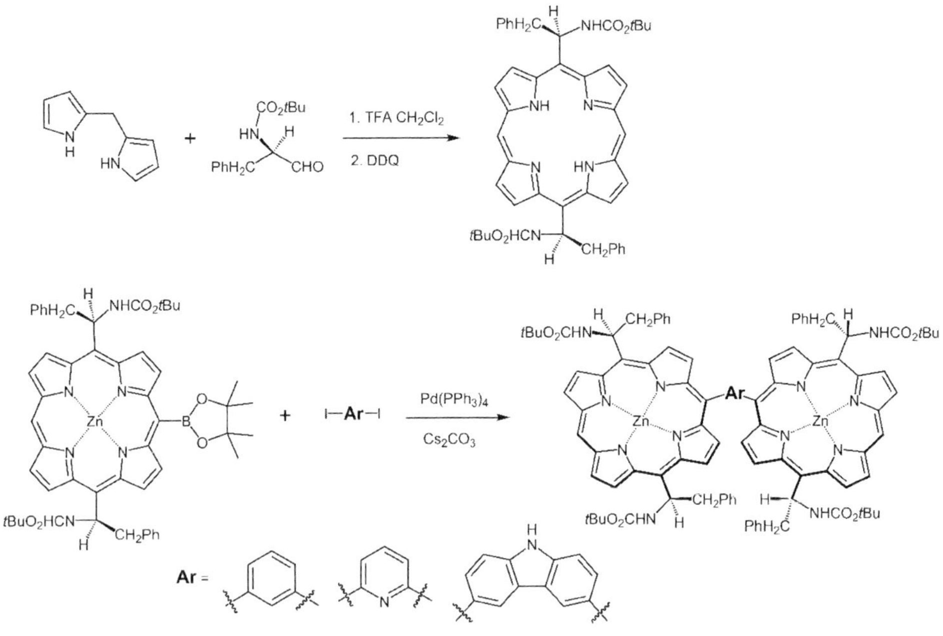

Scheme 2. Representative examples for the synthesis of a chiral porphyrin monomer from chiral aldehyde and the subsequent synthesis of a chiral porphyrin dimer *via* Pd-catalyzed coupling. Adapted from Ref. 7.

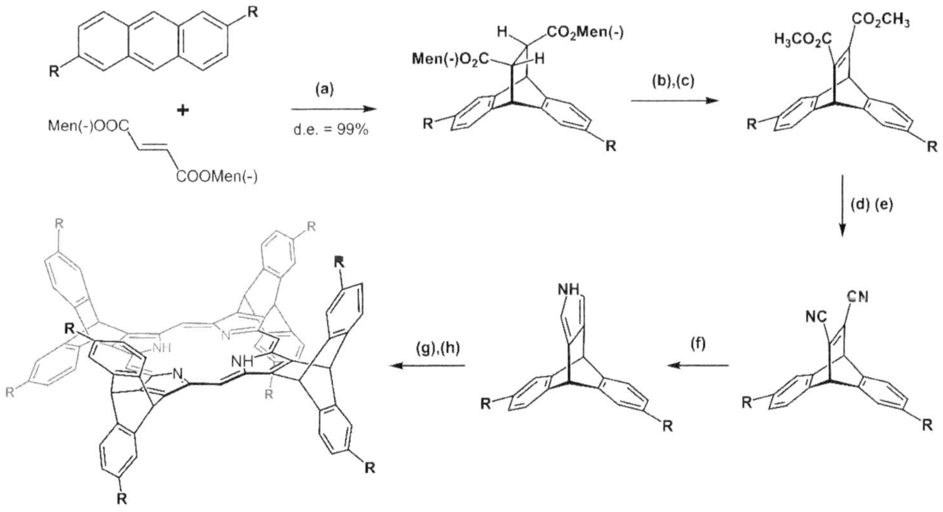

Scheme 3. Representative example for the syntheses of a chiral porphyrin monomer from a chiral pyrrole derivative (R= *tert*-Bu). (a) $AlCl_3$, –40°C; (b) (1) Ph-Se-Se-Ph, potassium *tert*-butoxide, (2) HCl; (c) MeOH, methanesulfonic acid, Δ; (d) NH_3, NaCN, Δ; (e) thionyl chloride; (f) DIBAH, CH_2Cl_2, –20°C; (g) HCHO, acetic acid; (h) DDQ. Adapted from Ref. 8.

B. Functionalization of Achiral Porphyrins with Chiral Molecules

Covalent attachment of chiral molecules to an achiral porphyrin unit represents the simplest and most common method for the synthesis of chiral porphyrins. Chiral porphyrin monomers, dimers and larger oligomers have been prepared by linking chiral molecules (including amino acids,[9,10] peptides,[11] nucleosides,[12] oligonucleotides,[13–18] sugars,[19,20] cyclodextrins[21] or steroids and other natural products[22]) to the *meso*- or *beta*-positions of an achiral porphyrin.

C. Chiral Resolution of Racemic Mixtures by HPLC and Capillary Electrophoresis

HPLC with chiral stationary phases has been used for resolution of chiral porphyrin monomers with C1, C2 or D4 symmetries (Figure 2) as well as chiral acyclic and cyclic dimers (Figure 3). Capillary electrophoresis with chiral selectors has been used for baseline-enantioseparation of chiral benzoporphyrin monomers. Representative examples are listed in the following sections.

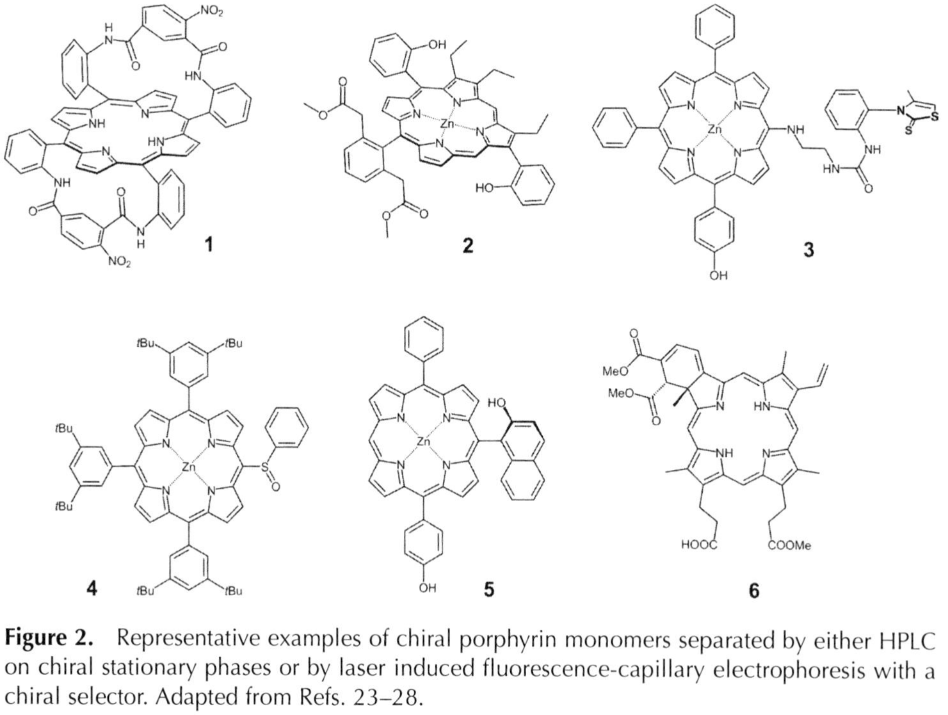

Figure 2. Representative examples of chiral porphyrin monomers separated by either HPLC on chiral stationary phases or by laser induced fluorescence-capillary electrophoresis with a chiral selector. Adapted from Refs. 23–28.

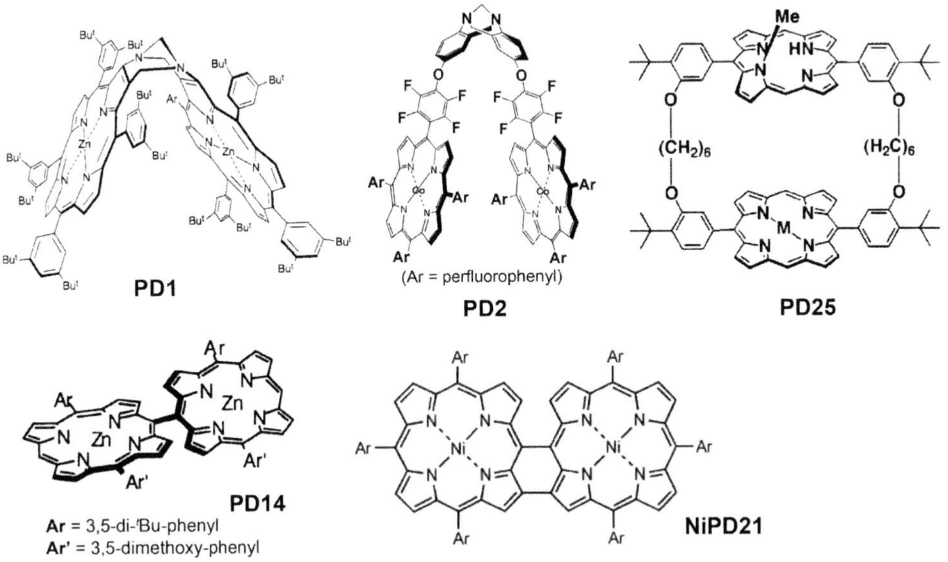

Figure 3. Representative examples of chiral porphyrin dimers separated by HPLC on chiral stationary phases. Adapted from Refs. 29–32.

1. *Porphyrin Monomers*

Chiral doubly bridged free-base tetraarylporphyrin **1** has been resolved by HPLC on a CHIRALCEL OD column (4.6 × 250 mm) using hexane:EtOH (1:1) as eluent.[23] The separation factor α and the resolution factor R were 2.98 and 1.98, respectively. Trifunctional chiral free base porphyrin **2** having C2 symmetry was resolved by HPLC on a YMC A-KO3 column (Yamamura Chemical Labs. Co., Ltd.) using hexane:$CHCl_3$:ethanol = 19:80:1 as the eluent.[24] Because only incomplete separation was achieved, several HPLC cycles were performed to attain the optically pure free base of (+)–**2**. Interestingly, attempts to resolve the zinc complex of **2** failed. Optically pure atropisomers of porphyrin **3** were obtained by chiral HPLC on a Chiralpak IC column using a mixture of hexane and ethanol (80:20) as the solvent system. At 25 °C, the two atropo-enantiomers were well-separated with retention times of R_{t1} (*aR*) = 16.97 min and R_{t2} (*aS*) = 21.93 min; the separation factor α and the resolution factor R were α = 1.35 and Rs = 2.71.[25]

Enantioseparation of (±)-*meso*-sulfinylporphyrin (chiral sulfoxide) **4** was performed by chiral HPLC using a semipreparative, chiral carbamate-modified polysaccharide DAICEL Chiralpak IA column with isocratic flow of hexane:2-propanol

(100:1).[26] Both enantiomers of the chiral porphyrin monomer **4** were isolated in high optical purities (99% ee for the 1st fraction, 97% ee for the 2nd fraction). A chiral, C1-symmerical *meso*-substituted ABCD-type porphyrin **5** was resolved by HPLC using isocratic flow rate (2 mL/min; 2-propanol:CH_2Cl_2:hexane = 1:50:50) on a chiral column (10 × 250 mm).[27] Unfortunately, the type of chiral column was not specified. Laser induced fluorescence-capillary electrophoresis (LIF-CE) have been successfully used to baseline-separate enantiomers of mono- and diacid derivatives of a chiral benzoporphyrin **6**.[28] Sodium chlorate was found to yield the best separation.

2. *Porphyrin Dimers*

Chiral Tröger's base bis-porphyrin **PD1** has been resolved on a chiral stationary phase of silica gel covalently functionalized with (*R*)-*N*-3,5-dinitrobenzoylphenylglycine (Analytical Jones APEX Chiral AP Column) using 0.5% propan-2-ol in light petroleum as the eluent.[29] **PD1** was also successfully resolved on a preparative scale by a simple flash chromatography on a silica gel pre-saturated with L-histidine benzyl ester, taking advantage of the strong enanatioselective binding of (+)–**PD1** and (–)–**PD1** bisporphyrins with L-histidine benzyl ester ($K = 1.1 \times 10^8$ dm^3/mol *vs.* 1.2×10^7 dm^3/mol).[30]

The preparative enantioseparations of a racemic mixture of the flexible chiral Tröger's base bis-porphyrin **PD2** was performed on HPLC using isocratic elution (14 mL/min; CH_2Cl_2:MeOH = 5:95) on a preparative chiral column Reprosil Chiral-NR (250 × 20 mm), achieving a separation selectivity of 1.16 and resolution of 0.83.[31] Optical resolution of chiral cyclic *N*-methylated porphyrin dimer **PD25** was achieved by preparative HPLC on a CHIRALPAK IA chiral stationary phase with hexane:CH_2Cl_2:Et_2NH (87.5:12.5:0.1) as the eluent.[32] Yoshida and Osuka used HPLC with a Sumichiral OA-3100 chiral column to separate enantiomers of the *meso–meso* linked porphyrin dimer **PD14**[33] while Muranaka *et al.* separated enantiomers of the *meso-meso* β–β doubly linked porphyrin dimer **NiPD21** (chiral fused bis-porphyrin).[34]

III. Chiral Porphyrin Monomers

Since the porphyrin ring is a symmetrical planar macrocycle it is an achiral compound. In order to obtain chiral porphyrins it is necessary to either introduce a chiral substituent by covalent bonds to the ligand (mainly in the *meso* positions) or by generating a porphyrin ring with an intrinsically chiral structure. However, in general, the expected CD signal is often weak since the magnetic transition moments of monomeric aromatic molecules are very small.

A. Porphyrins with Chiral Substituents in *meso* Positions

The most commonly used porphyrins for studying chiral systems are metalated with chiral substituents in the *meso* positions.

1. *Monosubstituted Porphyrins*

Kinoshita and coworkers exploited the coordination chemistry of the chiral Zn(II) *meso*-phenyl porphyrin **PM1** with a chiral pendant arm covalently attached at the ortho-position of one of the *meso*-phenyl groups.[35] As chiral pendant they used 2,6-pyridinedicarbonyl dichloride linked from one side to the 5-(2-aminopheny)-10,15,20-triphenylporphyrin and from the other side to (*R*)-1-(2-pyridyl)ethyl-amine (Figure 4). The resulting 2,6-diamidepyridine group is rigidly planar owing to hydrogen bonds between the amide nitrogen atom and the central pyridine nitrogen atom. A coordination of the terminal pyridine nitrogen atom to the central zinc induces rotation of the ortho-substituted *meso*-phenyl group, introducing

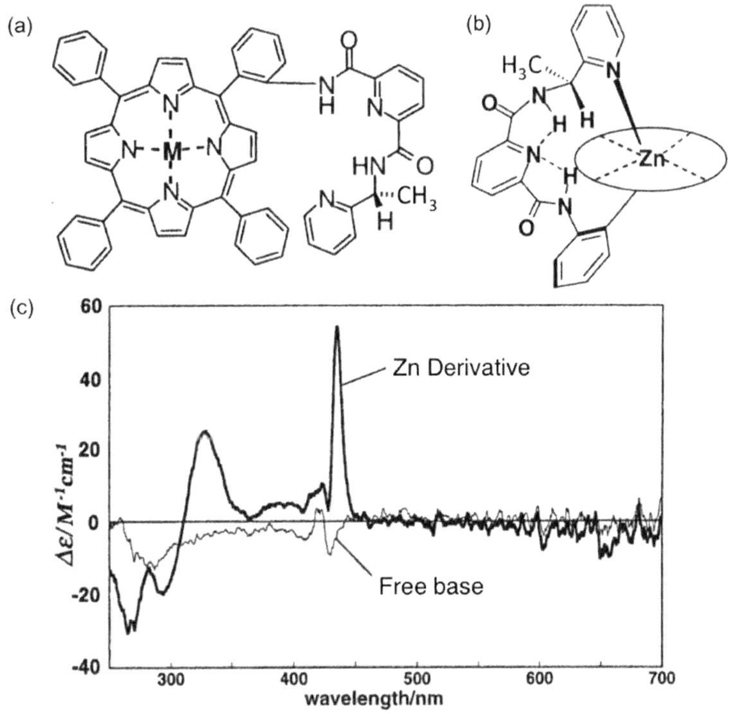

Figure 4. (a) Structure of chiral porphyrin **PM1**, (b) Schematic representation of the pendant arm of the Zn porphyrin derivative and (c) CD spectra of free base porphyrin **PM1** and Zn derivative. Adapted from Ref. 35.

chirality to the porphyrin ring (Figure 4). As a confirmation that the origin of the chirality could be mainly ascribed to coordination of the side arm, the authors obtained CD spectra of the free base and Zn metalated derivative. A significant Cotton effect appears around the Soret band of the Zn derivative and only a very weak signal is detected for the free base porphyrin (Figure 4).

Another efficient synthesis of optically active porphyrins has been reported by Tykwinski and coworkers who obtained several steroid–naphthoquinoline–Ni(II) petroporphyrin triads **PM2a–2e** (Figure 5).[36] These porphyrins were synthesized through a sequential one-pot multicomponent cyclocondensation reaction to construct the steroid and quinoline moieties, using the method of Lindsey for porphyrin synthesis[37] and metalation. These synthetic protocols were used to prepare porphyrin derivatives that represent both "continental"[38] and "archipelago"[39] models of asphaltene composition. Since it has been demonstrated that the naphthoquinoline moiety adopts a helical, homochiral conformation,[40,41] the authors explored whether helicity of the naphthoquinolines was maintained and even transferred to the Ni(II) porphyrin units. By CD spectroscopy, it is possible to note several

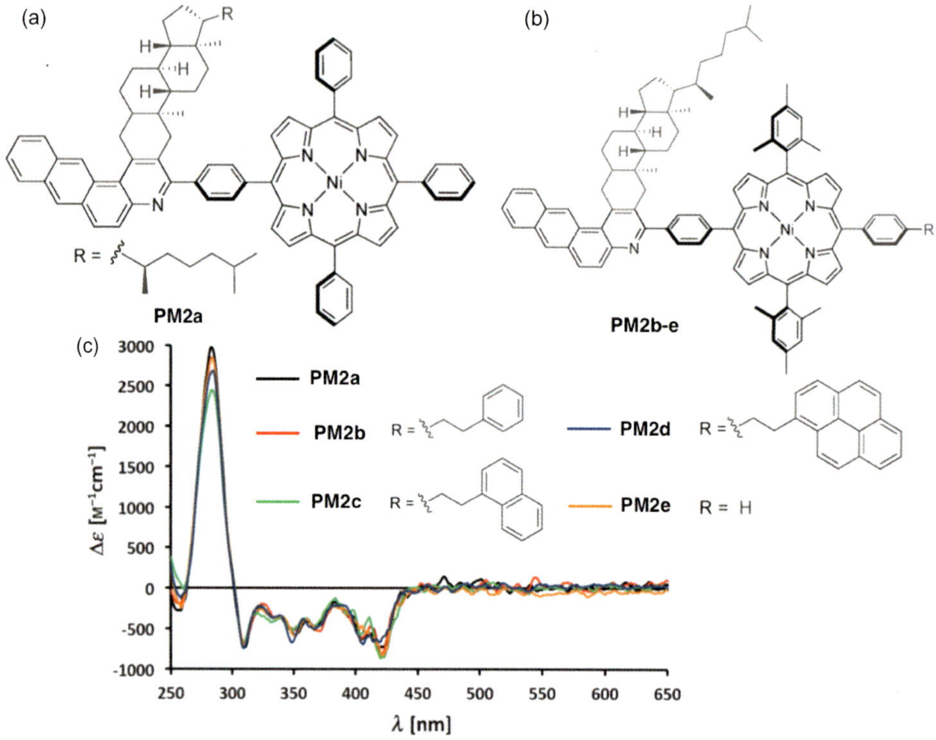

Figure 5. Structures of (a) steroid–quinoline–porphyrin triad (continental) **PM2a**, (b) steroid–naphthoquinoline–porphyrin triads (archipelago) **PM2b–2e** and (c) CD spectra of the chiral porphyrins **PM2a–2e**. Adapted from Ref. 36.

Cotton effects in the region from 280 nm to 370 nm according to the absorption of the naphthoquinoline units; however, the most interesting CD signals were observed at 405 nm and 420 nm which correspond to absorptions of the Soret band of the porphyrin units (Figure 5).

Monti and coworkers synthesized chiral manganese [5-(4-(carboxyphenyl-(*N*-L-proline)))-10-15-20-triphenylporphyrinyl] chloride **MnPM3** and investigated the interactions with micellar aggregates formed by either sodium dodecyl sulfate SDS or sodium *N*-dodecanoyl-(L)-prolinate, L-SDP (Figure 6).[42] They performed inclusion studies of porphyrins in micellar solutions (SDS and L-SDP) and found an inclusion of the chromophore in a hydrocarbon like region of the L-SDP micelle. In contrast, the inclusion of Mn porphyrin in SDS occurs with a consistent degree of aggregation. Indeed, by UV and CD spectroscopy the authors demonstrated that the presence of a chiral functionality on the periphery of the porphyrin macrocycle affects their aggregation mode within the biomembrane models. CD spectra of **MnPM3** in different media (SDS, L-SPD and H_2O/EtOH) show different CD signals. In particular, in the presence of L-SDP a negative coupled CD signal is detected; on the other hand, a positive bisignate couple is observed in SDS or in a solution of H_2O/EtOH (Figure 6). The authors ascribe this different

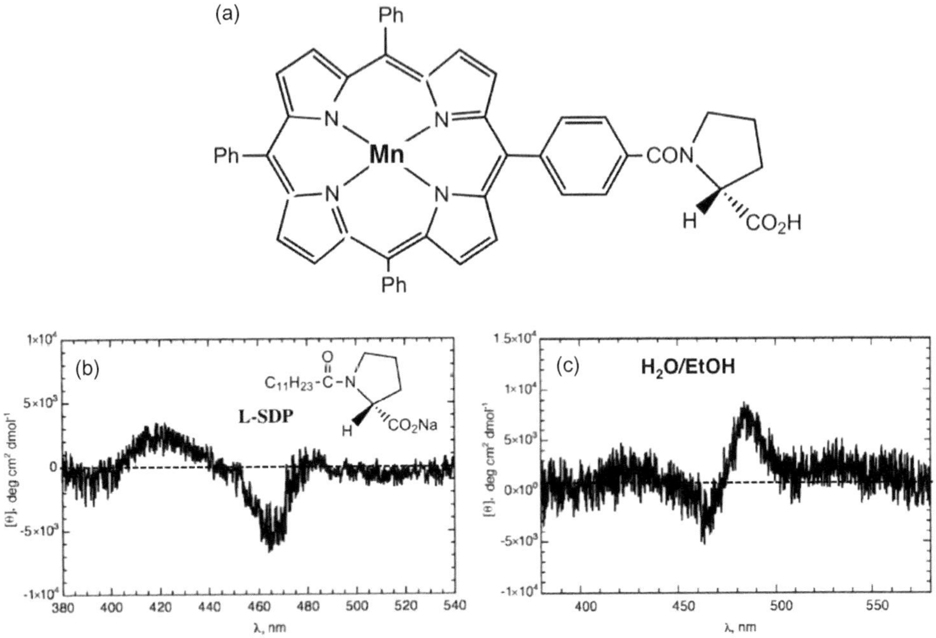

Figure 6. (a) Structure of manganese [5-(4-(carboxyphenyl-(N-L-proline)))-10-15-20-triphenylporphyrinyl] chloride **MnPM3** and CD spectra in the presence of (b) L-SDP (b) and (c) H_2O/EtOH solution. Adapted from Ref. 42.

behavior to the aggregation conditions; indeed, the sharpness of the relative Soret band (UV-vis) rules out the formation of porphyrin aggregates in the presence of L-SDP and the CD signal arises as a result of the induced CD (ICD) by the chiral biomembrane model. The presence of SDS or an H_2O/EtOH solution implies the formation of porphyrin chiral aggregates with the L-proline appended residues driving the helical sense of the self-assembling process.

Similar results have also been obtained using the naked L-proline porphyrin derivative **PM3**.[43] CD spectroscopy studies reveal that the inclusion of naked porphyrin in L–SDP micelles results in an intense, bisignate, positive band centered at ~440 nm. In contrast, in SDS there is an intense, coupled CD band with opposite (*i.e.*, negative) sign, centered at 450 nm, with intensity higher than that observed in L–SDP. These results suggest that the chiral recognition takes place in the hydrophobic region of the aggregates where a chiral environment is transmitted by the polar head groups to the hydrophobic chains.

Another example of how to design chiral porphyrin to tune the molecular orientation in self-assembled porphyrins was presented by Ihara *et al.*[44] They prepared a zinc(II) tetraphenylporphyrin **PM4** functionalized with an L-glutamide derivative as a self-assembling tool (Figure 7). In order to control the molecular ordering state of the **PM4** assembly, they investigated the interactions with

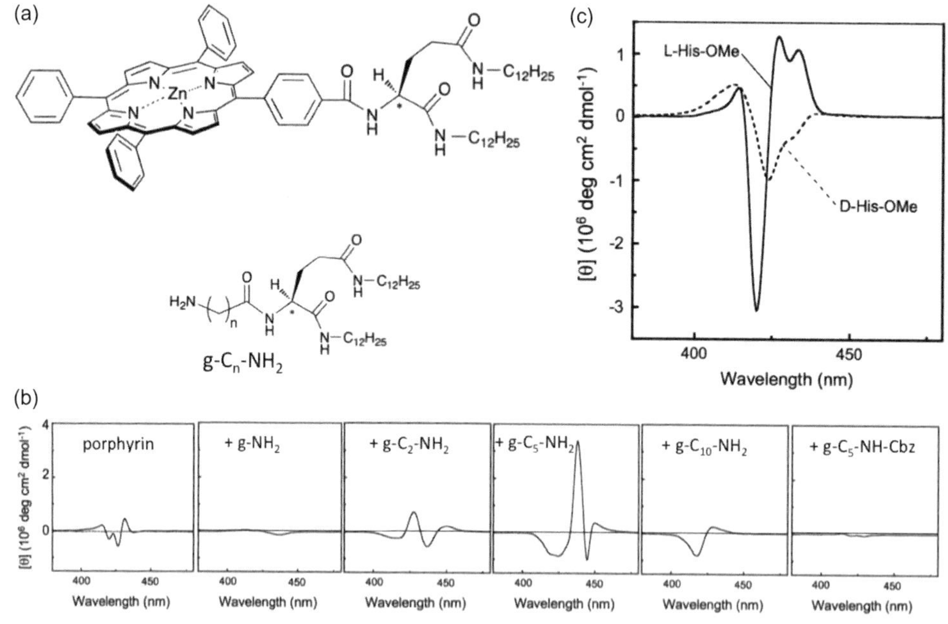

Figure 7. (a) Structure of zinc tetraphenylporphyrin L-glutamide **PM4**. CD spectra of **PM4** in the presence of (b) ω-aminoalkylated L-glutamide derivatives, and (c) L- and D-histidine methyl esters. Adapted from Ref. 44.

ω-aminoalkylated L-glutamide derivatives (g-C_n–NH_2) having different lengths of alkyl spacer. CD spectra show distinct Cotton effects around the Soret band in cyclohexane solutions of porphyrin glutamide derivative alone and in the presence of ω-aminoalkylated L-glutamide with C_0, C_2, C_5 and C_{10} alkyl chains (Figure 7). Therefore, one should consider that the spacer length of ω-aminoalkylated L-glutamide derivatives influences the molecular fitting for axial coordination on zinc. Moreover, to investigate the effect of the amino group, they prepared ω-aminoalkylated L-glutamide protected with *N*-benzyloxycarbonyl (g-C_5–NH–Cbz) in order to prevent coordination with Zn. They observed a remarkable reduction of CD intensity for **PM4** (Figure 7). These results suggest that the change of chiroptical signals, corresponding to different chirally stacked structures, is based on axial coordination of Zn with the amino group and is very sensitive to the spacer length of ω-aminoalkylated L-glutamide. **PM4** was then used for an enantioselective recognition of amino acids. Only with histidine methyl esters did the CD spectrum of the porphyrin show characteristic changes corresponding to the molecular chirality of histidine methyl ester (Figure 7).

Another example of chiral recognition of amino acid esters (AlaOMe, LeuOMe, ValOMe, PheOMe) in $CHCl_3$ by using chiral porphyrins was presented by Wen-Juan and coworkers.[45] In this work, the authors prepared a chiral Zn(II) tetraphenylporphyrin **PM5** carrying a L-BocTyr in the ortho-position of one of the *meso*-phenyl groups (Figure 8). By computational and experimental data (UV-vis titration and CD methods) the authors found that values of the association constants for the molecular recognition reactions, calculated for D-amino acids, were all higher than those obtained using L-amino acids ($K_D > K_L$). However, only the CD spectra obtained by the interactions of chiral porphyrin **PM5** with D- and L-alanine show a different shape. The splitting intensity of the ICD spectrum of D-$AlaOCH_3$ was larger than that of L-$AlaOCH_3$ (Figure 8). **PM5** with D-/L-$LeuOCH_3$ or D-/L-$ValOCH_3$ gave an unusual ICD spectrum and showed simple negative Cotton effect in the presence of both enantiomers. Theoretical studies were used to understand chiral recognition on a molecular level. The minimal energy conformation obtained by geometry optimizations shows that the distance between **PM5** and L-AlaOMe is larger than the distance with D-AlaOMe. The **PM5**-D-$AlaOCH_3$ complex has lower energy than the **PM5**-L-$AlaOCH_3$ system, indicating that the former was more stable than the latter.

2. *Di-Substituted Porphyrins*

An interesting example of difunctionalized porphyrin is represented by porphyrins with steric shielding on one side of the macrocycle plane, known

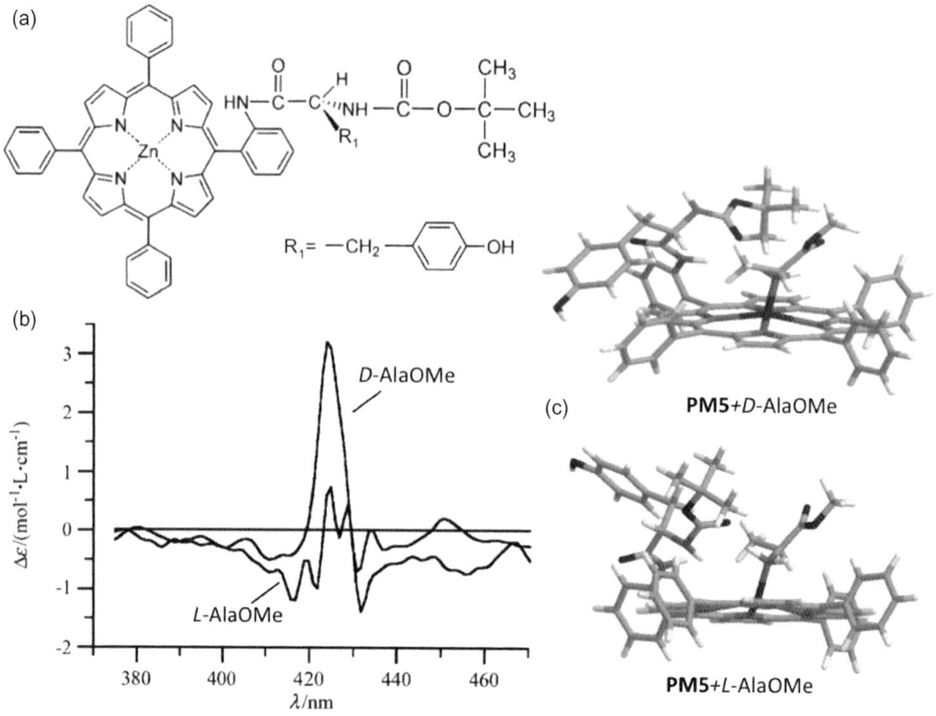

Figure 8. (a) Structure of zinc(II)tetraphenylporphyrin-L-BocTyr derivative **PM5**. (b) CD spectra of **PM5** in the presence of D-/L-AlaOMe. (c) The lowest energy conformations of **PM5** in the presence of D-AlaOCH$_3$ (upper) or L-AlaOCH$_3$ (down). Adapted from Ref. 45.

as basket-handle porphyrins (BHPs) **PM6**.[46,47] Bringmann and coworkers synthesized *via* a fast and efficient route both, chiral and achiral basket-handle porphyrins in significant amounts.[48,49] The synthetic approach to the **2HPM6** starts using a suitable *p*-xylene-linked dialdehyde, which reacts with dipyrromethane to yield **2HPM6** after a two-step, one-pot reaction consisting of a cyclocondensation and a subsequent oxidation of the initially formed porphyrinogen (Figure 9). If the substituents along the diagonal of the xylene unit of **2HPM6** are not identical (R ≠ R′), **2HPM6** then exits as a racemic mixture of two planar-chiral enantiomers. This method allows one to obtain, in few steps, a wide variety of stable **2HPM6** species which can be easily functionalized in order to improve solubility and stability. The authors also investigated the possibility of introducing a metal into the core of the **2HPM6** and successfully prepared Ni(II), Zn(II), Pd(II), Cu(II), In(III) and V(IV) derivatives (Figure 9). CD spectra were then recorded for all pairs of the metalated **MPM6** enantiomers resolved by HPLC on a chiral phase. The faster-eluting enantiomers always displayed a negative couplet within the

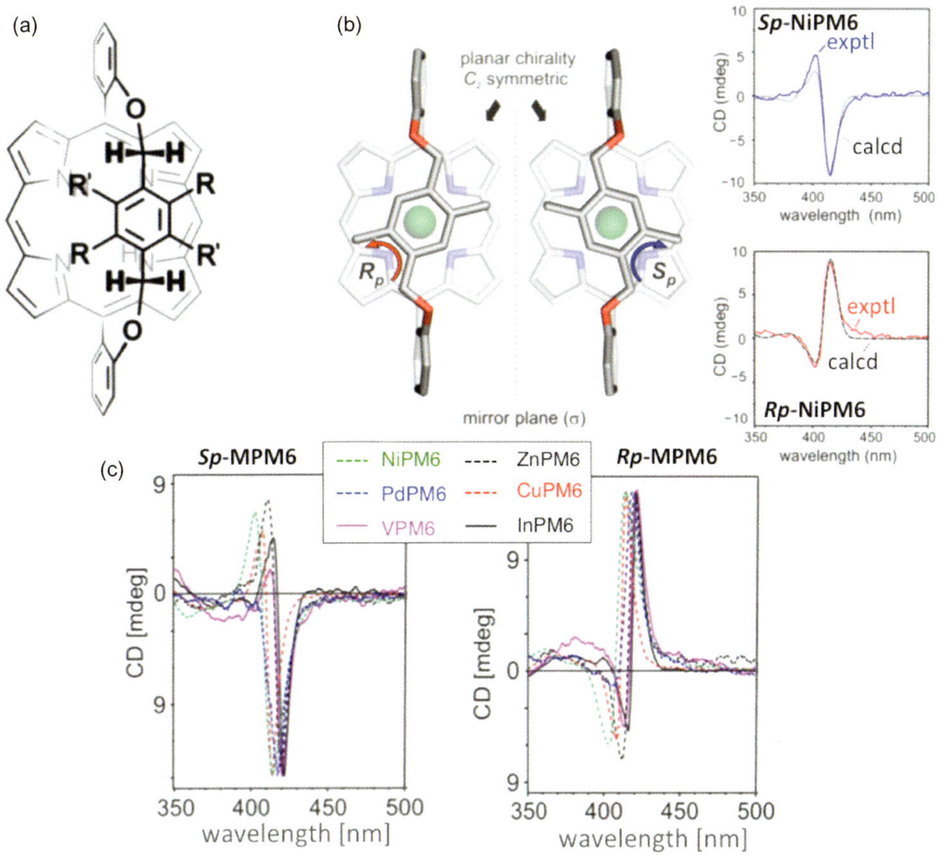

Figure 9. (a) Structure of metalated BHPs **PM6**, (b) comparison of calculated and experimental CD spectra of **NiPM6** and (c) experimental CD spectra of metalated S_p-**PM6** (left) and metalated R_p-**PM6**. Adapted from Refs. 46–49.

porphyrin absorption region, while the slower-eluting enantiomers showed a positive, mirror-image couplet (Figure 9). All ECD spectra were very similar; thus, the central metals do not influence the ECD spectra significantly. A conformational analysis (B97-D3/def2-TZVP) was performed only for the Ni(II) derivative **NiPM6** and assignment of the absolute configuration was done by comparison of the experimental CD spectra with results from quantum-chemical CD calculations using TDCAM-B3LYP (Figure 9). As a result, the faster-eluting enantiomer of **NiPM6** showing a negative ECD signal at ~420 nm, which was assigned as the S_p-configuration, while the slower-eluting enantiomer had a R_p-configuration. As ECD spectra of the metalated **MPM6** species all look similar, it is reasonable to extend the configuration assignment to all synthesized **MPM6**.

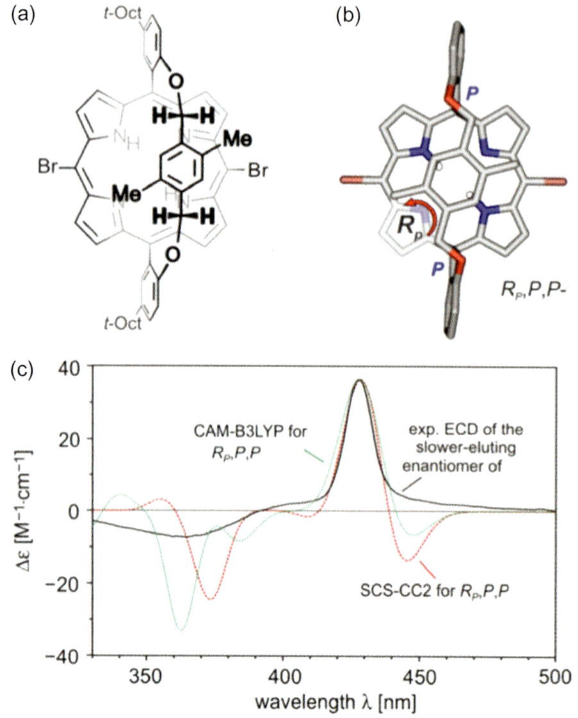

Figure 10. (a) Structure of free base **2HPM6**, (b) calculated structure of **2HPM6** with (R_p,*P*,*P*) configuration and (c) comparison of experimental and calculated CD spectra of (R_p,*P*,*P*) enantiomer of **2HPM6**. Adapted from Refs. 46–49.

Surprisingly, the CD spectrum of chiral free-base **2HPM6** with *tert*-octyl substituents is not comparable with the CD spectrum of metalated **NiPM6**. Indeed, the CD spectrum of **2HPM6** did not show a CD couplet in the Soret band region but two separate Cotton effects, positive and negative (Figure 10). The authors attributed this effect to NH tautomerism of the porphyrin macrocycle, restricted by the strap. A combination of the theoretical calculations and experimental NOE and COSY NMR data allowed the authors to establish that at room temperature only one tautomer was present for each enantiomer, having the NH-containing pyrrole unit positioned in proximity to the aromatic hydrogen of the xylene strap, while the other pyrrole clearly was located close to the methyl group of the strap (Figure 10). Finally, only the combination of high-quality experimental measurements (NMR, offline ECD) with high-level calculations (DLPNO-CCSD(T), SCS-CC2) permitted an unambiguous determination of the absolute configurations of the enantiomers of the free base **2HPM6**. These results showed that the enantiomer with a positive peak at ~430 nm did have the (R_P,*P*,*P*) configuration, while that with the negative signal had the (S_P,*M*,*M*) configuration (Figure 10).

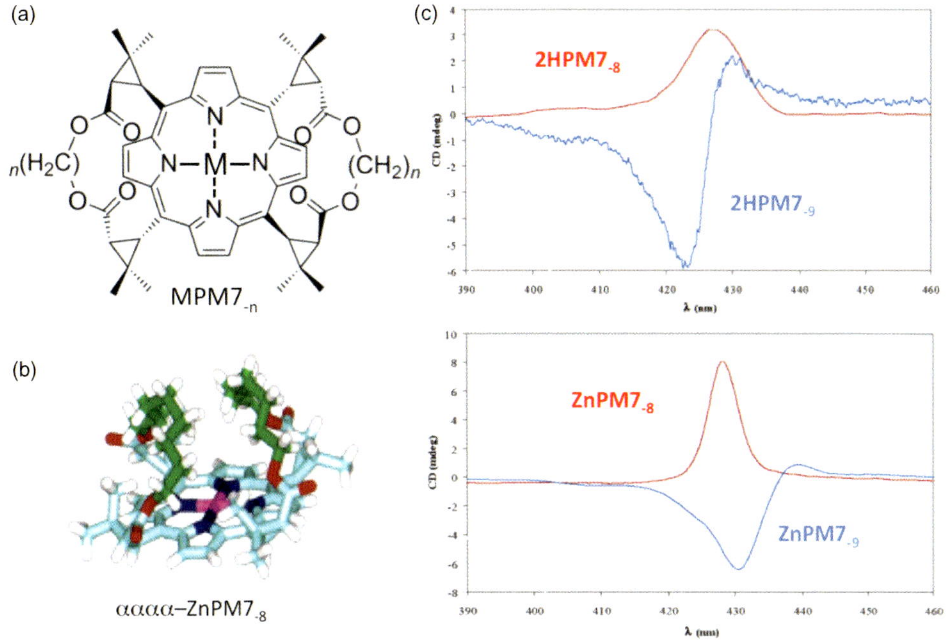

Figure 11. (a) Structure of **MPM7** derivative. (b) The lowest energy conformations of **ZnPM7$_{-8}$**. (c) CD spectra of **2HPM7$_{-8}$** and **2HPM7$_{-9}$** (up) and **ZnPM7$_{-8}$** and **2HPM7$_{-9}$** (down). Adapted from Refs. 50–52.

In 2001 Marchon, Gazeau and coworkers synthesized new "bridled" chiroporphyrins, **2HPM7**, in which two cyclopropyl were connected in adjacent *meso* positions by ester linkages of two *n*-methylene chains (Figure 11)[50,51] They demonstrated that the length of the methylene linker has a strong influence on the conformation of these molecules (**2HPM7-**n, n = 9–16). In particular when n = 9–16 methylene groups an $\alpha\beta\alpha\beta$ conformation is preferred; in contrast the **2HPM7$_{-8}$** with eight methylene bridles exhibits an $\alpha\alpha\alpha\alpha$ conformation. The authors then decided to combine CD spectroscopy and DFT calculations to characterize conformations of the Zn(II), Ni(II) and Cu(II) derivatives of **2HPM7** with eight **2HPM7$_{-8}$** or nine **2HPM7$_{-9}$** methylene groups (Figures 11 and 12).[52]

By DFT geometry optimizations, the αααα and αβαβ conformers were calculated to be the most stable form for the **ZnPM7$_{-8}$** and **NiPM7$_{-8}$** complexes, respectively, exhibiting a doming and a ruffling of the porphyrin macrocycle, as was also observed in the X-ray structures. The CD spectra were obtained for both the **MPM7$_{-8}$** and **MPM7$_{-9}$** porphyrins for all complexes (free-base and their square-planar nickel(II), copper(II) and zinc(II) derivatives; Figures 11 and 12). For **2HPM7$_{-8}$** and **ZnPM7$_{-8}$**, both exhibiting an $\alpha\alpha\alpha\alpha$ conformation, the CD spectra display a positive Cotton effect in the porphyrin absorption region. On the

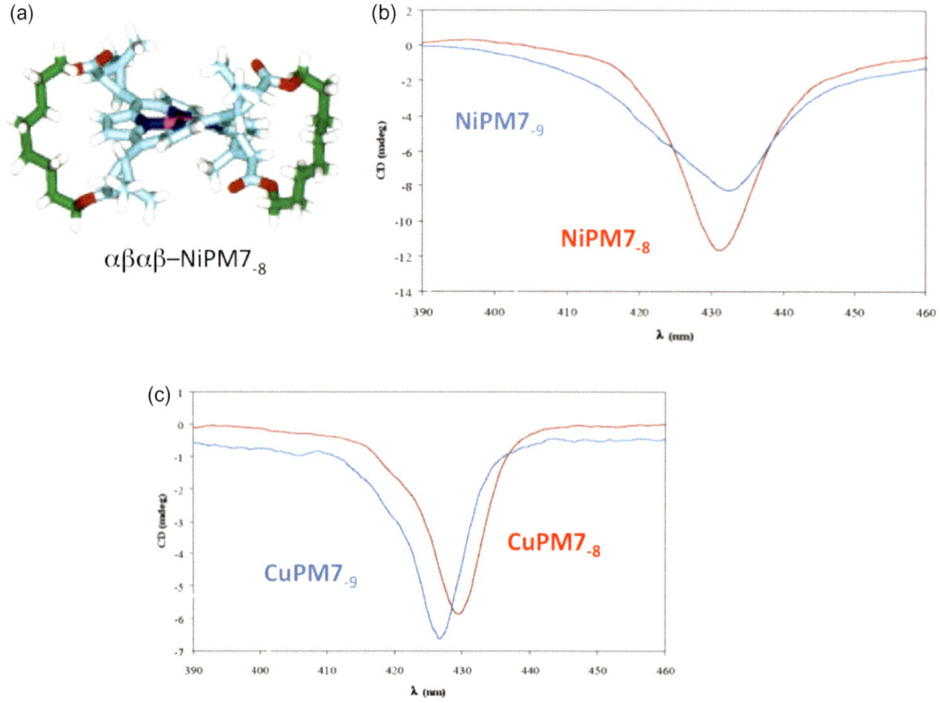

Figure 12. (a) The lowest energy conformations of **NiPM7**$_{-8}$. CD spectra of (b) **NiPM7**$_{-8}$ and **NiPM7**$_{-9}$ and (c) **CuPM7**$_{-8}$ and **CuPM7**$_{-9}$. Adapted from Ref. 52.

contrary, the CD spectra of both **2HPM7**$_{-9}$ and **ZnPM7**$_{-9}$, which have a $\alpha\beta\alpha\beta$ conformation, show a negative Cotton effect with a weak positive signal.

Thus, the sign and shape of the CD spectrum seems to be correlated with the molecular conformation. Indeed the $\alpha\beta\alpha\beta$ nickel complexes **NiPM7**$_{-8}$ and **NiPM7**$_{-9}$ both display an intense negative Cotton effect in the Soret region (Figure 12). Since both copper(II) complexes of Cu **CuPM7-*n*** (n = 8–9) display an intense negative Cotton effect (Figure 12), one can conclude that they possess an αβαβ conformation. Calculations performed on the basis of the crystal structures indicate that the CD spectra seem to arise from several competitive mechanisms of optical activity owing to a combination of both the inherent chirality of the porphyrin and extrinsic contributions of their chiral environment.

However, asymmetric syntheses of metalloporpyrins suffer from low yields for synthesis of both enantiomers. In order to eliminate this problem, Dubois and coworkers obtained the two enantiomers of "bridled" chiroporphyrins **2HPM8-*n*** by kinetic resolution using Amano PS lipase.[53] This time the authors linked two chiral *meso*-cyclohexyl substituents (instead of cyclopropyl, seen in **PM7**) with diester *n*-methylene, for the synthesis of the two enantiomers of the bridled (*R,S*)- or (*S,R*)-**2HPM8-*n*** (Figure 13). They obtained yields ranging from 3% for the shortest

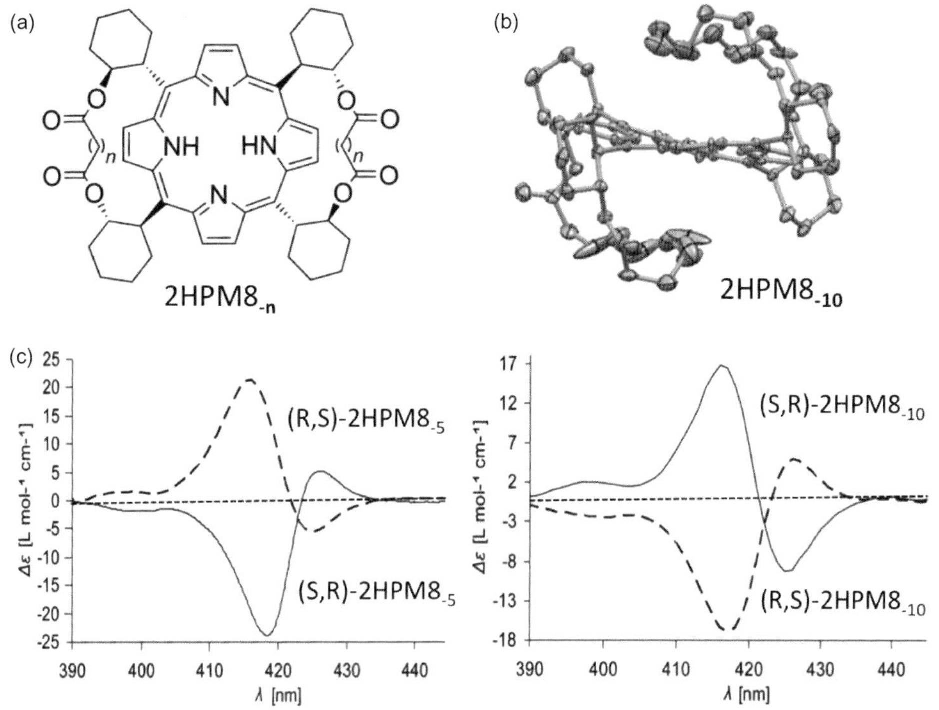

Figure 13. (a) Structure of **2HPM8**$_{-n}$, (b) X-ray structure of (*S,R*)-**2HPM8**$_{-10}$ and (c) CD spectra of two enantiomers of **2HPM8**$_{-5}$ (left) and **2HPM8**$_{-10}$ (right). Adapted from Ref. 53.

bridles (**2HPM8**$_{-5}$, $n = 5$) to 15% for the longer ones ($n = 8$ or 10). As reported for the previous bridled chiroporphyrins, NMR spectroscopy suggested that the conformation of **2HPM8-*n*** porphyrins is dependent on the bridle length. Indeed the porphyrin with the shortest straps, **2HPM8**$_{-5}$, exhibits an $(\alpha\alpha)(\alpha\alpha)$ conformation; in contrast, porphyrins with longer bridles (n = 8, 10) exist as mixtures of atropisomers. Nevertheless, it has been found that $(\alpha\alpha)(\beta\beta)$ atropisomer is the predominant species in the mixture and therefore the more stable atropisomer for **2HPM8**$_{-10}$ according to the crystal structure (Figure 13). CD spectra of the two enantiomers present mirror-image CD signals. In particular, negative Cotton effects are observed for $(\alpha\alpha)(\alpha\alpha)$ (*S,R*)- **2HPM8**$_{-5}$, whilst a positive signal is detected for $(\alpha\alpha)(\beta\beta)$ (*R,S*)-**2HPM8**$_{-5}$ (Figure 13). Even in the case of **2HPM8**$_{-10}$, mirror image CD spectra are observed for the two enantiomers.

3. *Tetrasubstituted Porphyrins*

One of the most interesting applications of chiral porphyrins is their use as catalysts for asymmetric synthesis. Significantly, structural information of a catalyst takes

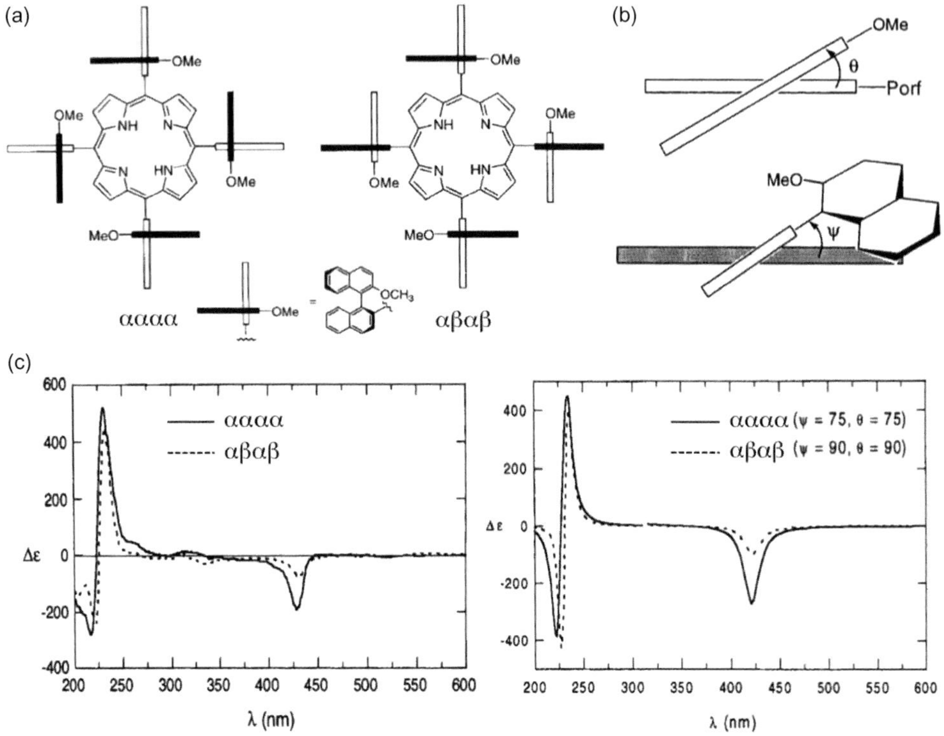

Figure 14. (a) Two atropisomers of **PM9**. (b) Definition of dihedral angles. (c) Experimental (left) and calculated (right) CD spectra for the **PM9** atropoisomers. Adapted from Ref. 54.

on great importance in order to rationalize the observed asymmetric induction, especially in view of the development of more efficient catalytic compounds.

With this in mind, Di Bari *et al.* analyzed the solution conformation of two intrinsically chiral atropisomeric *meso*-tetrakis[(aS)-28-methoxy-1,18-binaphth-2-yl]porphyrins, **PM9** (Figure 14), used as catalytic precursors in asymmetric synthesis, after analysis of data from a combination of experimental (^{1}H-NMR ROESY, UV-vis and CD) and computational (semiempirical structure optimization, DeVoe's coupled oscillators calculations) methods.[54]

One of the most prominent structural characteristics of the two **PM9** species is that they are endowed with multifold rotational symmetry: the (5α, 10α, 15α, 20α) isomer is C4, and the (5α, 10β, 15α, 20β) isomer is D2. This is a desirable property in catalysis, since it reduces the number of nonequivalent stereochemical pathways for approach of reactant and substrate to the catalyst.[55]

Upon measuring the CD spectra of the two atropoisomers (Figure 14), two main signals can be detected: one positive exciton couplet centered around 225 nm, due to the (S)-binaphthyls 1Bb absorption,[56,57] and a negative Cotton effect in the Soret band region (around 430 nm).[58] With a detailed comparison of

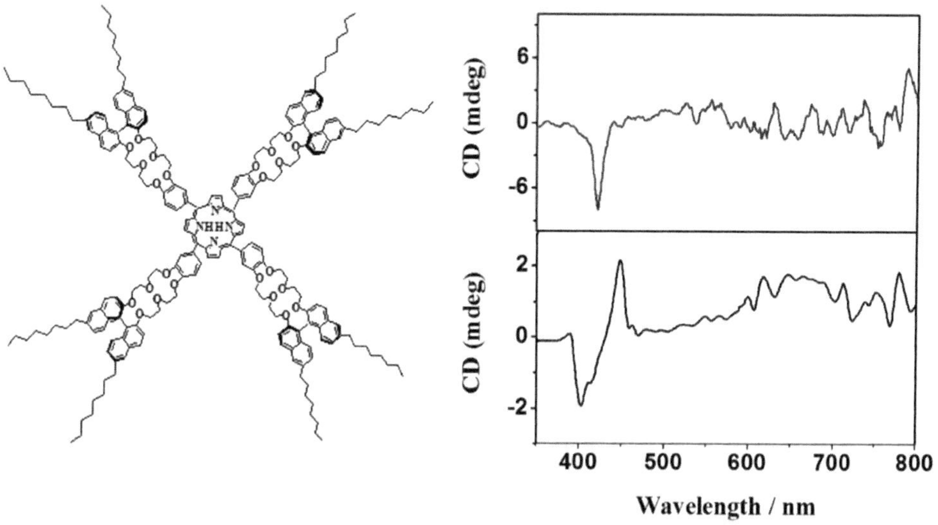

Figure 15. Structure of (*S*)-**PM10** (left) and CD spectra of the porphyrin in solution ($CHCl_3$, up) and LB film (bottom). Adapted from Ref. 59.

simulated and experimental spectra, the authors found the best match of the dihedral angles ($\theta = 75°$, $\psi = 75°$ for $\alpha\alpha\alpha\alpha$) and ($\theta = 90°$, $\psi = 90°$ for $\alpha\beta\alpha\beta$) for the two atropoisomers (Figure 14).

Lu *et al.* designed and synthesized (*S*)-*meso*-5,10,15,20-tetra-[(11,12:13,14-di(60-octyl-10,20-naphtho)-4,7,10,15,18,21-hexaoxacyclo-eicosaphenyl]-porphyrin **PM10** in order to prepare high quality monolayer and/or multilayer porphyrin films (Figure 15).[59] The authors confirm the transfer of chiral information from binaphthyl moieties to the porphyrin core by the presence of negative Cotton effect in the Soret absorption region centered at 424 nm (Figure 15). According to the binaphthyl linked metal-free porphyrin systems reported previously, appearance of a negative ICD signal in the Soret absorption regions indicates a right-handed configuration for the porphyrin derivative.[60,61] Moreover when Langmuir–Blodgett (LB) films were studied, a bisignate Cotton effect was observed in the CD spectrum of the LB film of the porphyrin, revealing that chirality is maintained in the LB film (Figure 15).

With the aim to improve the catalytic efficiency of a carbene ruthenium porphyrin already tested as a catalyst in asymmetric cyclopropanation,[62] Miyamoto and coworkers in 2008 obtained both enantiomers of carbene Ru porphyrin **PM11** by resolving a racemic mixture (Figure 16).[63] The authors characterized these porphyrins by X-ray diffraction and CD spectroscopy, determining the absolute configurations (Figure 16). The two enantiomers of **PM11** displayed mirror image CD spectra in the 300–700 nm range. Moreover, through the cyclopropanation of styrene with methyl diazoacetate, the enantiomeric porphyrins yielded 1-methyl-2-phenylcyclopropanecarboxylate with *trans/cis* ratios of 3:1 and turnovers of up

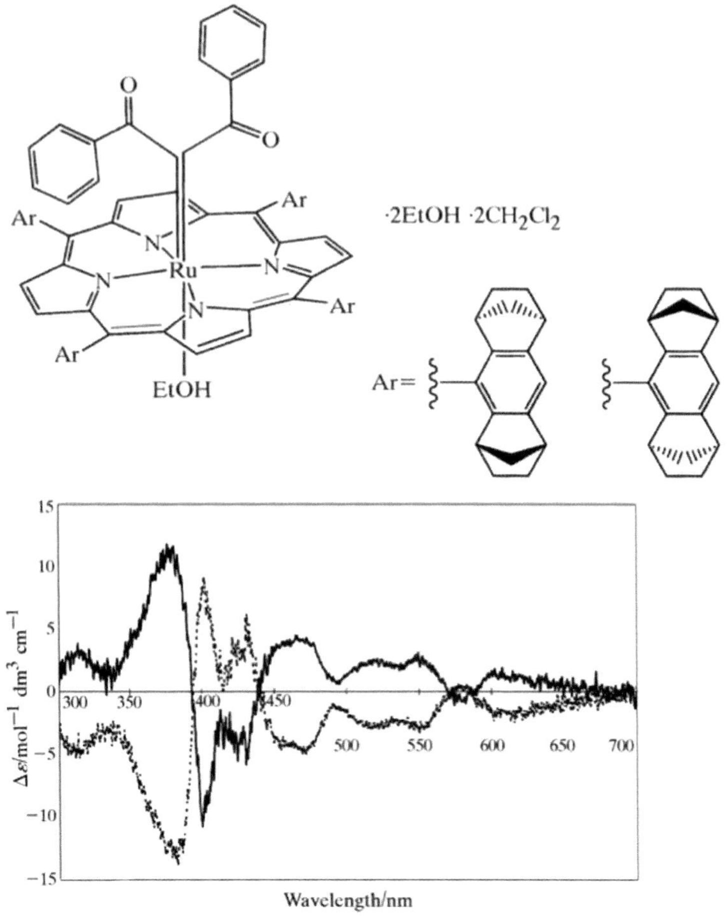

Figure 16. Structures of two enantiomers of carbene Ru porphyrin **PM11**. CD spectra of the two enantiomers of **PM11**. Adapted from Ref. 63.

to 3×10^4. In addition, **PM11** allowed enantioselective synthesis of the (*1S,2S*)-1-methyl-2-phenylcyclopropanecarboxylate in 75% ee, whereas its enantiomer gave the (1R, 2R) enantiomer in 72% ee.

Zhang and coworkers investigated the role of peripheral substituents of porphyrin chromophores in an attempt to provide new insight into chiral information transfer for synthetic conjugated systems at molecular and also supermolecular levels.[64] They noted that even after installing optically active molecules onto the periphery of the porphyrin molecule, it is not always possible to detect a CD signal in the whole porphyrin absorption range, due to the weak perturbation induced by the peripheral substituents with stereocenters.[65] Therefore these authors designed and synthesized a (*R*)-enantiomer of optically active metal-free porphyrin and its

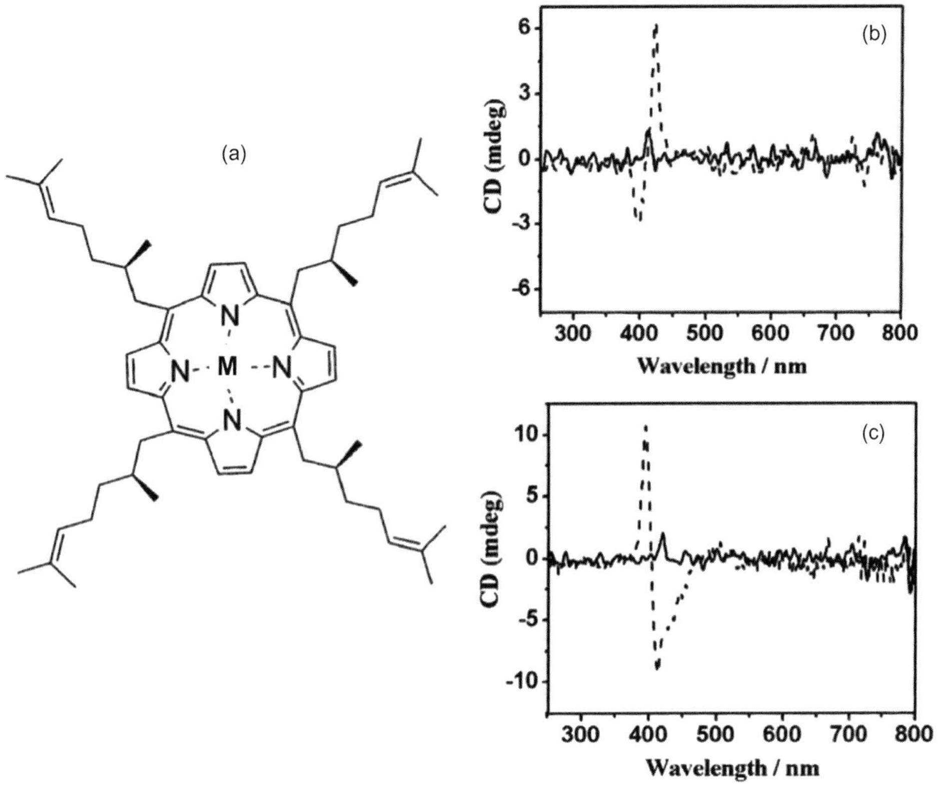

Figure 17. (a) Structure of citronellal porphyrin **MPM12** and CD spectra of (b) **2HPM12** and (c) **ZnHPM12**. **MPM12** in methanol (solid lines) and in 1:1 methanol/water mixture (dashed lines). Adapted from Ref. 66.

zinc derivative, bearing four chiral citronellal units linked directly onto the *meso* positions of the porphyrin ring (**PM12**; Figure 17), comparing the chiral features with a similar compound N,N′,N″-tris{21H,23H-5-p-aminophenyl-10,15,20-tris-[p-(S)-3,7-dimethyloctoxy-phenyl]porphyrin} 1,3,5-tricarboxamid.[66] Indeed, no signal was observed in the CD spectrum of the porphyrin used as comparison, indicating the lack of effective asymmetrical perturbation of the chiral citronellal side chains due to the relatively longer distance between the chiral citronellal side chains and the porphyrin core. In contrast, **2HPM12** and **ZnHPM12** displayed positive CD signals in the Soret absorption regions (Figure 17). This result indicates that the short distance between the chiral citronellal side chains and the porphyrin chromophore allows effective chiral information transfer. Dissolving the porphyrins in a 1:1 methanol/water mixture, the authors induced aggregation of the **PM12** porphyrins. As a result, upon measuring the CD spectra of the two com-

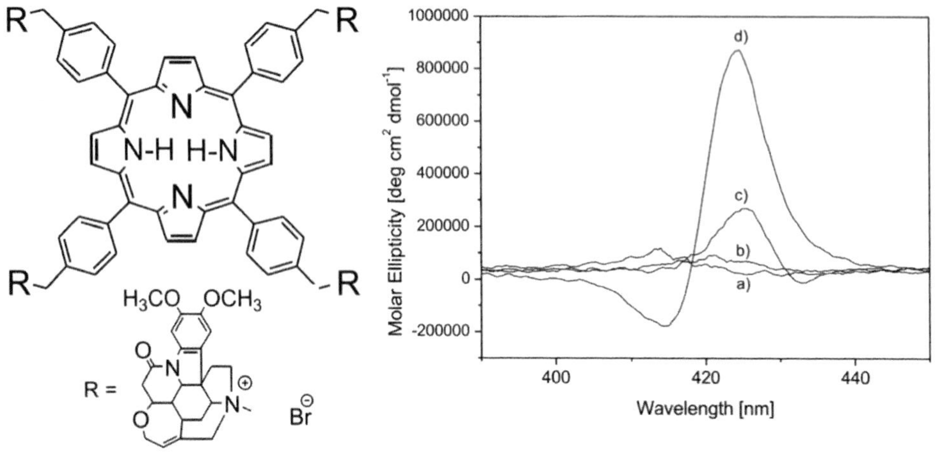

Figure 18. Structure of brucine porphyrin **PM13** and CD spectra of **PM13** in different solvents: (a) methanol, (b) dimethylsulfoxide, (c) isopropyl alcohol and (d) water. Adapted from Ref. 67.

pounds in a methanol/water (1:1) solvent mixture, positive and negative bisignate Cotton effects were detected: [399(–)/424(+)] for the free base and [411(–)/393(+)] for the Zn derivative, Figure 17).

The possibility of immobilizing the chirality by low-molecular weight organic compounds called gelators inspired Kral *et al.*[67] who have designed, synthesized and characterized (^{1}H NMR, Raman, infrared, electronic and vibrational CD spectroscopies) a novel porphyrin **PM13** with four brucine moieties in the *meso* positions, which forms an organogel with methanol even at extremely low concentration (Figure 18). The authors measured CD spectra of **PM13** in various solvents (Figure 18). Whereas the CD signals of the porphyrin are almost undetectable in methanol or dimethylsulfoxide, a weak CD signal was observed in isopropyl alcohol and a large CD signal was observed in water (Figure 18). Such spectral patterns imply the formation of chiral aggregates in this solvent.

B. Chiral Self-Assembly of Chiral Porphyrins

The design of chiral supramolecular multiporphyrin arrays displaying directional energy and electron transfer is a promising strategy for a broad range of potential applications in areas such as chlorophyll mimic systems and photovoltaic cells,[68] sensors,[69] optoelectronics,[70] nanoelectronic devices[71] and also nonlinear optical materials.[72] For this purpose, an optimal mesoscopic organization of the constituent donor and acceptor chromophores is essential for achieving efficient energy or charge-transfer processes. Therefore, CD spectroscopy is a useful technique to investigate the arrangement adopted by assembled porphyrins. One of the first examples of chiral

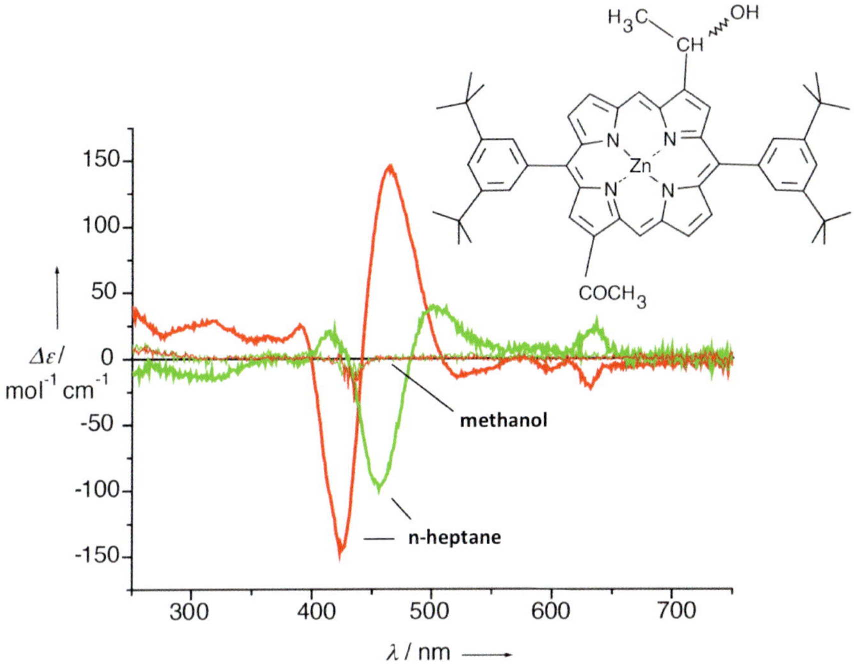

Figure 19. Structure of **PM14**. CD spectra of the enantiomerically enriched (*R*)- and (*S*)-**PM14** after self-assembly in *n*-heptane (thick red and green lines) and after addition of methanol which disrupts the assemblies (thin red and green lines overlapping the 0 line). Adapted from Ref. 73.

self-assembled structures originated by the aggregation of chiral porphyrins was reported by Balaban *et al.* in 2003.[73] The authors synthesized a new chiral Zn(II) porphyrin **PM14** with different substituents in the β positions (acetyl and hydroxyethyl) and two 3,5-di-*tert*-butylphenyl groups in the 5- and 15-*meso* positions (Figure 19).

With both enantiomers of **PM14**, self-assembly was induced in *n*-heptane and the CD spectra of both fractions were recorded at room temperature. Intense bisignate CD signals are observed in the porphyrin Soret band absorption region. Notably, the two fractions show mirror-imaged CD signals, indicating that the different configurations (R or S) of the hydroxethyl group induce the different chirality in the self-assembled species (Figure 19). When methanol was added, a complete disassembly of the porphyrin structures was detected as evidenced by the disappearance of the intense CD signals due to the formation of the monomeric **PM14**-methanol adducts. Indeed, methanol competes for the zinc ligation and totally disrupts the hydrogen bonding network between Zn–OH and the carbonyl group in conjugation with the porphyrin ring. This indicates that the observed CD signal does not represent the chirality of the monomeric porphyrins, but the supramolecular helical assemblies (M or P) of the two enantiomers.

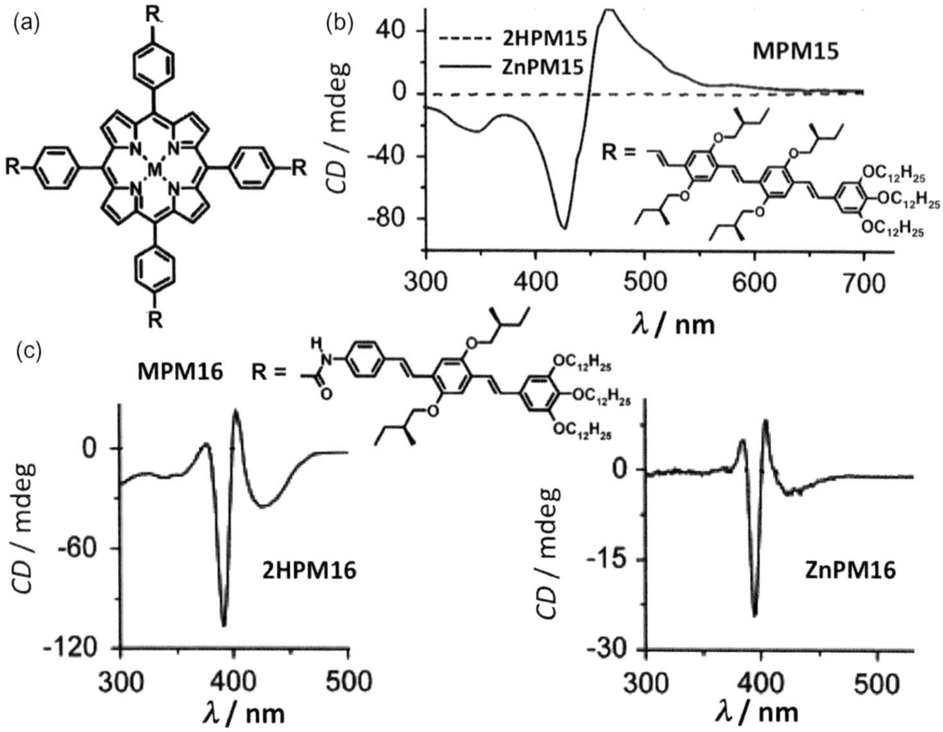

Figure 20. (a) Structures of porphyrins **MPM15** and **MPM16**, (b) CD spectra in methylcyclohexane of **2HPM15** (dashed line) and **ZnPM15** (solid line) and (c) CD spectra in methylcyclohexane of **2HPM16** (left) and **ZnPM16** (right). Adapted from Ref. 74.

A comparative study on the aggregation behavior of oligo(*p*-phenylenevinylene)-appended porphyrins containing all *trans*-vinylene **PM15** or amide **PM16** linkages was presented by Meijer and coworkers (Figure 20)[74] who demonstrated that the type of supramolecular arrangement obtained in organic solvents strongly depends on the nature of the covalent connection. In particular, in the case of all trans-vinylene linkages, a J-type intermolecular packing is obtained and the assemblies are only of moderate stability. However, upon dissolving the **2HPM15** in methylcyclohexane, the π–π stacking of the conjugated backbones results in a bisignate Cotton effect corresponding to the UV-vis absorption. For the Zn derivative **ZnPM15**, the absence of a Cotton effect indicates that aggregation occurs in an achiral fashion or without preferential helicity (Figure 20). Conversely, the supramolecular structures obtained from the amide-linked **PM16** display an H-type stacking arrangement of enhanced stability as a consequence of intermolecular hydrogen bonding along the stack direction, favorably interlocking the stacked building blocks. Both free base **2HPM16** and the Zn derivative **ZnPM16** systems display supramolecular chirality in the assembled state as observed from their similar CD spectra (Figure 20).

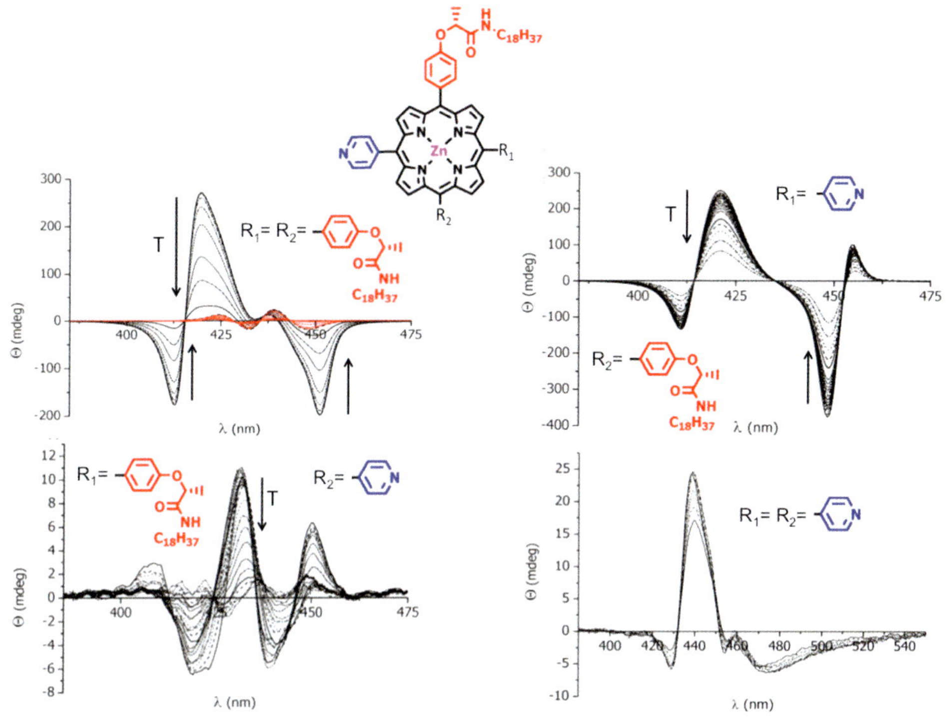

Figure 21. Structures of **ZnPM17a–17d** and variable temperature CD spectra. Adapted from Ref. 75.

Amabilino and coworkers[75] exploited the coordination of zinc porphyrins with axial nitrogen derivative ligands such as pyridyl groups and described the formation of well-organized macromolecules. In particular, they designed and synthesized several *meso*-substituted Zn porphyrins **ZnPM17a–17d** having different numbers (from 1 to 3) and positions of the pyridyl and phenyl alkyl amide (4-(R)-2-*N*-octadecylamidoethyloxyphenyl) groups (Figure 21), in order to study the influence of the proportion of stereogenic centers in the side groups connected to the chromophore ring in their optical activity and the important role of pyridyl groups in the self-assembly of these chiral macrocycles.[75] This arrangement allows for the self-assembly of these chromophores by more than one interaction. Indeed, two different coordinating atoms (the pyridyl nitrogen atom or the carbonyl oxygen atom) can in principle bind the metal ion; furthermore, hydrogen-bonding between amide groups can be the main noncovalent interaction to form an aggregate. Solutions of **ZnPM17a–7d** in methylcyclohexane/3% $CHCl_3$ were prepared (at a concentration of 5 μM), and CD spectra were recorded over a range of temperatures from 338 to 263 K after a heating and cooling cycle to ensure that equilibrium conditions applied (Figure 21). The temperature dependence of CD spectra for compounds with only one pyridyl group shows two clearly distinct states of self-assembly. A weak Cotton effect arises at the position of the band associated with the pyridyl-Zn(II) porphyrin complex, suggesting a perpendicular disposition of the two metalloporphyrins and a thermodynamically stable aggregate in the range of temperatures from 300.5 to 338 K. When the temperature was decreased, a highly intense positive Cotton effect and a huge negative CD signal indicate extension of the interchromophore interactions. The CD spectra of Zn porphyrin having in 10, 20-*meso* positions two pyridyl groups were simpler than those of the metalloporphyrin with only one pyridyl group, presumably because of the higher symmetry in the former compound (Figure 21). Thus the CD spectra of Zn-porphyrin with two pyridyl groups in the 15, 20-*meso* positions show two bisignate signals of low intensity. This pattern is similar to that of a compound with only one pyridyl group above 300 K that the authors associated to a cyclic tetramer. In contrast, the CD spectra of a compound with only one stereogenic center in its structure and three pyridyl groups reflects an environment of the molecules that is different from the homologs principally because the solubility of the compound is too low in the solvent used. Therefore, the assembly of these compounds is extremely sensitive to the position of substituents.

The solvent effect on the aggregation of chiral porphyrins has been studied by Monti *et al.* for a mono-*meso* substituted porphyrin with a (L)-prolinium moiety **PM18** (Figure 22).[76] They analyzed the aggregation process of **PM18** in mixed water–ethanol solvent mixtures. In the composition range of 100% to 50% ethanol (v:v) the macrocycle is in the monomeric form. The self-aggregation process starts as soon as the water proportion is increased. CD spectroscopy reveals that aggregation promotes the formation of large chiral suprastructures depending on the composition of the solvent (Figure 22). Indeed, the aggregation induced in 75% H_2O results in the formation of porphyrin suprastructures featuring signals when are 1–2 orders of magnitude larger than those obtained at a higher water proportion (90% H_2O). The authors, by kinetic studies, ascribe this effect to the different rates of aggregation processes. In water–ethanol 90:10 (v:v), for example, the aggregation of **PM18** is complete within the time of mixing and results in the formation of aggregates with a poorly defined structure. On the contrary, at higher ethanol proportions (*i.e.* 75% H_2O), a slower aggregation process occurs, resulting in the formation of J-type aggregates of **PM18**.

Starting from these interesting results, Monti *et al.* decided to induce the aggregation of (L)-prolinium cationic porphyrin **PM18** in the presence of chiral aggregates of the anionic L-proline porphyrin derivative **PM19** in water/ethanol 75–25 (v:v) mixtures (Figure 22).[77] This experiment promotes an efficient hetero **PM18–PM19** self-aggregation process to form J-type aggregates, as evidenced by UV-vis and CD spectra (Figure 22). Interestingly, a remarkable amplification of

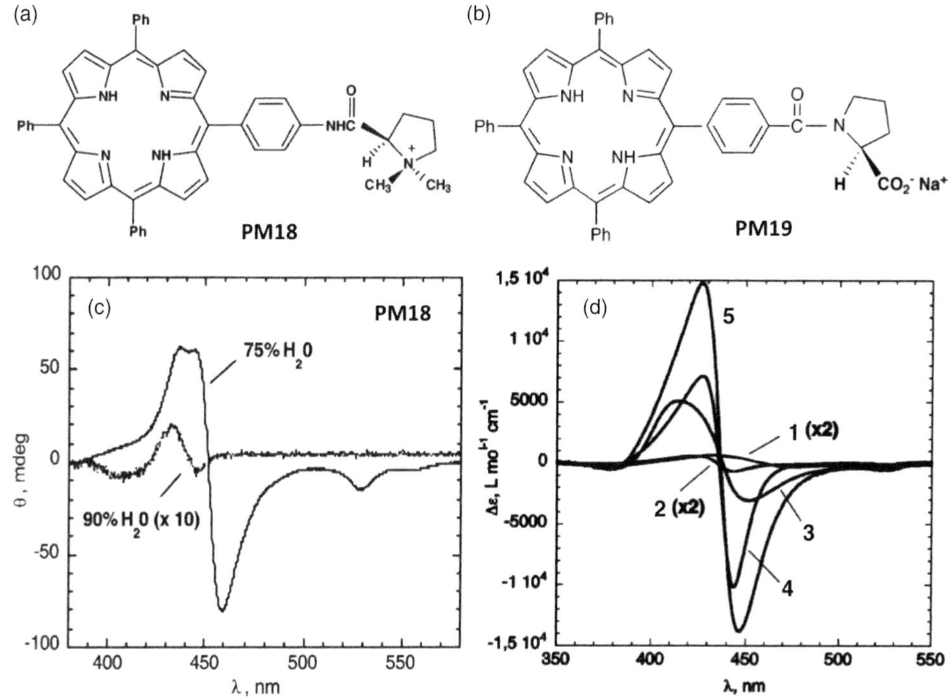

Figure 22. Structures of (a) cationic mono-*meso* substituted porphyrin with (L)-prolinium moiety **PM18,** (b) anionic L-prolin porphyrin derivative **PM19**, (c) CD spectra of cationic **PM18** in different solvent compositions and (d) CD spectra in H_2O/EtOH (75/25 v/v) of: **1** cationic **PM18**, **2** anionic **PM19** alone and **3**, **4**, **5** at different concentrations (5 μM, 10 μM, 20 μM) of cationic **PM18** in the presence of anionic **PM19** (5 μM). Adapted from Ref. 76.

the CD intensity is observed compared with what is detected for cationic **PM18** alone. This should indicate a strong template effect exerted by the presence of the anionic assemblies, onset by electrostatic forces.

Ihara and coworkers examined the association behavior of L-glutamide-functionalized tetraphenylporphyrin **PM20** (Figure 23) in different solvents in order to promote the self-assembly.[78] In particular they found that the porphyrin can be dissolved in various kinds of organic solvents such as acetonitrile, methanol, ethanol, DMF, THF, chloroform and benzene. However, the addition of cyclohexane converted the solution of **PM20** to a gel state. Therefore the authors decided to dissolve **PM20** in a cyclohexane–THF (20:1) mixture. Under these condition, the TEM analysis showed well-developed fibrous networks. CD spectroscopy revealed a tetrasignate CD signal in the 375–460 nm region upon increasing the concentration of **PM20** (Figure 23). A plot of the CD signal at 417 nm *vs.* concentration of **PM20** indicated that the critical aggregation concentration (cac) of **PM20** is around 0.1 mM in a cyclohexane–THF (20:1) mixed system at 10 °C.

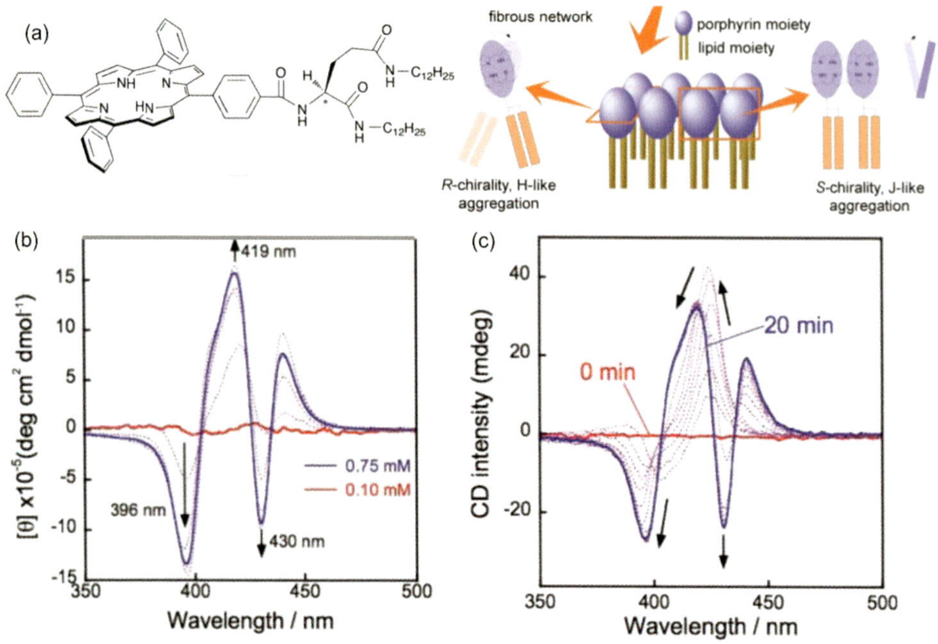

Figure 23. (a) Structure of L-glutamide-functionalized tetraphenylporphyrin **PM20** and schematic representation of gel state. (b) CD spectra of **PM20** in a cyclohexane–THF (20:1) mixture at 10 °C (from 0.1 mM to 0.75 mM). (c) Time-course of CD spectral change of **PM20** (0.25 mM) in a cyclohexane–THF (20:1) mixture at 10 °C. Adapted from Ref. 78.

Detailed investigation of the CD spectrum of a 0.2 mM solution of **PM20** in a cyclohexane–THF (20:1) mixture was performed in order to describe the complex Cotton effects observed. A time-course of the CD reveals hierarchical spectral changes. The CD pattern centered at 398 nm can be assigned to hypsochromically shifted H-like aggregates of **PM20** and this is also assigned to right-handed (R-) chirality, while another centered at 428 nm is assigned to the bathochromically shifted J-like aggregates and left-handed (S-) chirality according to Symonyi *et al.*'s explanation (Figure 23).[79]

Zhang and coworkers analyzed the effect of solvent on the morphology and chirality of aggregates of optically active Zn porphyrin–pentapeptide (glycinyl–alanyl–glycinyl–alanyl–glycine) conjugate **PM21** (Figure 24).[80] By electronic absorption, CD, IR, TEM, SEM and X-ray diffraction techniques they demonstrated that the self-assembly of **PM21** leads to the formation of nanofibers with right-handed helical arrangement in a nonpolar THF/*n*-hexane (1:3) system and nanotubes with a left-handed helical arrangement in the polar THF/water (1:3) system. In particular, the CD spectrum of **PM21** aggregates formed in THF/*n*-hexane (1:3) shows a positive bisignated Cotton effect [424(–) /448(+)], with the

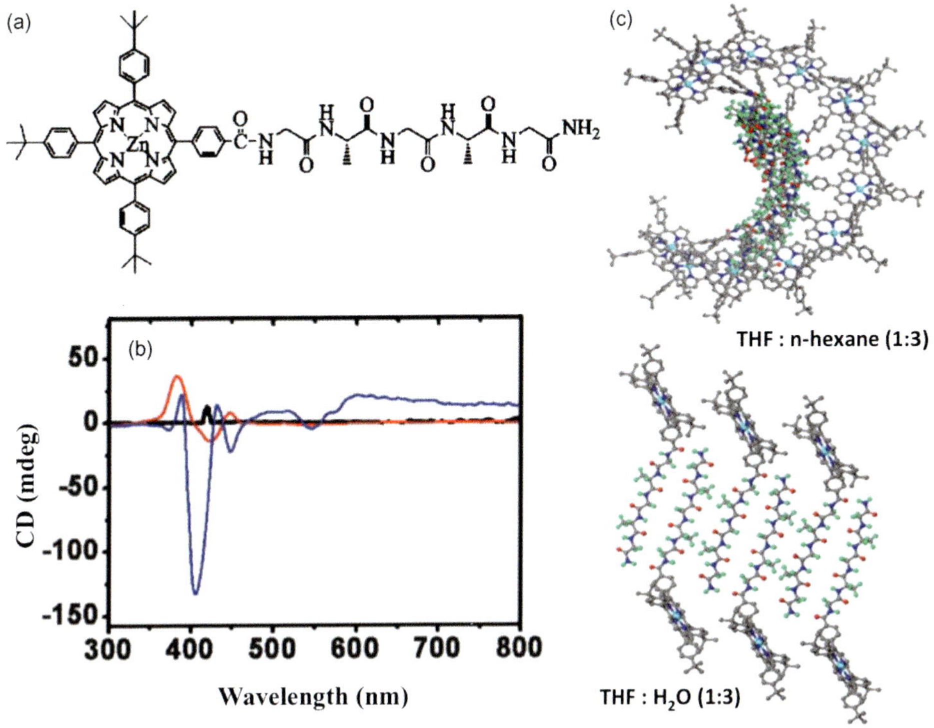

Figure 24. (a) Structure of Zn porphyrin–pentapeptide (glycinyl–alanyl–glycinyl–alanyl–glycine) conjugate **PM21**. (b) CD spectra of **PM21** (black line), its nanostructures fabricated from THF/*n*-hexane (1:3) dispersed in *n*-hexane (red line) and its nanostructures fabricated from THF/water (1:3) dispersed in water (blueline). (c) Schematic illustration of a proposed mechanism for the formation of **PM21** nanostructures. Adapted from Ref. 80.

crossover at 439 nm (Figure 24). On the other hand, two negative bisignated Cotton effects [405(–)/388(+), 448(–)/432(+)], with the crossovers at 395 nm and 443 nm were observed in the CD spectrum of **PM21** aggregates formed in THF/water (1:3) (Figure 24). According to the exciton chirality method[81] the given sign of coupling and the direction of dipole moment suggest a right-handed helical arrangement of **PM21** in aggregates fabricated in *n*-hexane and a left-handed helical arrangement of **PM21** in aggregates fabricated in water. As clearly revealed previously, peptides usually employ α-helix and antiparallel β-sheet secondary conformations in nonpolar and polar solvent systems, respectively.[82,83] As a result, when the **PM21** conjugate starts to aggregate in a nonpolar THF/*n*-hexane (1:3) system, the hydrophilic pentapeptide chain forms the interior of the helix, while the hydrophobic porphyrin chromophore constitutes the periphery of the helix resulting in the right-handed helix in a stack of porphyrin chromophores. On the other hand, the pentapeptide chain adopts an antiparallel β-sheet secondary conformation in the polar THF/water (1:3) system, which induces formation of a dimeric supramolecular bilayer structure depending on the intermolecular hydrogen bonding interaction between the pentapeptide chains of neighboring porphyrins. Then the interactions between dimeric building blocks lead the formation of nanotubes with a left-handed helical arrangement (Figure 24).

Exploiting the potential to form organogels of tetraphenylporphyrin **PM22** bearing three tails with stereogenic carbon followed by an achiral chain with an amide group (Figure 25), Casado and coworkers studied the conformation of aggregates by means of CD and VCD.[84] Gels were prepared with an increasing concentration of **PM22** in methylcyclohexane followed by a heating the solution to boiling and by rapid cooling to room temperature and then analyzed by CD for the (*R*,*R*,*R*) enantiomer of **PM22**. At low concentration (100 mM) the CD spectrum shows a bisignate feature centered at 400 nm and mirror-like behavior for the two enantiomers of **PM22** (Figure 25). To gain further insight into the electronic structure, they performed molecular modeling calculations building a molecular stack of 74 (R,R,R) **PM22** disposed helicoidally in a counter-clockwise disposition. Calculations predicted the appearance of absorption bands with CD activity in the

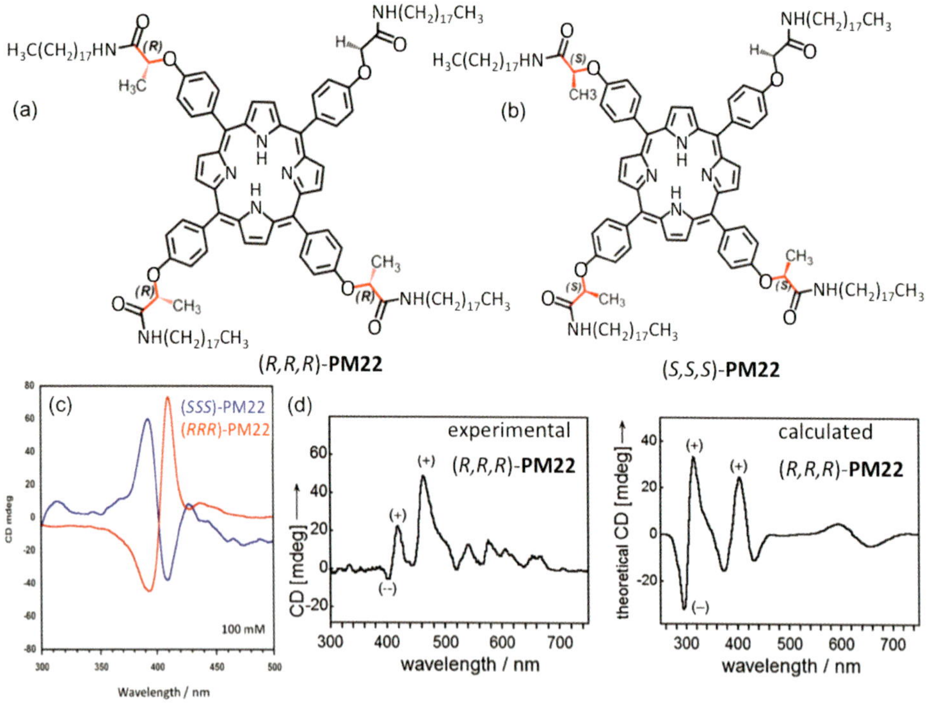

Figure 25. Structures of (a) (*R*,*R*,*R*)-**PM22** and (b) (*S*,*S*,*S*)-**PM22**, (c) CD spectra of the two enantiomers of **PM22** at 100 mM in methyl-cyclohexane; (d) experimental (left) and calculated (right) CD spectra of (*R*,*R*,*R*)-**PM22**. Adapted from Ref. 84.

400–800 nm region, in good agreement with the experimental CD spectrum recorded for (*R,R,R*)-**PM22** (Figure 25). This indicates that the *R,R,R*-enantiomer aggregates in the organo-gel in a counter-clockwise orientation and with left-handed helix. On the other hand, the (*S,S,S*)-**PM22** should yield a clockwise orientation and right-handed helix.

Another example of chiral assembly of chiral porphyrins was presented by Scolaro and coworkers using **PM23** (*meso*-5,10,15,20-tetra-(2,6-dimethylheptyl) porphyrin; Figure 26)[85] a porphyrin bearing, in the *meso* positions, four (*R*) or (*S*) hydrogenated citronellal units. According to previous reports,[59] CD spectra recorded on the two enantiomeric forms in the non-aggregating solvent (CH_2Cl_2) show no induced CD signals in the porphyrin absorption region, indicating the absence of any asymmetrical perturbation from the chiral alkyl side chains on the aromatic ring (Figure 26). On the contrary, in line with the hydrophobic properties of the alkyl chains present on the periphery of the macrocycle, changes were observed in the spectroscopic behavior of **PM23** when an acetone/water mixture was used as solvent. In particular, the Soret band of the porphyrin, centered at 415 nm in the neat acetone, appears to split into two broad components centered

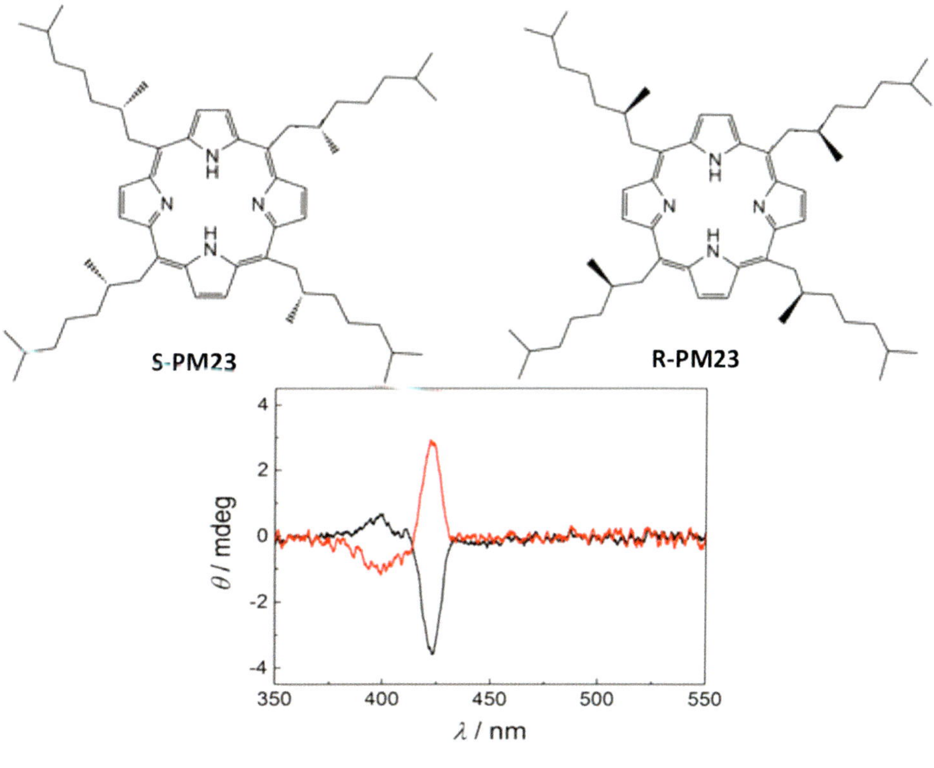

Figure 26. Structure of two enantiomeric forms of **PM23** and their CD spectra in 40/60 v/v acetone/water. Adapted from Ref. 86.

at 404 nm and 420 nm, respectively. According to exciton theory of McRae and Kasha,[86] the splitting of the Soret band is ascribable to an intermolecular stacking interaction among the porphyrin rings, revealing H- and J-type molecular aggregates. Typical CD spectra performed on such porphyrin aggregates show signals that are clearly correlated with the enantiomorphism of the stereocenter present on the alkyl chains in the periphery of the macrocycle, with negative and positive bisignate Cotton effects for (*S*)- and (*R*), respectively (Figure 26).

Meijer and Shenning investigated the chiral memory effect at the supramolecular level by the selective depolymerization of porphyrin aggregates.[87] They first induced coaggregation of the chiral and achiral metalloporphyrins in methylcyclohexane and after treatment with the Lewis base quinuclidine (QND), leading to depolymerization, observed retention of the initial chirality. The same authors synthesized zinc and copper *meso*-tetraphenylporphyrins functionalized with chiral amide ***c*-MPM24** and achiral amide ***a*-MPM24**

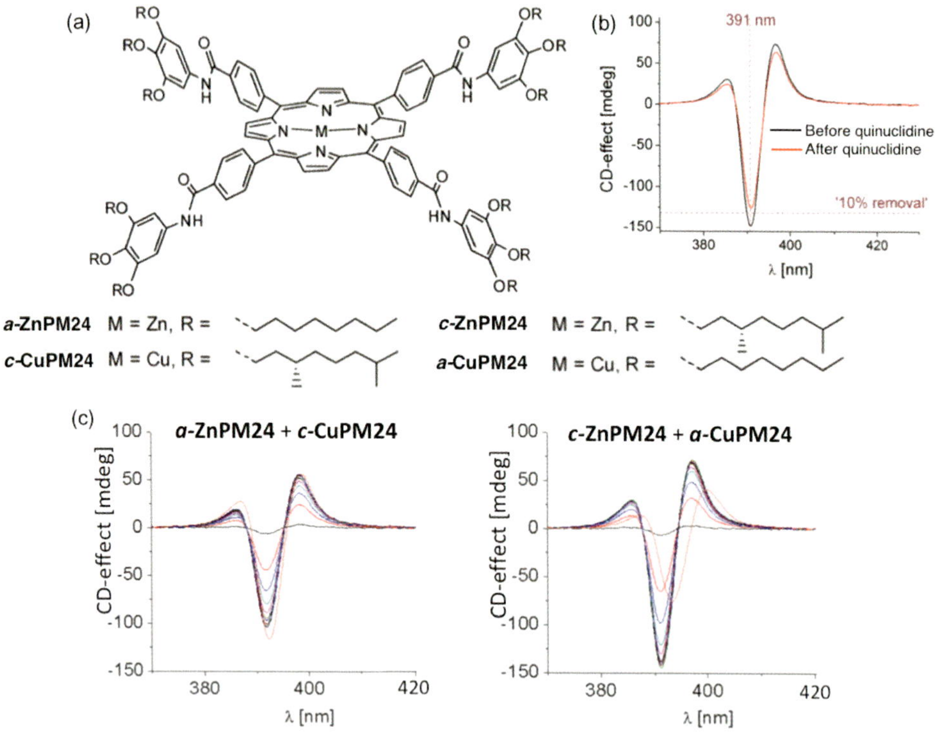

Figure 27. (a) Structure of chiral ***c*-MPM24** and achiral ***a*-MPM24**, (b) CD spectra of mixed chiral ***c*-ZnPM24** and achiral ***a*-CuPM24** before and after extraction with QND and (c) CD spectra of mixed achiral ***a*-ZnPM24** and chiral ***c*-CuPM24** (left) and mixed chiral ***c*-ZnPM24** and achiral ***a*-CuPM24** (right). Adapted from Ref 88.

(Figure 27), being familiar with different expected behavior of the two metalloporphyrins[88] in the presence of a Lewis base indeed axial ligation occurs predominantly with Zn-porphyrin monomers in the presence of the Lewis base quinuclidine, leading to depolymerization. The authors first studied the system with a chiral Cu-porphyrin and an achiral Zn-porphyrin showing that according to the "sergeant-and-soldiers" principle, increasing the fraction of chiral porphyrin yields a strong amplification of the helical aggregate with a single handedness (Figure 27). Similar results are observed in the reverse experiment using chiral Zn-porphyrin and achiral Cu-porphyrin (Figure 27). This latter system was used for the selective removal of chiral Zn-porphyrin by axial ligation with Lewis base QND. The helicity observed in the remaining "soldier" supramolecular polymer indicates a chiral memory effect that is obtained by removing only the chiral building block from the helical amplified stacks after self-assembly (Figure 27).

Noteworthy, the remarkable inertness of these systems at room temperature prompted the authors to investigate the possibility of releasing and restoring the chiral memory by a series of heating and cooling experiments on solutions of imprinted chiral stacks containing achiral ***a*-CuPM24**. In particular, in the presence of the Lewis base QND and after removal of the chiral ***c*-ZnPM24** from the coaggregate, heating and subsequent cooling from 60 °C produced a partial recovery of the chiral memory. At the relatively high temperature of 60 °C, the conformational inertness of the remaining assemblies was sufficient for these "achiral ***a*-CuPM24** seeds" to maintain their imprinted chirality, allowing enantiospecific self-assembly of achiral ***a*-CuPM24** upon cooling.

Another example of switching chiral information obtained by self-assembling and disassembling of chiral porphyrins has been reported by Balaban and coworkers.[89] They prepared two enantiomers of *meso*-tetraphenylporphyrin dipeptide conjugates **PM25** by covalently linking to the C-terminus of diphenylalanine (Figure 28), demonstrating that these porphyrins self-assemble to fibrils, platelets or nanospheres depending on the solvent composition. The fibrils show the most interesting properties owing to the possibility to switch the exhibited chirality by changing the polarity of the solvent. By modifying the solvent conditions, they observed a different morphology of the **PM25** aggregates, in particular in dichloromethane, which is a "good solvent," as the porphyrin conjugates are in monomeric form. Then "injecting" into a dry nonpolar solvent such as *n*-heptane, they observed broadened, red-shifted and split Soret bands typical of porphyrinic J-aggregates. With a 1:1 ratio of dichloromethane to *n*-heptane, they observed spheres with diameters ranging from 400 nm to 600 nm coexisting with fibrils. They recorded the formation of micrometer long fibrils in the case when the dichloromethane: *n*-heptane ratio was 3:7, then with a ratio of 2:8, larger platelets

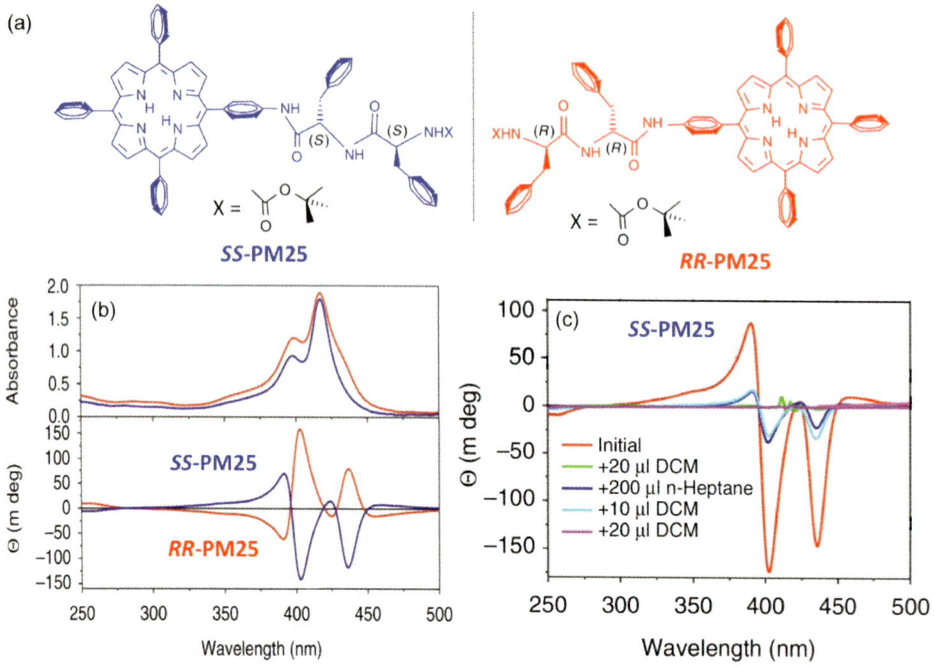

Figure 28. (a) Structures of enantiomers of *meso*-tetraphenylporphyrin dipeptide **PM25**, (b) UV-vis and CD spectra of (*R,R*) (red curve) and (*S,S*) (blue curve) self-assembled **PM25** in dry *n*-heptane and (c) CD spectra of (*S,S*)-**PM25** when changing the solvent composition as indicated in the inset. Adapted from Ref. 89.

appeared. After self-assembly of **PM25**, very intense and multiple Cotton effects prove that the porphyrin chromophores are excitonically coupled with almost equally strong negative Cotton effects at 403 nm and 437 nm (Figure 28). By adding small aliquots of dichloromethane, the solvated self-assemblies change their morphology gradually, until an almost silent ECD spectrum is obtained (Figure 28). Remarkably, further addition of *n*-heptane restores the exciton couplets.

IV. Chiral Porphyrin Dimers

A. Indirectly Linked Chiral Porphyrin Dimers: Porphyrin Dimers with a Chiral Bridge

Bisignate CDs with a positive and a negative component are typically observed as a result of exciton coupling in systems of two chirally-oriented chromophores. For systems with two identical chromophores, a conservative couplet with rotational strengths of equal magnitude is predicted. The bisignate couplet has been attributed to the through-space exciton coupling of the Soret band

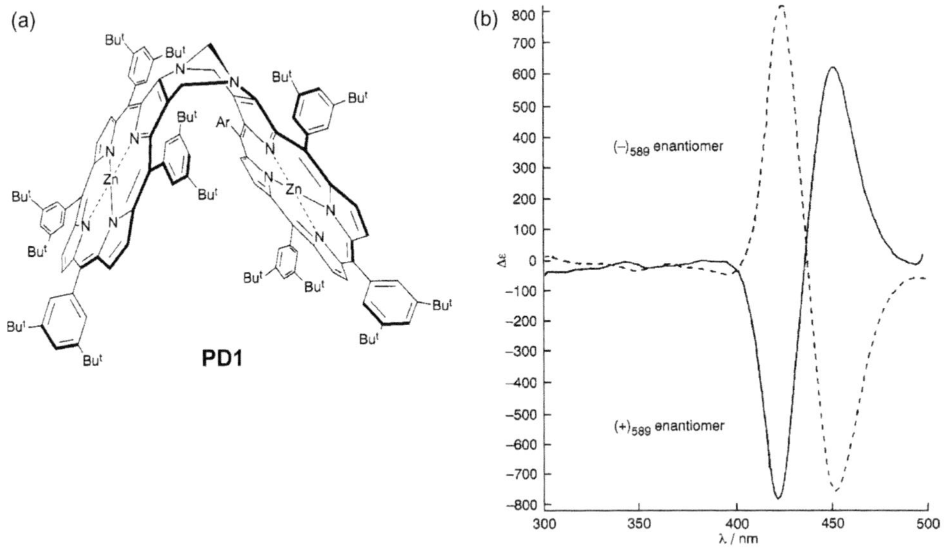

Figure 29. (a) Structure of chiral Zn(II) bisporphyrins **PD1** and (b) Soret band CD spectra of the enantiomers of Zn(II) bisporphyrin **PD1** in $CHCl_3$. Adapted from Ref. 29 with permission of the Royal Society of Chemistry.

transitions of two porphyrins, specifically coupling between B transitions of 5,15 direction).

In 1995, Crossley *et al.* reported the synthesis and chiroptical properties of chiral bisporphyrin **PD1** with a C2 symmetry (a Troger's base analogue, Figure 29a) prepared by chiral resolution by HPLC.[29] The (–)-enantiomer was obtained in 100% e.e. and the (+)-enantiomer with ≥98% e.e ($[\alpha]_D = -2000$; $CHCl_3$). The resolved enantiomers (+)-**PD1** and (–)-**PD1** displayed extremely high values of specific rotation $[\alpha]_D = +/- 2000$ ($CHCl_3$) consistent with their helicity. Their CD spectra showed mirror-image CD couplets in the Soret band region centered around 420 nm and 452 nm with very high molar ellipticities ($[\theta]_{425} = 2.7 \times 10^6$ deg cm^2 $dmol^{-1}$) (Figure 29b). Based on the exciton chirality method developed by Nakanishi, Harada and coworkers,[5,90,91] these authors proposed that the splitting was due to the chiral exciton coupling of the two identical chromophores oriented clockwise or counterclockwise. They used the signs of the bisignate CD signals to assign the absolute configuration of (–)-**PD1** as being M-enantiomer.

In 2014, a flexible chiral Tröger's base cobalt-bis-porphyrin derivative **PD2** (Figure 30) was explored by Dolensky and coworkers as a potential enantioselective receptor of amino acid methyl esters.[31] Bis-porphyrin **PD2** was synthesized as a racemic mixture and separated into enantiomers by chiral HPLC. CD spectra revealed bisignate mirror-image CD couplets in the porphyrin Soret band region

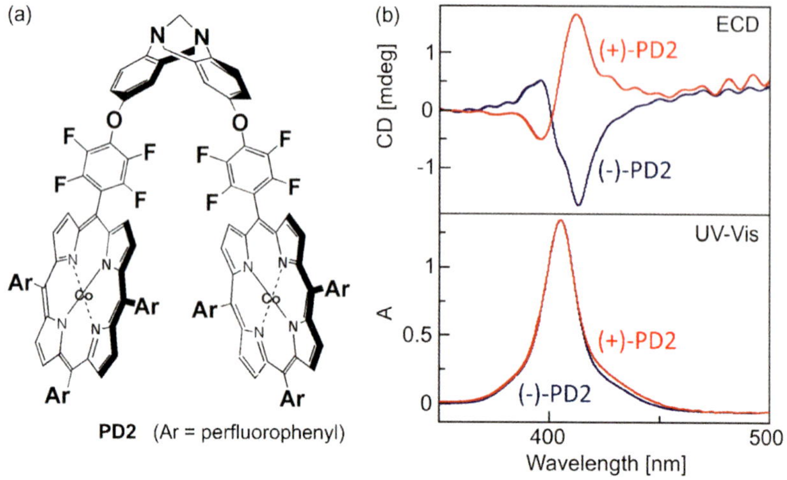

Figure 30. (a) Structure of the chiral bis-Co-porphyrin **PD2** and (b) CD and UV-vis spectra of (+)-**PD2** and (–)-**PD2** in CH_2Cl_2. Adapted from Ref. 31 with permission of Wiley-VCH.

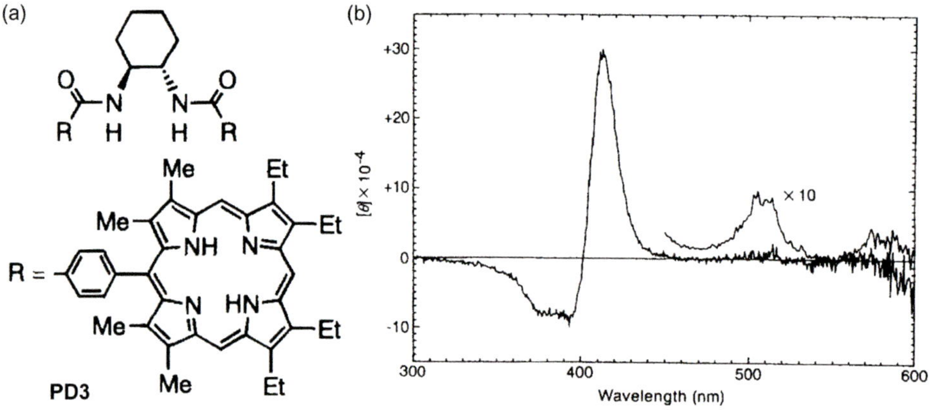

Figure 31. (a) Structure and (b) CD spectra of chiral porphyrin dimer (1*S*,2*S*)-**PD3** in $CHCl_3$. Adapted from Ref. 92 with permission of Elsevier.

for (+) and (–) enantiomers with a positive CD band at ~412 nm and a negative CD band at ~395 nm for (+)-K (Figure 30). The absolute configuration of the **PD2** enantiomers has not been determined.

The chiroptical characteristics of chiral (1*S*, 2*S*) porphyrin dimer **PD3** with C2 symmetry prepared by covalently linking two achiral porphyrin units *via* a (1*S*,2*S*)-diaminocyclohexane chiral bridge was reported in 1995 by Ema, Utaka, and coworkers (Figure 31).[92] These authors used an aliphatic chiral linker to avoid possible electronic interactions between the chiral linker and the porphyrins that could impact the CD spectrum of the porphyrin dimer. The CD spectrum of (1*S*,2*S*)-**PD3** showed a strong bisignate CD in the Soret region ([θ] = +29 × 10^4 at 412 nm; [0] = –8.5 × 10^4 at 392 nm) that was attributed to the chiral exciton coupling between the electric transition moments of chirally oriented porphyrins. Using molecular modeling, the authors estimated the average center-to-center distance between the two porphyrin moieties to be around 16 Å. The observed CD signal was stronger than the CD observed for naturally-occurring hemoproteins. The authors therefore concluded that the exciton coupled CD can be stronger than the CD, resulting from the presence of chiral amino acids at the porphyrin periphery.

In 1995 and 1996, Nakanishi and coworkers explored chiroptical properties of long-range interacting chromophores (up to 50 Å) in a series of chiral steroid-, dimeric steroid- and brevetoxin B-linked porphyrin dimers **PD4** (Figure 32a).[93,94] Porphyrin dimers were synthesized by covalently linking two 5-(*p*-carboxyphenyl)-10,15,20-triphenylporphin carboxylic acids with C2 symmetry to various chiral dihydroxy- and amino-hydroxy steroid derivatives *via* EDC/DMAP coupling. A strong exciton couplet CD in the Soret band region ($\Delta\varepsilon$ up to +111/–77) was observed for all steroid-porphyrin dimers, with an interchromophoric porphyrin-porphyrin distance up to 24 Å as observed also in brevetoxin B which presents an interchromophoric distance of ~50 Å. Based on comparison of the CD spectra for chiral porphyrin dimer **PD4** (Figure 32b) with chiral bis-benzoate derivatives having an identical chiral bridge, the authors proposed that the electric transition

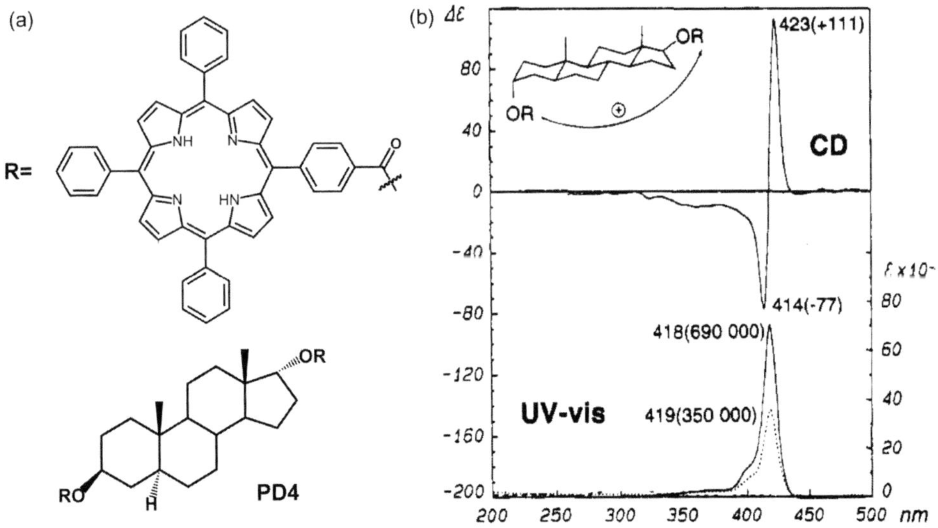

Figure 32. (a) Structure and (b) CD and UV-vis spectra of porphyrin dimer **PD4** linked *via* a chiral steroid bridge. Adapted from Refs. 93 and 94 with permission of the American Chemical Society.

moment in tetraphenylporphyrin runs in the C-5/C-15 direction. The authors concluded that the exciton coupled CD between the two porphyrin chromophores attached to a chiral linker could be used for conformational analysis of biopolymers.

Fleischhauer, Woody, Berova and coworkers later reported a theoretical analysis of the porphyrin–porphyrin exciton interaction in CD spectra of dimeric tetraarylporphyrins.[95–97] These authors demonstrated that coupled-oscillator DeVoe calculations provide fairly accurate predictions of the exciton interactions between Soret transitions of chirally oriented porphyrins. On the other hand, the circular oscillator calculations represent the only allowed approach for simulating the CD spectra of conformationally or configurationally locked porphyrin dimers. Here the application of the exciton chirality approach is unjustified, since the Soret transition cannot be depicted in terms of a single linear dipole. Figure 33 shows a comparison of calculated and experimental UV-vis and CD spectra of Zn(II) bisporphyrin **PD1** reported by Crossley and coworkers in 1995,[29] and the Zn(II) complex of the chiral porphyrin dimer **PD5** with a chiral cyclophane as spacer reported by Ema and coworkers in 1997.[98]

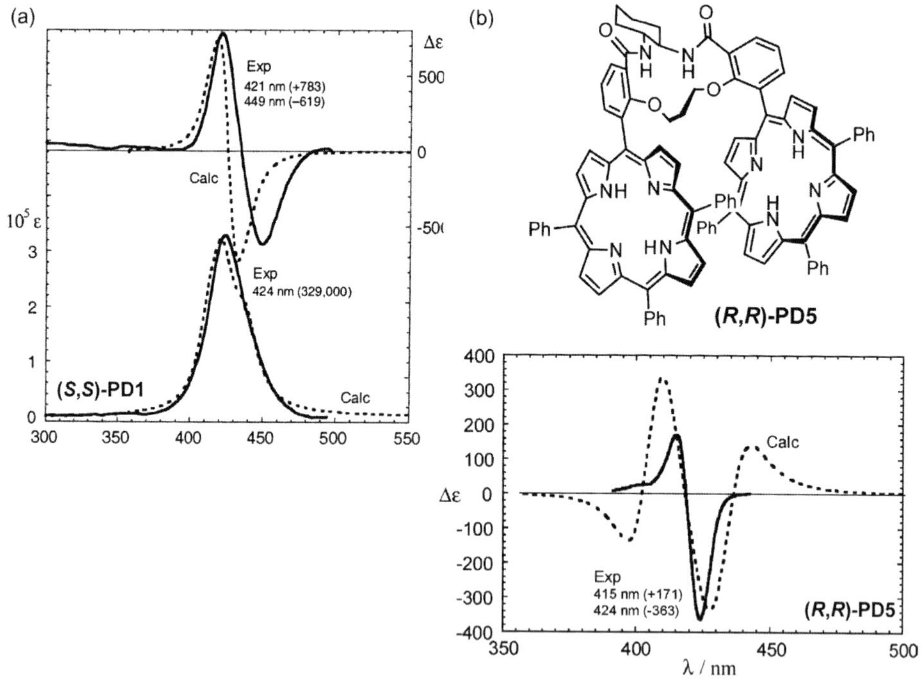

Figure 33. (a) Theoretical and experimental UV-vis and CD spectra of Zn(II) bisporphyrin **PD1**. (b) Structure, theoretical and experimental UV-vis and CD spectra of porphyrin dimer **PD5**. Adapted from Refs. 29 and 98, with permission of the American Chemical Society.

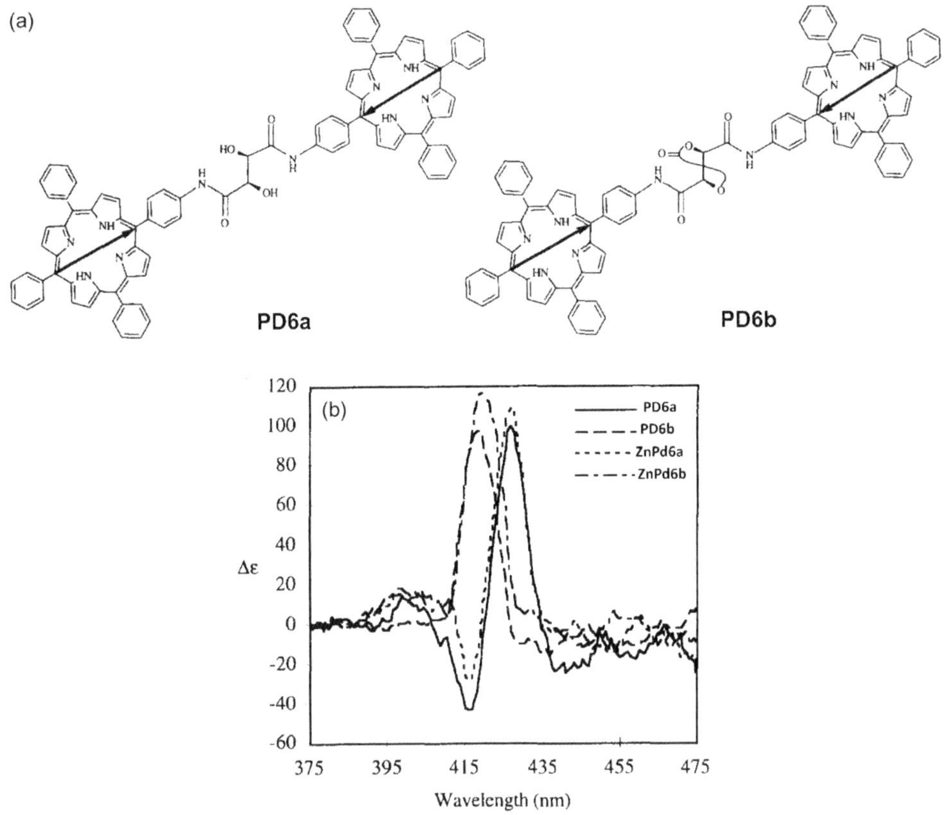

Figure 34. (a) Structures of free-base porphyrin dimers **PD6a** and **PD6b** and (b) CD spectra of free-base porphyrin dimers **PD6a** and **PD6b** and their zinc derivatives. Adapted from Ref. 99 with permission of Elsevier.

In 1996, Shachter and coworkers reported the synthesis and circular dichroism of two tartrate-linked free-base dimers **PD6** and zinc(II)-porphyrin dimers **ZnPD6** (Figure 34).[99] The **PD6** dimer was prepared by reaction of 5-(4-aminophenyl)-10,15,20-(triphenyl)porphyrin with 1*R*,2*R*-disuccinimidyltartrate. Porphyrin dimers **PD6a** and **ZnPD6a** displayed a bisignate CD in the porphyrin Soret region, consistent with exciton coupling of the porphyrin transitions. **PD6a** showed a positive CD band at 427 nm ($\Delta\varepsilon = +100$) and a negative CD band at 416 nm ($\Delta\varepsilon = -42$) (Figure 34b). Surprisingly, the bisignate couplet of the zinc derivative **ZnPD6a** was found at exactly the same wavelengths (427 nm and 416 nm). The appearance of a non-conservative couplet with a stronger positive CD band and a weaker negative CD band was tentatively attributed to interactions of the Soret transitions with other high-energy transitions. The couplet amplitudes ($A = \Delta\varepsilon_{427} - \Delta\varepsilon_{416} \approx +142$ and $+139$) for **PD6a** and **ZnPD6a** were smaller than those of previously reported rigidly linked porphyrin dimers, likely due to larger conformational freedom of the former. Interestingly, conformational restriction of the tartaric acid bridge in the form of carbonate **PD6b** resulted in disappearance of the bisignate CD signal and a positive CD band was observed instead. Explanation for this behavior was not provided.

BINOLs represent an important class of chiral structural motifs and several porphyrin dimers with BINOL bridges have been reported. In 1997, Hayashi, Ogoshi and coworkers reported synthesis and circular dichroism spectra of semi-flexible chiral porphyrin dimers **PD7**. The dimers were prepared as single enantiomers by linking achiral porphyrins to (*S*)- or (*R*)-binaphthyl derivatives (Figure 35).[6] The **PD7** enantiomers displayed mirror-image CD couplets in the Soret band region (only the CD spectrum of the (*S*)-enantiomer has been shown; Figure 35). The authors suggested chiral exciton coupling between stereo-oriented porphyrin chromophores to rationalize the observed porphyrin dimer CD spectra.[5,90,91] The intensity and wavelength of the CD couplet was shown to be sensitive to the addition of amine and diamines such as *n*-octylamine (10) or 1,8-diaminooctane. Hayashi and coworkers later explored a biphenyl version of **PD7**.[100]

Jiang, Bian and coworkers have synthesized chiral (*R*)- and (*S*)-BINOL-Zn(II) porphyrin dimers **PD8a–8c** and explored their chiroptical properties.[101] The authors varied the length and the structure (ether or ester) of the linker between the 2- and 2′-hydroxy groups of the (*R*)- and (*S*)-BINOL and two identical

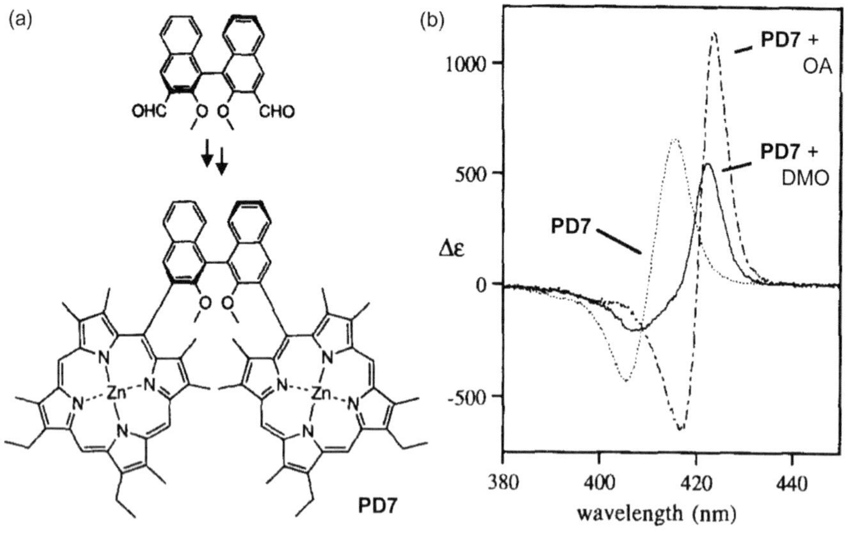

Figure 35. (a) Structures of chiral porphyrin **PD7** and (*S*)-2,2′-dimethoxy-[1,1′-binaphthalene]-3,3′-dicarbaldehyde. (b) CD spectra of chiral porphyrin dimer **PD7** alone, and in the presence of *n*-octylamine (OA) or 1,8-diaminooctane (DAO). Adapted from Ref. 6 with permission of Elsevier.

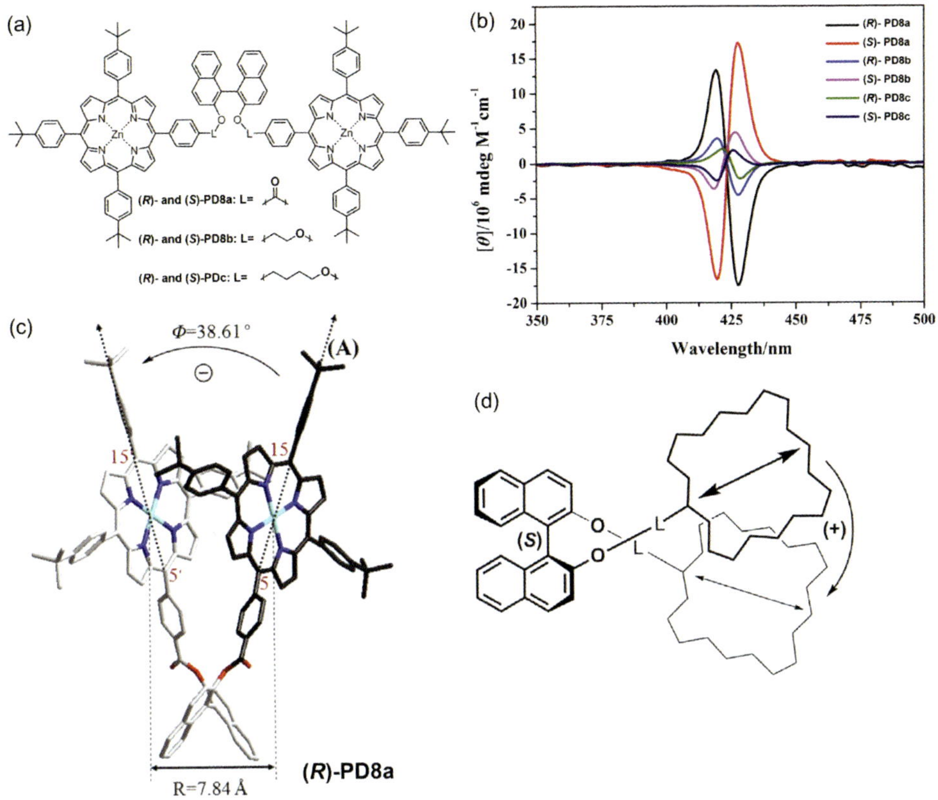

Figure 36. (a) Structure and (b) CD spectra of the chiral porphyrin dimers (*R*)- and (*S*)-**PD8a–8c** in $CHCl_3$ and (c) optimized molecular structure of (*R*)-**PD8a** by DFT method at the B97D/LANL2DZ level. The torsion angle ϕ is the angle of C15-C5-C5′-C15′; the interchromophoric distance is the Zn–Zn distance in Å. (d) Schematic representation of a BINOL induced a chiral twist in the porphyrin dimer. Adapted from Ref. 101 with permission of Elsevier.

Zn(II)-tetraphenylporphyrin derivatives to prepare three pairs of chiral Zn(II) porphyrin dimers **PD8a–8c** with different flexibilities as well as different porphyrin–porphyrin distances (Figure 36). All six **PD8** dimers displayed a bisignate CD signal in the Soret band region with mirror-image spectra for each pair of enantiomers (Figure 36b and Table 1). The sign of the CD couplet correlated with the chirality of the chiral BINOL bridge with (*R*)-BINOL linker giving rise to a counterclockwise twist between the porphyrins and (*S*)-BINOL to a clockwise twist as determined by the exciton chirality method. A decrease of the CD intensity was observed in **PD8a** to **PD8c** as the length and flexibility of the linker between porphyrin and BINOL increased. DFT molecular modeling on **PD8a** confirmed the chiral twist predicted by the exciton chirality method.

Table 1. CD data of porphyrin dimers **PD8a–8c** (1.0×10^{-6} M solution in $CHCl_3$).

Dimer	λ ($\Delta\varepsilon$) nm(mol^{-1}.L.cm^{-1})[a]		A_{CD} (mol^{-1}.L.cm^{-1})[b]
(*R*)-**PD8a**	428 (−531)	420 (+403)	−934
(*S*)-**PD8a**	428 (+525)	420 (−506)	+1031
(*R*)-**PD8b**	428 (−139)	420 (+109)	−248
(*S*)-**PD8b**	427 (+133)	419 (−112)	+245
(*R*)-**PD8c**	428 (−70)	422 (+64)	−134
(*S*)-**PD8c**	426 (+58)	420 (−76)	+134

[a]λ: peak or trough wavelength in nm; molar extinction coefficients $\Delta\varepsilon = [\theta]/32982$; [b]total amplitude $A_{CD} = \Delta_{\varepsilon 1} - \Delta_{\varepsilon 2}$ in mol^{-1}.L.cm^{-1}.

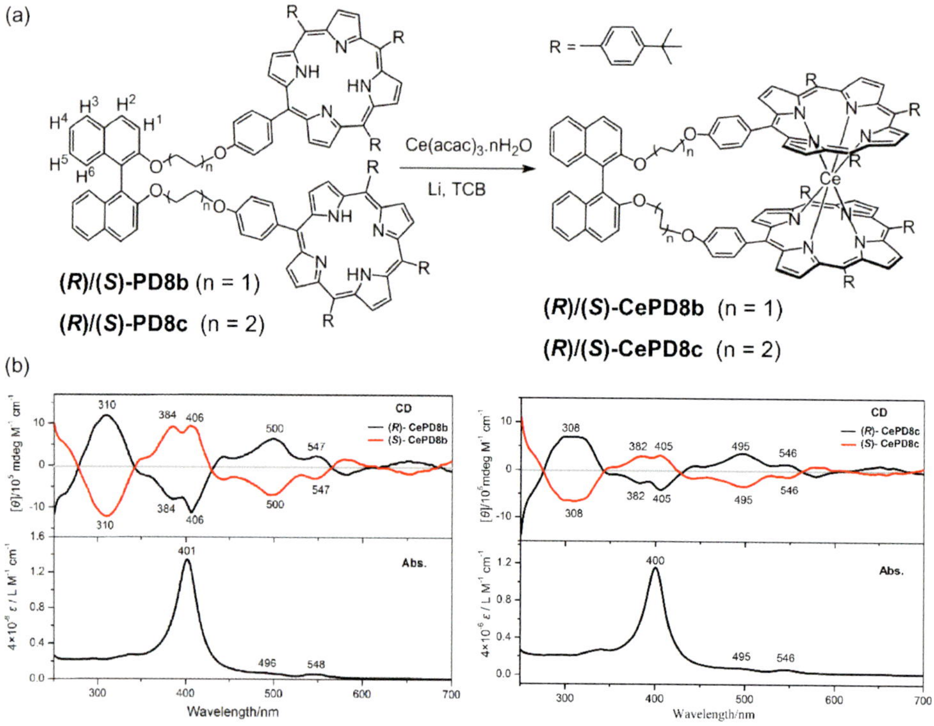

Figure 37. (a) Synthesis and structure and (b) CD and UV-vis spectra of chiral double-decker porphyrin dimers **CePD8b** (left) and **CePD8c** (right) in $CHCl_3$. Adapted from Ref. 102 with permission of Elsevier.

Bian and coworkers next prepared chiral cerium double-decker porphyrin dimers **CePD8b,8c** from chiral dimers **PD8b,8c** (Figure 37a).[102] **CePD8b,8c** displayed mirror-image CD spectra with a negative sign in the Soret band (380–410 nm) for (*R*)-enantiomers and a positive sign in the Soret band for (*S*)-enantiomers

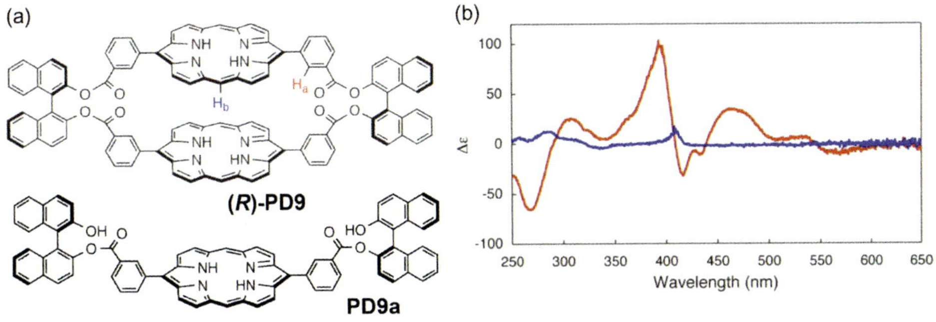

Figure 38. (a) Structure of the chiral porphyrin dimer (*R*)-**PD9** and chiral monomer **PD9a**. (b) CD spectrum of (*R*)-**PD9** (red curve) and **PD9a** (blue curve) in $CHCl_3$. Adapted from Refs. 103 and 104 with permission of the Royal Society of Chemistry.

(Figure 37b). The CD spectra confirmed that the C2-symmetry of a BINOL bridge was transcribed into the helical twist of the double-decker porphyrin dimer. The **CePD8b** dimer with a shorter linker between BINOL and porphyrin units possessed more intense CD spectra than the **CePD8c** with a longer linker. The authors did not elaborate on the differences in the CD spectra of **PD8b,8c** and **CePD8b,8c**.

In 2011 and 2012, Ema, Sakai and coworkers reported a BINOL containing chiral cyclic free-base porphyrin dimer (*R*)-**PD9** (Figure 38a) possessing a macrocyclic cavity for intercalation of achiral and chiral aromatic guests.[103,104] The porphyrin dimer (*R*)-**PD9** was prepared as a pure enantiomer by using (*R*)-BINOL and achiral 5,15-substituted porphyrin. The (*R*)-**PD9** CD spectrum displayed a complex pattern in the 600–250 nm region (Figure 38b). On the other hand, the CD spectrum of the chiral porphyrin monomer **PD9a** showed only weak induced CD with a positive CD band in the Soret band region (~410 nm) confirming that the CD signal of (*R*)-**PD9** originated from coupling between two porphyrin moieties.

MacMillan and Molinski reported CD spectra of chiral porphyrin dimers **PD10a–10c** with chiral C2 symmetric enantiomeric acylic diols of different lengths as chiral bridges between two tetraphenylporphyrins aligned in highly uniform, unilamellar DCP liposomes of narrow size distribution (∅ = 26 ± 5.1 nm).[105] The CD spectrum of (*R,R*)-**PD10a** showed a conservative (+/–) bisignate signal with a positive CD band at 430 nm ($\Delta\varepsilon = +27$) and a negative CD band at 413 nm ($\Delta\varepsilon = -25$) (Figure 39b). The (*S,S*)-**PD10a** enantiomer displayed a mirror-image CD profile (413 nm, $\Delta\varepsilon = -34$; 429 nm, $\Delta\varepsilon = +31$). CD spectra of (*R,R*)-**PD10c** and (*R,R*)-**PD10d** with larger interchromophoric distances between the porphyrins than in the **PD10a** displayed (+/–) bisignate couplets of weaker intensity (Figure 39c). It appeared that the intensity of the CD couplet decreased linearly with the interchromophoric distance.

In 2005, Arai and coworkers reported the synthesis and CD characteristics of the chiral porphyrin trimer **PD11**.[106] Two zinc(II) porphyrins where linked to the

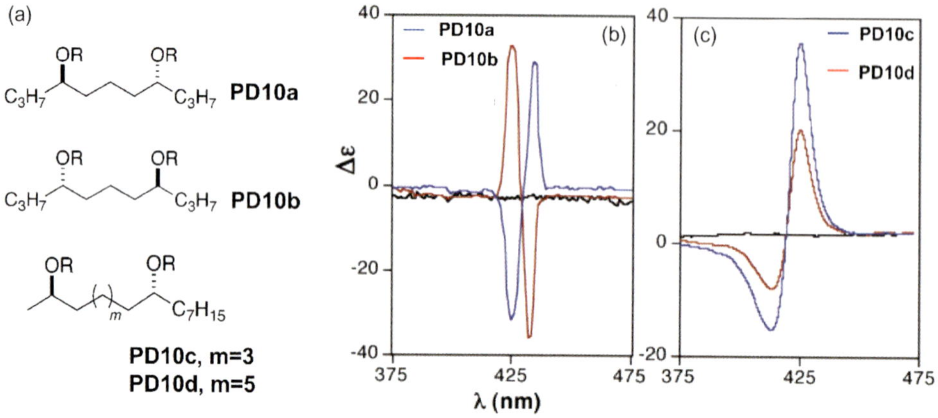

Figure 39. (a) Structures of chiral porphyrin dimers **PD10a–10d** (*R* = tetraphenylporphyrin). (b) CD spectra of (*R,R*)- **PD10a** (blue curve) and (*S,S*)-**PD10b** (red curve) in DPC liposomes. (c) CD spectra of (*R,R*) enantiomers of **PD10c** (blue curve) and **PD10d** (red curve) in DPC liposomes. Adapted from Ref. 105 with permission of the American Chemical Society.

5,15-aryl groups of a free-base porphyrin template by achiral glycine (–NH–CH_2–CO–) spacers. Chiral L-leucine benzyl esters were attached to the free base porphyrin template from the opposite side of the zinc-porphyrins as depicted in Figure 40a. CD spectra of **PD11** displayed a (+/–) bisignate couplet with a positive CD band 438 nm ($\Delta\varepsilon$ = +638) and a negative CD band at 428 nm ($\Delta\varepsilon$ = –872) together with a small CD band at 413 nm ($\Delta\varepsilon$ = +106) (Figure 40b). The authors rationalized the bisignate CD by exciton-coupled Cotton effects between two closely located zinc porphyrins locked in a chiral right-twisted orientation due to the effect of the chiral L-leucine benzyl esters tails. The correlation between the chirality of **PD11** and the L-leucine benzyl ester tail could have not been made. Interestingly, zinc derivative of **ZnPD11** displayed a complex (–/+/–/+/–/+) CD spectrum in CH_2Cl_2 probably as a result of aggregation [λ/nm ($\Delta\varepsilon/M^{-}1cm^{-1}$) = 443 (–63.9), 439 (+86.3), 430 (–280), 425 (+403), 419 (–134), 412 (+54.7)]. A change of CH_2Cl_2 to $CHCl_3$ to toluene caused decrease of the Soret band CD signal from 100% to 50% and 10% respectively. Addition of toluene to the CH_2Cl_2 caused linear decrease of the **PD11** CD signal intensity in the Soret band region (Figure 40c). The authors pointed out that the CH_2Cl_2, $CHCl_3$, and toluene solvent effect the dipole moments (μ/C = 5.2, 3.8, 1.0), the relative permitivities (ε_r = 8.93, 4.81, 2.38), or the Kamlet–Taft π^* parameters (π^* = 0.82, 0.58, 0.54), though the solvent effect on the CD may be more complex.

In 2006 and 2011, Takeuchi and coworkers reported chiral double-decker porphyrins **PD12a** having one chiral substituent per double-decker[107] and **PD12b** bearing two chiral substituents per double decker[108] for enantioselective recognition of dicarboxylic acid (Figure 41). The CD spectra of both **PD12a** and **PD12b**

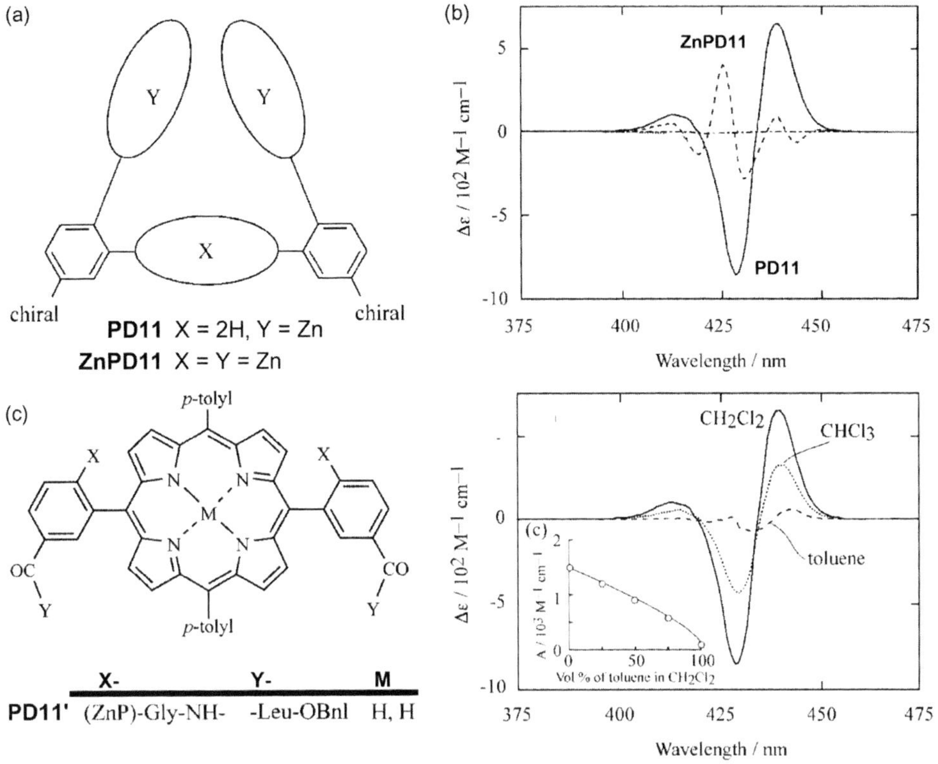

Figure 40. (a) Structure of chiral porphyrin trimers **PD11** and **ZnPD11**, (b) CD spectra of **PD11** and **ZnPD11** in CH_2Cl_2 and (c) CD spectra of **PD11** in CH_2Cl_2, $CHCl_3$ and toluene (Inset: CD signal as a function of toluene concentration in CH_2Cl_2. Adapted from Ref. 106 with permission of Wiley-VCH.

displayed a bisignate signal with negative CD bands around 400 nm and positive bands around 310 nm, with the latter attributed to the chirality of the pyridyl groups (Figures 41c, 41d). The addition of excess of triethylamine to **PD12a** or **PD12b** caused disruption of the intermolecular H-bonds and disappearance and significant decrease of the positive and the negative CD bands respectively. The CD spectra also suggested suitable preorganization of **PD12** for the chiral dicarboxylic acid guest.

Canary and coworkers communicated the chelation-enhanced CD of tripodal porphyrin dimer **PD13** (Figure 42).[109] **PD13** contained an *N,N,N,O*-coordination environment of previously characterized tripodal ligands and was prepared by covalently attaching two achiral tetraphenylporphyrins to the amino group of a chiral aminoalcohol. Despite being chiral, **PD13** did not display any CD due to its conformational flexibility. Complexation with Cu(II) in the presence of NH_4SCN led to formation of the **CuPD13** dimer with locked conformation and

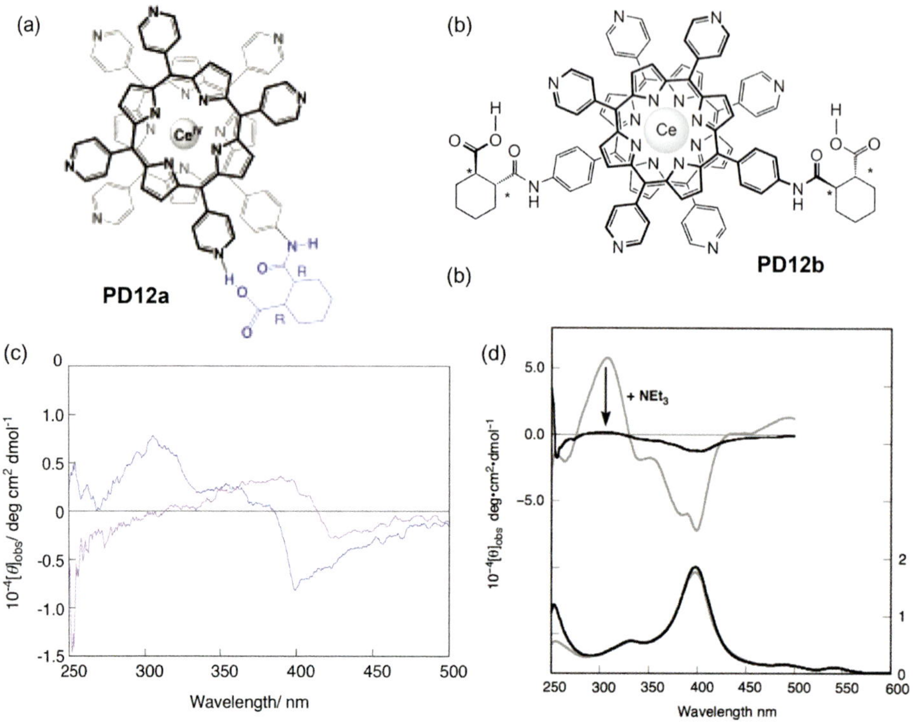

Figure 41. Structure of the chiral double-decker porphyrins (a) **PD12a** and (b) **PD12b**. (c) CD spectra of **PD12a**. Adapted from Ref. 105 with permission of the American Chemical Society. (d) CD and UV spectra of **PD12b** in tetrachloroethane-tetrahydrofuran 30:1 (v/v). Adapted from Ref. 108 with permission of Taylor & Francis.

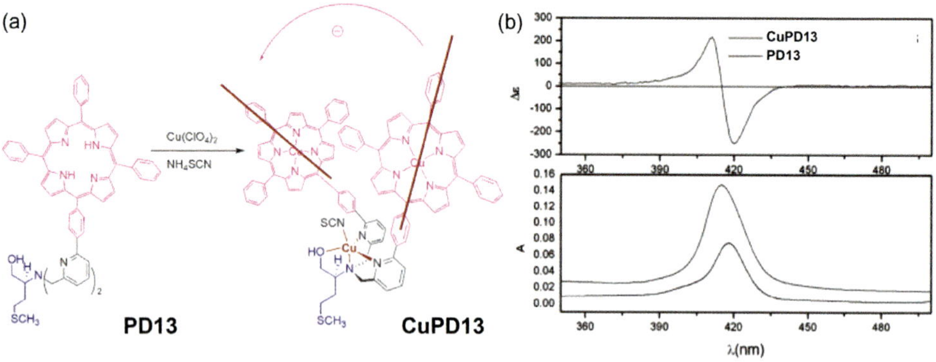

Figure 42. (a) Structure and (b) CD (top) and UV-vis (bottom) spectra of chiral porphyrin dimer **PD13** and its copper(II) complex **CuPD13**. Adapted from Ref. 109 with permission of the American Chemical Society.

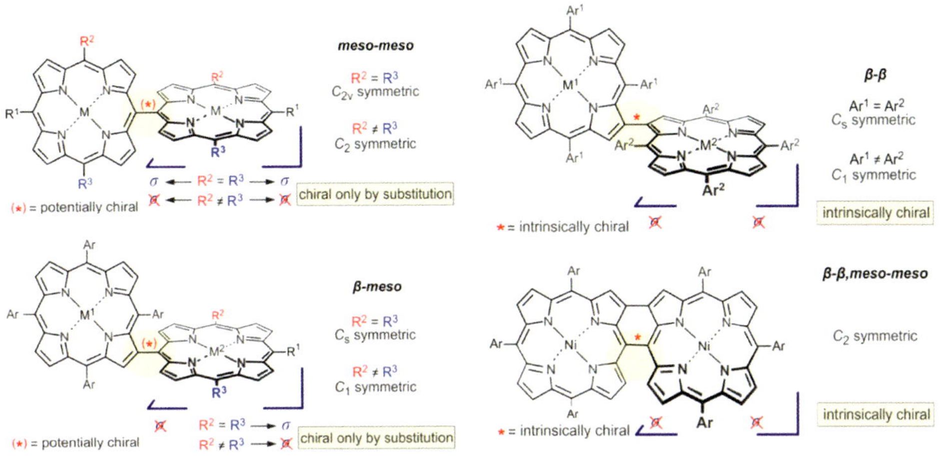

Figure 43. Origin of chirality in directly linked and in fused porphyrin dimers. Reproduced from Ref. 110 with permission of the American Chemical Society.

the appearance of a strong negative bisignate exciton couplet with a negative CD band at 420 nm and a positive CD band at 411 nm (Figure 42b). The exciton couplet CD spectra also suggested free rotation of porphyrins along their 5,15 axis and a counterclockwise twist between porphyrins. Complexation with Co(II), Ni(II), or Ca(II) gave rise to a much weaker bisignate CD signal and was explained by possible less optimal orientation of the porphyrin units. Reduction of **CuPD13** with ascorbic acid caused disappearance of the Soret bisignate CD, while consecutive addition of Cu(I)PF_6 followed by oxidation by air resulted in reappearance of the CD signal. Overall, the chiroptical properties of the **PD13** dimer are strongly dependent on the metal ion and its oxidation state and could be used in chiroptical on/off switch devices.

B. Directly Linked Chiral Porphyrin Dimers: *meso–meso, meso–β* and *β–β* Porphyrin Dimers

Porphyrin dimers with two porphyrin moieties linked together *via meso–meso*, *meso–β* and *β–β* positions were reported in the late 1990s. They displayed interesting optical and electrochemical properties.

1. *meso–meso Connected Chiral Dimers*

Meso–meso porphyrin dimers that possess a large rotational barrier along the *meso–meso* C–C bond and carry different aryl groups at 5- and 15-positions are chiral and

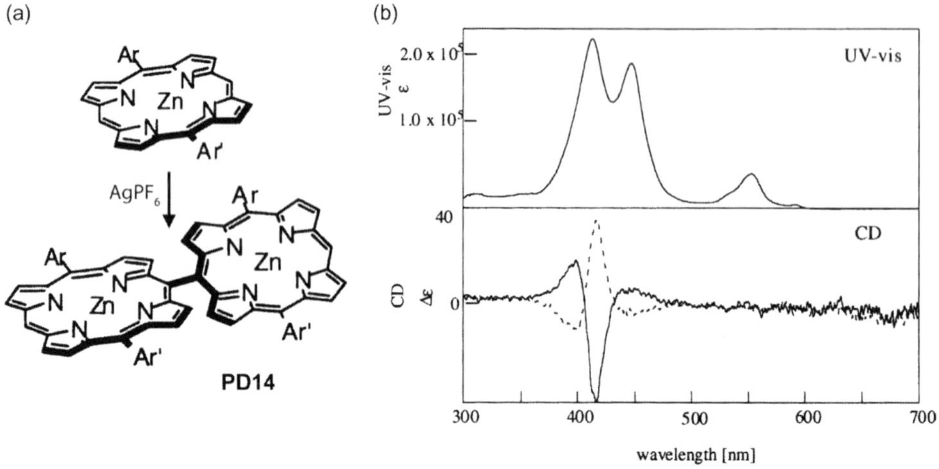

Figure 44. (a) Structure and (b) UV-vis (top) and CD (bottom) spectra of chiral *meso–meso* porphyrin dimer **PD14**. Ar = 3,5-di-*tert*-butylphenyl; Ar′ = 2,5-di-methoxyphenyl. Adapted from Ref. 33 with permission of Elsevier.

exist as a pair of enantiomers. The first chiral, enantio-pure *meso–meso* linked porphyrin dimer was reported by Yoshida and Osuka in 2000.[33] The Ag(I)-promoted oxidative *meso–meso* coupling of Zn(II) 5-(3,5-di-*tert*-butylphenyl)-15-(2,6-dimethoxy)phenylporphyrin yielded dimer **PD14** (Figure 44), trimer and higher oligomers that were separated by size exclusion chromatography. 1H NMR data revealed restricted rotation along three bonds: the *meso–meso*, the *meso*-(2,6-dimethoxyphenyl), and the *meso*-(3,5-di-*tert*-butylphenyl). The enantiomers of **PD14** were prepared by chiral resolution of racemic mixture by HPLC on a chiral preparative column and their UV-vis and CD spectra are shown in Figure 44b. The split Soret band was explained by exciton coupling theory with two degenerate transition dipole moments Bx and By that are parallel and perpendicular to the long axis of the *meso–meso* coupled dimer **PD14**. The coupling of two Bx components led to the red-shifted Soret band at 448 nm while the orthogonal geometry cancelled all other dipole-dipole interactions thus giving rise to the Soret band at 413 nm. The authors interpreted the trisignate CD spectrum in the Soret region as an overlap of two bisignate couplets with opposite signs: a strong bisignate couplet at the unshifted Soret band (413 nm) and a weak couplet at the red-shifted Soret band (448 nm). The molar ellipticity of **PD14** was reported to be $[\theta]_{413} = -1.3 \times 10^5$ deg.cm^{-1}.dm^3.mol^{-1}.

In 2003, Osuka, Kim and coworkers prepared a series of the *meso–meso*-linked strapped porphyrin dimers strapped with a dioxymethylene chain (O–$(CH_2)_n$–O) of various length (Figure 45a).[111] Pure enantiomers obtained by chiral HPLC on an analytical Sumichiral OA4400 column were stable and did not racemize at room temperature. The enantiomers of **PD17** (n = 10) displayed mirror image

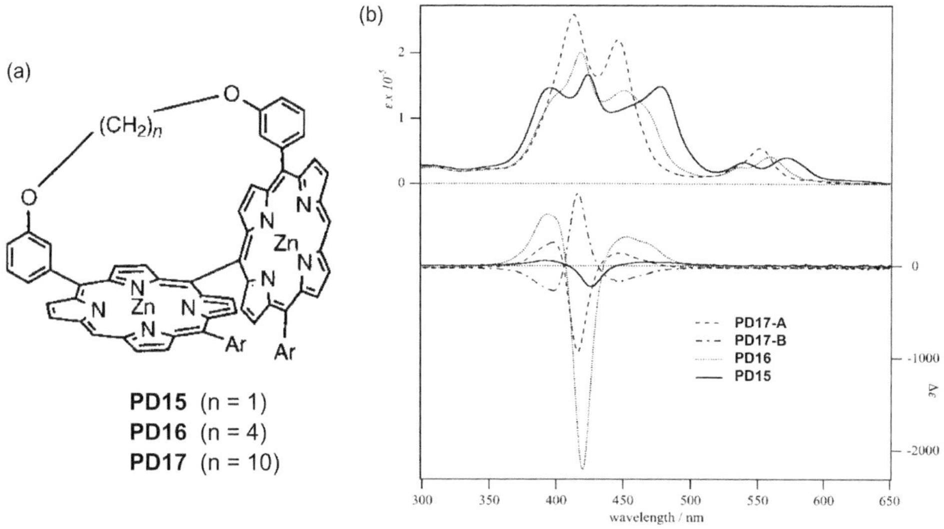

Figure 45. (a) Structure of chiral *meso–meso* porphyrin dimers **PD15** (n = 1), **PD16** (n = 4), **PD17** (n = 10). Ar = 3,5-di-*tert*-butylphenyl and (b) UV-vis (top) and CD spectra (bottom) of **PD17** (A: fast eluting enantiomer, B: slow eluting enantiomer), **PD16** and **PD15**. Adapted from Ref. 111 with permission of Wiley-VCH.

trisignate CD spectra in the Soret band region (–/+/– and +/–/+) with CD bands at 446 nm, 416 nm and 398 nm (Figure 45b). No CD signal has been observed for **PD17** in the Q band region. Similar trisignate CD profiles with CD bands have also been observed for **PD16** (n = 4; CD bands at 394 nm 420 nm and 454 nm) **PD15** (n = 1; CD bands at 392 nm, 425 nm and 470 nm). Decreasing the chain-length of the strap resulted in decreased dihedral angle and increased distortion of porphyrins as suggested by MM2 calculations and ^{1}H NMR spectra. These structural changes had a pronounced effect on the position and intensity of the trisignate CD signal with **PD16** (n = 4) displaying the strongest CD signal. The authors interpreted the trisignate CD signal as an overlap of two split Cotton effects with opposite sign; one at ~400 nm and the second one at ~420–440 nm.

In 2009, Zheng and coworkers prepared chiral *meso–meso* linked strapped porphyrin dimer **PD18** by intermolecular diastereoselective *meso–meso* coupling using enantiomers of the binaphyl linked porphyrin dimer **PD19** as starting material (Figure 46a).[112] (*SR*)- and (*SS*)-**PD18** diastereomers were prepared in a 2.2:1 ratio by intermolecular coupling of (*S*)-**PD19** at low temperature (–45 °C) and purified by flash chromatography on silica gel. Similar diastereoselectivity was observed for (*RR*)- and (*RS*)-**PD18** prepared from the enantiomeric (*R*)-**PD19**. The CD showed a trisignate signal in the Soret region with CD bands at ~460 nm, ~425 nm and ~398 nm (Figures 46b, 46c). The CD patterns were proposed to arise from the

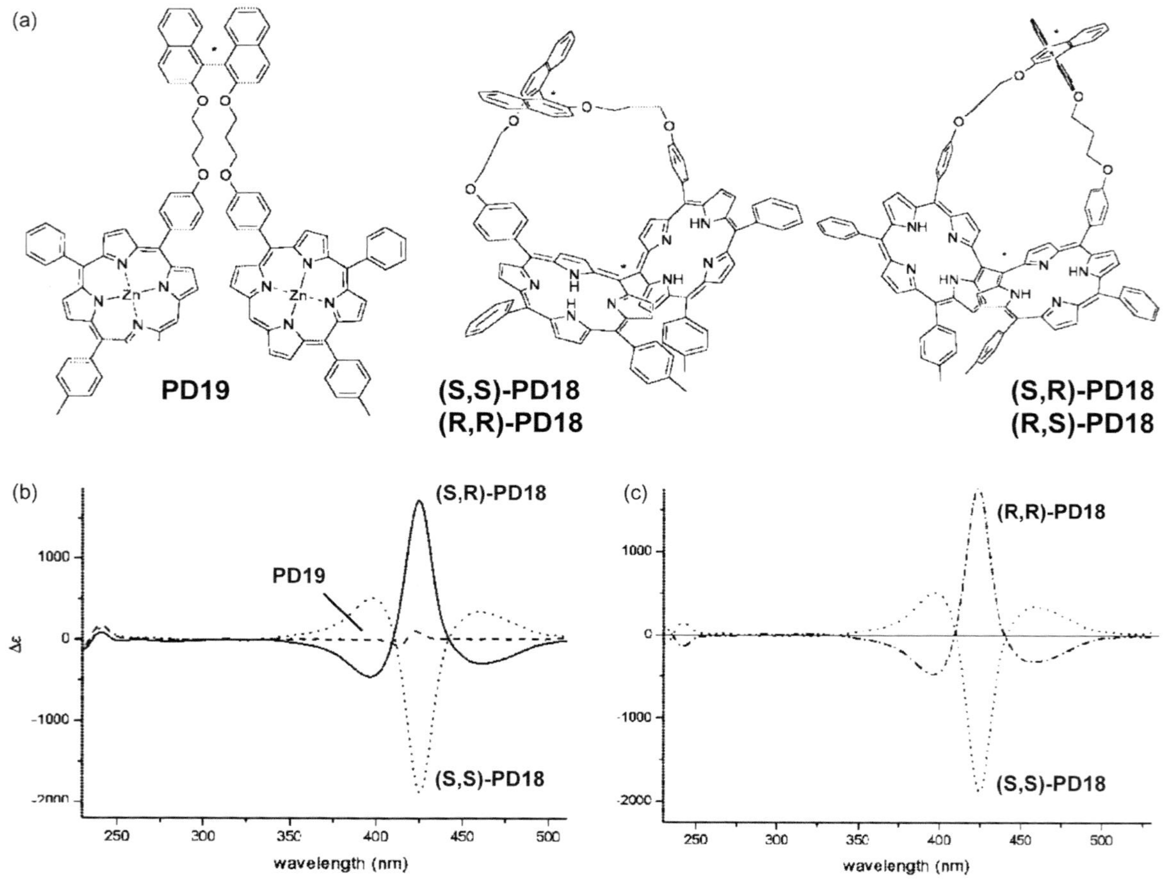

Figure 46. (a) Structure of **PD19** and (*SS*)-, (*RR*)-, (*SR*)- and (*RS*)-**PD18**. (b) CD spectra of **PD19** (dashed line), (*SR*)-**PD18** (full line), and (*SS*)-**PD18** (dotted line). (c) The CD spectra of (*SS*)-**PD18** (dotted line) and (*RR*)-**PD18** (dashed-dotted line). Adapted from Ref. 112 with permission of the American Chemical Society.

coupling of the B_x and B_y transition dipole moments of the chirally oriented porphyrin units. The CD spectra were used to identify the four diastereomers, relying on previously published results and the CD signal of the binaphtyl unit ~240 nm.

2. *β–β* Connected Chiral Dimers

Combination of useful properties of axial chirality and electronic, photonic and structural properties of porphyrin in rotationally hindered directly *β–β*-linked porphyrin dimers could lead to novel chiral catalysts, chiroptical and electronic devices, chiral reporter groups, or novel materials with pronounced chiral, structural and chiroptical properties. Bringmann and coworkers prepared the first intrinsically axially chiral *β–β* linked Zn(II) porphyrin dimer **ZnPD20** by palladium catalyzed coupling of *β*-bromo-porphyrin with *β*-porphyrin boronic ester (Figure 47a).[113,114] The pure enantiomers were obtained by chiral resolution using chiral HPLC of racemic **ZnPD20** on Chiralcel OD-H or Chirex 3010 columns. CD spectra of **ZnPD20** enantiomers showed mirror-image profiles with CD bands at ~445 nm and

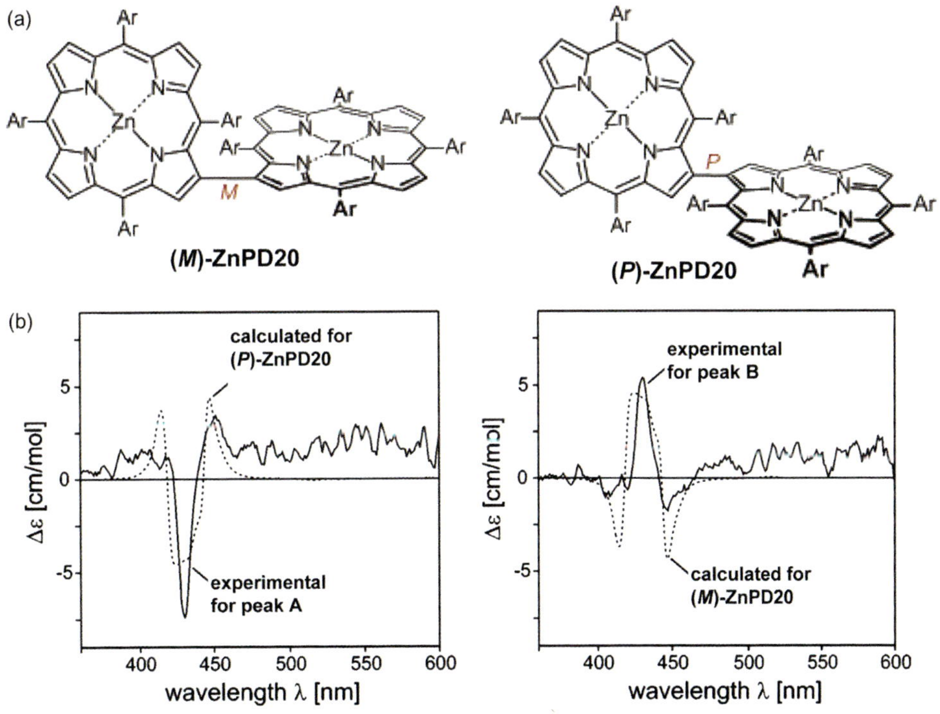

Figure 47. (a) Structure of (*M*)- and (*P*)-**ZnPD20**. Ar = phenyl. (b) Experimental and calculated CD spectra of (*M*)- and (*P*)-**ZnPD20**. Adapted from Refs. 113 and 114 with permission of the American Chemical Society.

~430 nm (Figure 47b). The absolute configuration was determined by the exciton chirality method assigning (*P*)-configuration to the faster eluting atropoisomer and (M)-configuration to the slower eluting atropoisomer. The empirical results were corroborated by quantum chemical CD calculations (Figure 47b). Configurational stability studies using HPLC analysis showed no interconversion of **ZnPD20** in 24 h at room temperature, suggesting that the porphyrin–porphyrin axis is rotationally stable. On the other hand, the enantiomeric purity of **Pd(II)ZnPD20** decreased to 82% ee and the ee of **Cu(II)ZnPD20** dropped to 4%. The barrier of racemization thus depended on central metals with Zn(II) having the highest (ΔG = 115 kJ/mol) and free-base the lowest energy barrier.

3. *meso–meso β–β* Fused and *meso–β* Connected Chiral Dimers

Osuka, Kobayashi and coworkers have prepared a chiral *meso–meso* β–β doubly linked porphyrin dimer **NiPD21** (chiral fused bis-porphyrin) and explored its chiroptical properties (Figure 48a).[34,115] The X-ray structure revealed a helically twisted conformation and the perturbed UV-vis absorption spectrum indicated π conjugation in the system. The enantiomers of **NiPD21** were resolved by chiral HPLC on a Sumichiral OA3100 preparative column. The CD spectra of enantiomers of **NiPD21** displayed complex mirror image profiles with CD bands at around 725 nm, 550 nm, 500 nm, 415 nm and 380 nm (Figure 48b). The racemization was studied with the help of variable temperature CD spectra at 333–363 K and the following activation parameters were obtained:

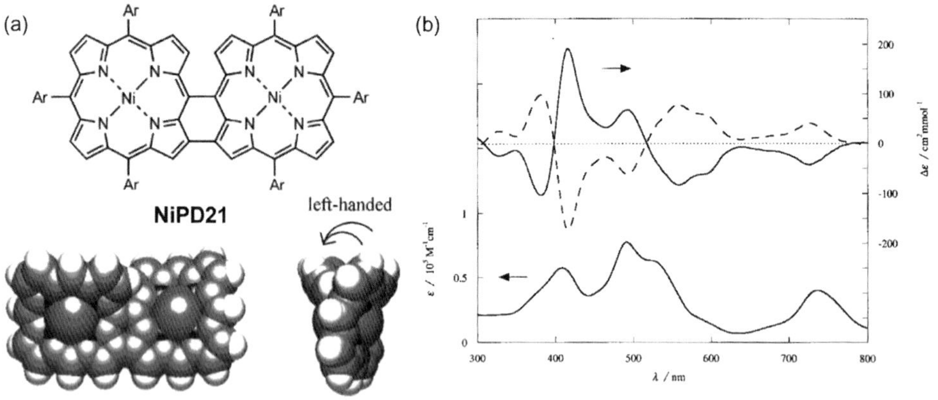

Figure 48. (a) Structure of *meso–meso* β–β doubly linked porphyrin dimer **NiPD21**. Ar = 3,5-di-*tert*-butylphenyl. Adapted from Refs. 114 and 115 with permission of Wiley-VCH. (b) CD (top) and UV-vis absorption spectra (bottom) of the first eluting (solid curve) and the second eluting (dashed curve) enantiomers of **NiPD21**. Adapted from Ref. 34 with permission of the American Chemical Society.

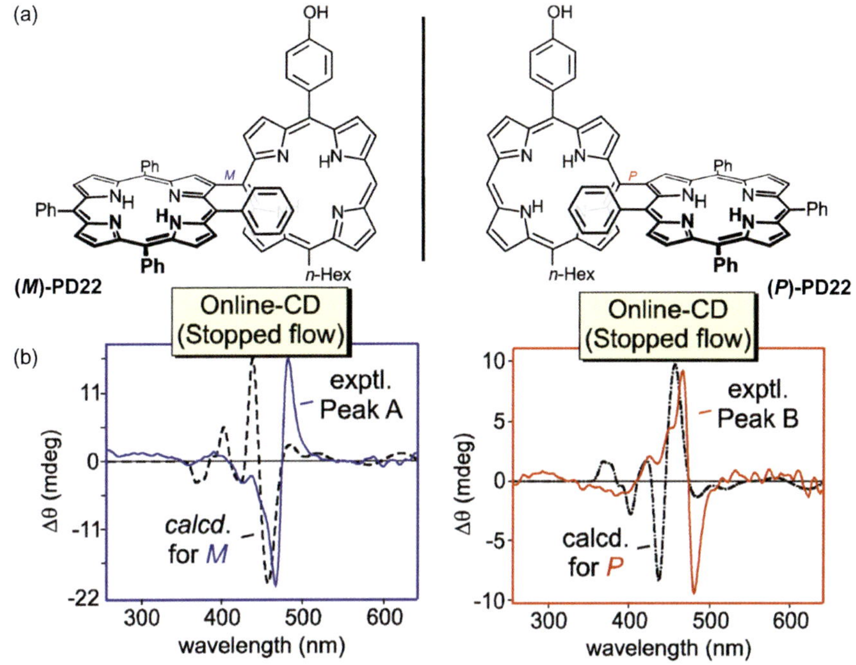

Figure 49. (a) Structure of (*M*) and (*P*) enantiomers of *meso-β*-linked porphyrin dimer **PD22**. (b) Experimental and calculated CD spectra of (*M*)- and (*P*)-**PD22**. Adapted from Ref. 116 with permission of the American Chemical Society.

$\Delta G^{\ddagger} = 111$ kJ/mol, $\Delta H^{\ddagger} = 101$ kJ/mol, $\Delta S^{\ddagger} = -30.4$ J/mol.K, $\Delta E = 104$ kJ/mol. An absolute configuration assignment was achieved by comparison of experimental and time-dependent DFT (TDDFT) calculated optical and chiroptical spectra. Importantly, the assignment by the empirical exciton chirality method yielded an incorrect answer and confirmed that the **NiPD21** dimer could not be treated as a fused chiral system with two independent porphyrin moieties.

In 2009, Senge, Bringmann, and coworkers reported axially chiral *meso–β*-linked porphyrin dimer **PD22** prepared by Suzuki-type coupling of a *meso*-bromoporphyrin with a *β*-porphyrin boronic ester (Figure 49a).[116] Pure enantiomers of the highly asymmetric (C_1-symmetry) **PD22** were obtained by chiral HPLC on a Chirex 3010 column (Phenomenex). Their CD spectra showed bisignate mirror image spectra with CD bands around 485 nm and 465 nm. The first eluting atropo-enantiomer of **PD22** had a positive CD couplet, whereas the second had a negative CD couplet. The absolute configuration could not be assigned by the empirical exciton chirality method,[5,89,117] and quantum chemical CD calculations must have been used, clearly showing the difficulties for stereochemical elucidation of molecules containing multiple chromophores. Comparison of experimental and calculated CD spectra (Figure 49b) allowed the M-configuration to be assigned to the slower eluting atropo-enantiomer.

In 2016, Gehrold, Bruhn and Bringmann, reported experimental and theoretical (chir)optical characteristics of helically chiral *meso–meso β–β* doubly linked basket-handle porphyrin dimers including nickel derivative **NiPD23**. The unique structure of basket-handle porphyrins reduced their symmetry and allowed the differentiation of the two sides of the porphyrin macrocycle. **NiPD23** was prepared by oxidation of enantiopure *meso–meso* linked basket-handle porphyrin dimer **NiPD24** (Figure 50a).[110] Enantiomers of **NiPD24** were prepared by HPLC using Diacel Chiralpak IA (4.6 × 250 mm; 3 μm) as a chiral phase. The CD spectra of **NiPD23** enantiomers displayed a complex, mirror-image profile (Figure 50b). The absolute configuration of **NiPD23** was determined by comparison of the experimental CD spectra with CD calculated by Osuka and coworkers using TD B3LYP methods.[34] The calculated CD spectra showed a very good fit and allowed assignment of the (*P*)-chirality to the fast eluting peak of **NiPD23**.

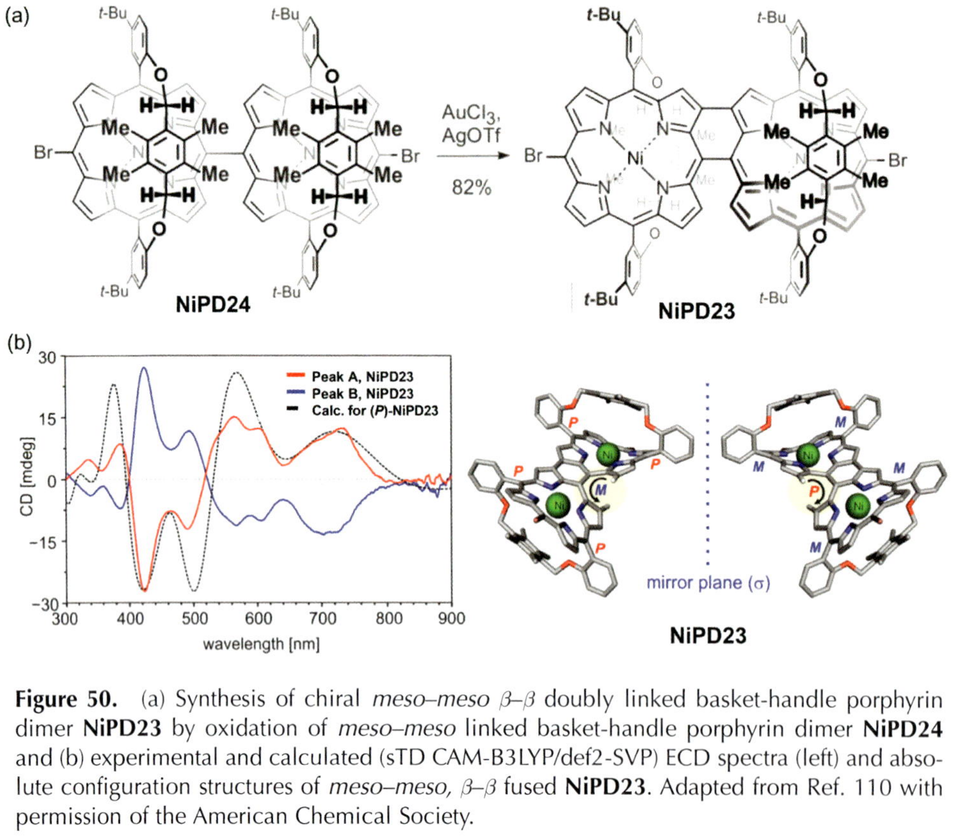

Figure 50. (a) Synthesis of chiral *meso–meso β–β* doubly linked basket-handle porphyrin dimer **NiPD23** by oxidation of *meso–meso* linked basket-handle porphyrin dimer **NiPD24** and (b) experimental and calculated (sTD CAM-B3LYP/def2-SVP) ECD spectra (left) and absolute configuration structures of *meso–meso, β–β* fused **NiPD23**. Adapted from Ref. 110 with permission of the American Chemical Society.

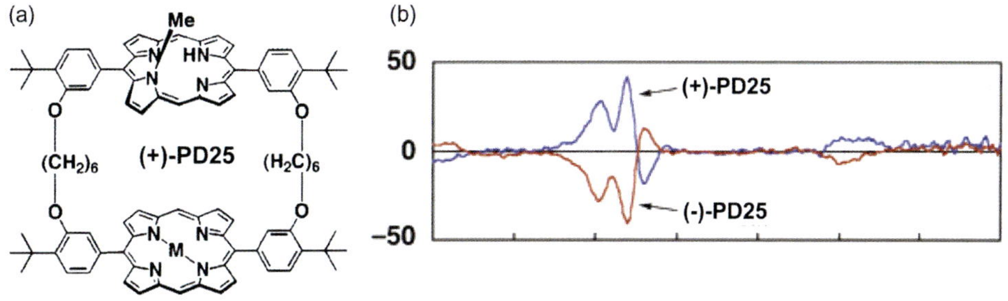

Figure 51. (a) Structure of the chiral bis-porphyrin hosts (+)-**PD25** (M = CH3Rh). (b) CD spectra of the (–)-**PD25** and (+)-**PD25** bis-porphyrins. Adapted from Refs. 32 and 118 with permission of the American Chemical Society.

C. Porphyrin Dimers Containing Intrinsically Chiral Porphyrin Monomers

Synthesis and CD spectra of chiral cyclic porphyrin dimer **PD25** with an embedded C1-symmetric chiral *N*-substituted porphyrin were reported in 2006 by Aida and coworkers (Figure 51a).[32,118] The chiral dimer was prepared as a racemic mixture and pure enantiomers were obtained by preparative chiral HPLC on CHIRALPAK IA with hexane/CH_2Cl_2/Et_2NH as eluent. The CD spectra of (+)-**PD25** and (–)-**PD25** displayed mirror-image signals with a trisignate feature in the Soret band region (CD bands at ~430, 420 and 405 nm) and a single CD band in the Q-band region (~550 nm) (Figure 51b).

V. Applications of Chiral Porphyrin Monomers and Dimers: Molecular Recognition

Applications of chiral porphyrins and metalloporphyrins lie in molecular recognition and optical resolution of chiral carbon nanomaterials and in stereoselective synthesis, especially epoxidation and cyclopropanation, of alkenes. Chiral recognition is a very active area of the host–guest chemistry field due to practical real-life applications including biosensing, determination of absolute configuration, enantiomeric excess, and enantiomer resolution as well as model systems to enhance our understanding of biomolecular recognition in biological and biochemical processes. Porphyrins and metalloporphyrins are attractive components of host systems due to their well-developed chemistry and readability of peripheral functionalization of the porphyrin core, while the interaction with guest molecules can be monitored by a variety of spectroscopic techniques. The use of chiral porphyrins allows recognition of chiral molecular targets, including amines, alcohols, amino acids, hydroxy-carboxylic acids and their derivatives. Due to their extended aromatic surface, chiral porphyrins have also been used successfully for recognition and sensing of chiral nanotubes and fullerenes.

A. Recognition and Sensing of Chiral Organic Molecules

The majority of chiral recognition studies by chiral porphyrin hosts have focused on amino acids (especially esters) as biotargets, given their biological importance, availability as both enantiomers, and presence of suitable functional groups for multipoint recognition. To achieve strong intermolecular interactions and good recognition, organic solvents have often been used as media. For the purpose of this section, chiral (metallo)porphyrin hosts were divided into two groups, chiral porphyrin monomers and chiral porphyrin dimers.

1. *Chiral Porphyrin Monomers*

In 1993 and 1995, Kuroda, Ogoshi, Mizutani and coworkers showed chiral recognition of amino acid methyl esters by (i) chiral doubly bridged free-base tetraarylporphyrins **PM26**[119] and **PM27**,[23] (ii) chiral zinc(II) porphyrins **PM28** with two functional arms (5,15),[120] and (iii) chiral zinc(II) porphyrin **PM29** with three arms (5,10,15) (Figure 52a).[24] The porphyrins were prepared as racemic mixtures and resolved by chiral HPLC. Only the CD spectra of enantiomers of

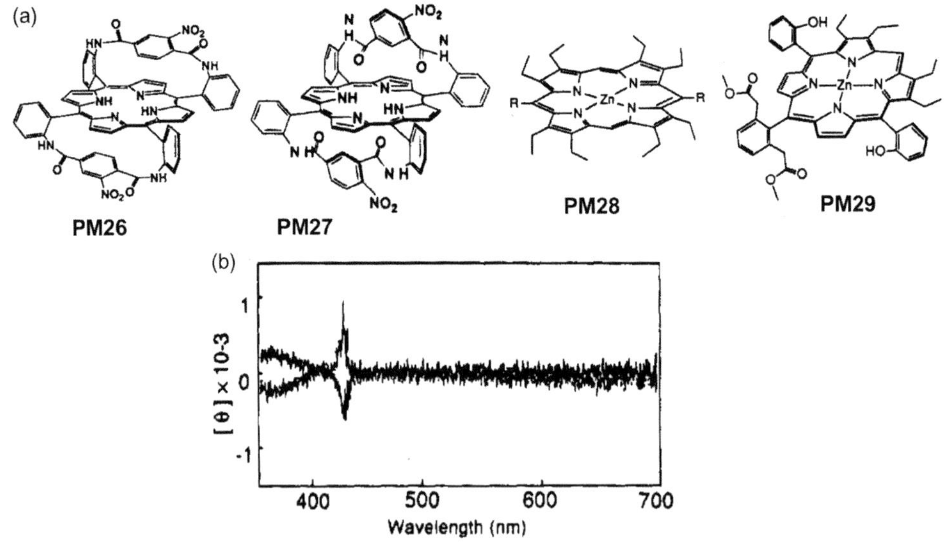

Figure 52. (a) Structure of chiral porphyrins **PM26**, **PM27**, **PM28** and **PM29** and (b) CD spectra of enantiomers of chiral porphyrin monomer **PM27**. Adapted from Refs. 23 and 24 with permission of the American Chemical Society.

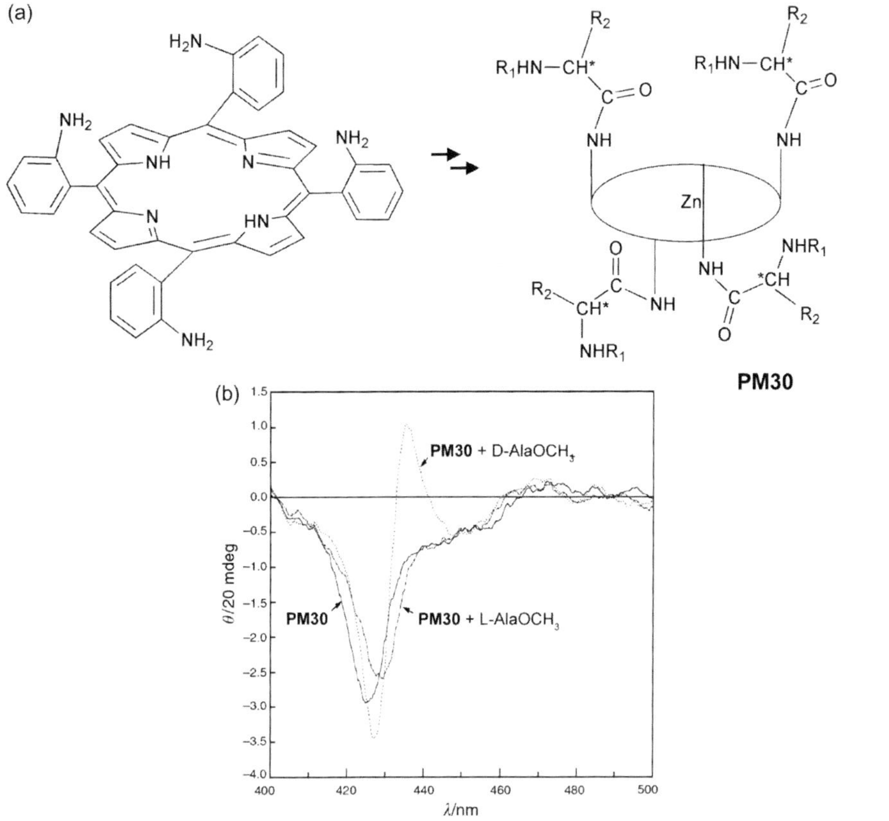

Figure 53. (a) Structures of tetraaminiphenylporphyrin and chiral porphyrin **PM30**. (b) CD spectra of chiral porphyrin **PM30** and its complexes with D-AlaOCH and L-AlaOCH in dichloromethane. Adapted from Ref. 121 with permission of the Royal Society of Chemistry.

the chiral porphyrin **PM27** have been shown with CD bands centered at ~429 nm (Figure 52b).[23]

Using a chiral α,β,α,β-tetraphenyl-substituted Zn(II) porphyrin **PM30** with C2 symmetry prepared from tetraaminiphenylporphyrin (Figure 53a), Zhu and coworkers reported the chiral recognition of amino acid methyl esters in 2001.[121] The chiral Zn(II) porphyrin **PM30** displayed single negative CD band at ~425 nm in the CD spectra (Figure 53b). Complexes of Zn(II) porphyrin **PM30** with L-enantiomers of methyl esters of amino acids exhibited negative CD bands very similar in shape and position to that of Zn(II) porphyrin **PM30**. On the other hand, addition of D-enantiomers of amino acid methyl esters gave rise to bisignate CD signals with positive CD bands at ~435 nm and negative at ~428 nm (Figure 53b). UV-vis spectrophotometric titration experiments showed enantioselectivities up to 21.5 for enantiomers of phenylalanine methyl ester.

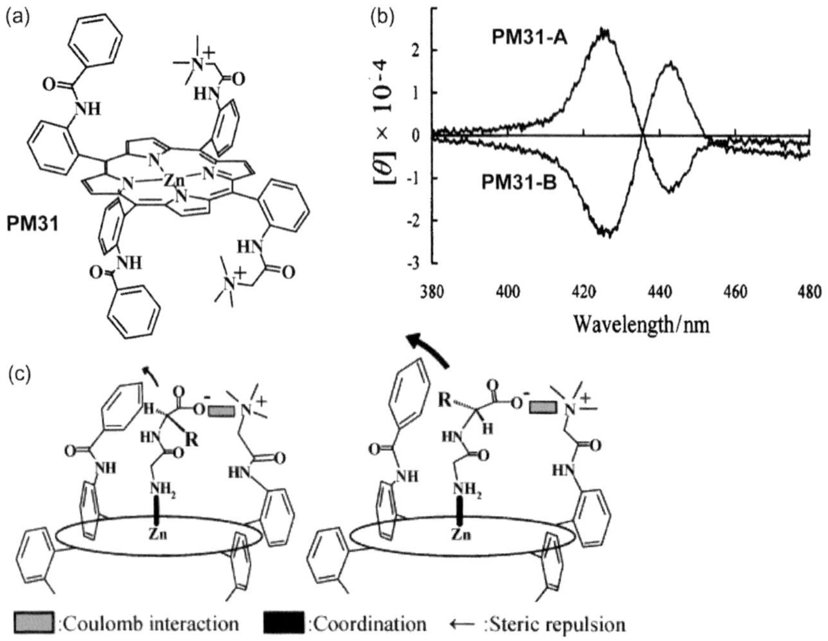

Figure 54. (a) Structure of chiral water-soluble zinc(II)-porphyrin **PM31**. (b) CD spectra of HPLC separated enantiomers of **PM31** in water (pH = 10.4, I = 0.02). (c) Proposed model of chiral recognition. Arrows indicate the steric hindrance between the **PM31** host and amino acid guest. Adapted from Ref. 122 with permission of the American Chemical Society.

In 2004, Imai and coworkers reported stereoselective recognition of amino acids and dipeptides by chiral, water-soluble zinc(II)-porphyrin **PM31** with a C2 symmetry (Figure 54a).[122] Porphyrin **PM31** was prepared as a racemic mixture and enantio-separation was performed using chiral HPLC. The enantiomers displayed mirror-image bisignate couplets in the Soret region at ~425 nm and ~442 nm (Figure 54b). The L-enantiomers of amino acids and dipeptides displayed chiral selectivity towards two enantiomers of **PM31** ranging from 1.2 to 3.3 with the highest value reported for L-Leu and Gly-L-Trp dipeptide. The authors proposed a three-point binding mode relying on electrostatic and coordination interactions together with steric repulsion to explain the observed stereoselectivity (Figure 54c). CD data of the porphyrin-amino acid complexes have not been reported.

Bürgi, Kräutler and coworkers explored a series of biconcave D4-symmetric (**Zn(II)-PM32**, **dioxo-Ru(VI)-PM32**) and C4-symmetric (**pyridino-(CO)-Ru(II)-PM32** and **iodo-Rh(III)-PM32**) metalloporphyrins with a rigid framework (Figure 55a) as chiral NMR shift agents for chiral molecules and chiral nonpolar fullerene adducts.[8,123] The free-base porphyrin was prepared as a single enanatiomer relying

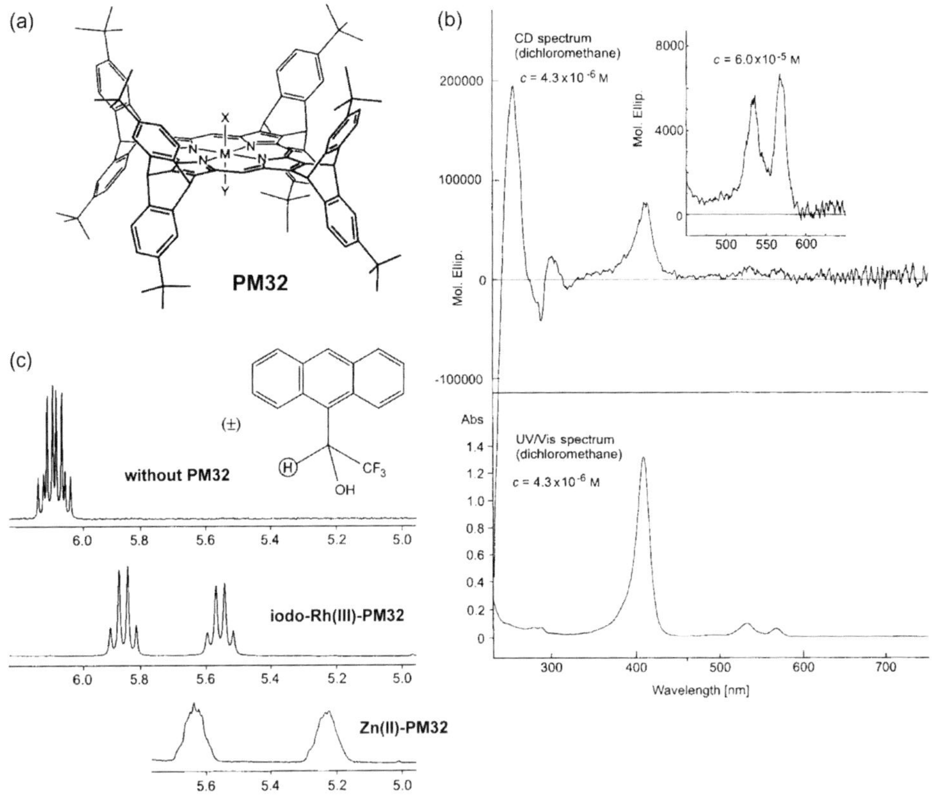

Figure 55. (a) Structure of metalloporphyrins **PM32**. (b) UV-vis and CD spectra of Zn-**PM32** in CH_2Cl_2. (c) ^{1}H NMR shift experiments in C_6D_6 of racemic 1-(9-anthryl)-2,2,2-trifluoro ethanol in the absence (top) and presence (middle and bottom) of chiral bis-metalloporphyrins iodo-Rh(III)-**PM32** and Zn(II)-**PM32**. Adapted from Refs. 8 and 123 with permission of Wiley-VCH.

on a highly diastereo-discriminating Diels–Alder reaction with fumaric acid di-(–) menthyl ester used as a chiral auxiliary.[8] The CD spectrum of **Zn(II)-PM32** has positive CD bands at 408 nm, 300 nm and 248 nm and negative CD bands at 317 nm and 283 nm (Figure 55b).

The diamagnetic **Zn(II)-PM32** and **iodo-Rh(III)-PM32** porphyrins turned out to be the most versatile NMR shift reagents, allowing detection of a broad range of chiral molecular substrates including chiral amines (α-phenylethylamine), alcohols (1-(9-anthryl)-2,2,2-trifluoroethanol), carboxylic acids (α-methoxyphenylacetic acid) and esters (menthyl acetate). On the other hand, paramagnetic **Co(II)-PM32** porphyrin proved to be a suitable chiral shift agent for dinitrile (2,6-di-*tert*-butylanthracene) and several chiral nonpolar fullerene derivatives. CD data of the porphyrin complexes with chiral molecules or fullerenes were not shown.

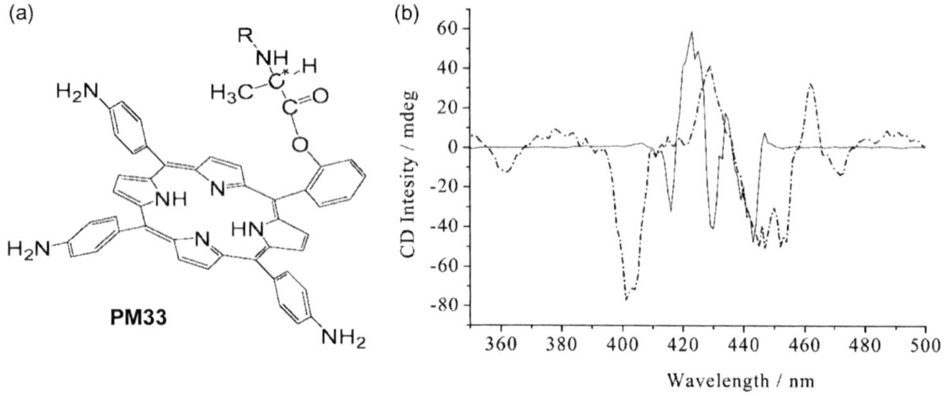

Figure 56. (a) Structure of porphyrin **PM33** (R = Boc). (b) CD spectra of porphyrin **PM33** in $CHCl_3$ (solid line) and the **PM33**/SBA-15 material (dashed-dotted line, solid state) at 298 K. Adapted from Ref. 124 with permission of Wiley-VCH.

Wang and coworkers have explored the amino acid recognition abilities of the solid composite material made by incorporation of chiral porphyrin **PM33** (Figure 56a) into mesoporous molecular sieves SBA-15.[124] Incorporation of the porphyrin **PM33** into the SBA-15 caused a significant change in its CD spectrum in comparison to **PM33** in solution (Figure 56b). The authors rationalized the change in the CD signal by intermolecular interaction of **PM33** with the OH groups in the channels of the SBA-15. UV-vis and CD spectroscopic studies showed higher affinity of the **PM33**:SBA-15 mesoporous composite material for the D-alanine methyl ester than the L-alanine methyl ester (data not shown).

2. *Chiral Porphyrin Dimers*

Crossley and coworkers have reported enantioselective recognition of methyl and benzyl esters of histidine and lysine amino acids by a Troger's base analog chiral porphyrin dimer **PD1** (Figure 29a).[29] Enantiomers of bisporphyrin **PD1** were prepared by a chiral resolution with a chiral HPLC. The resolved enantiomers (+)-**PD1** and (–)-**PD1** possessed a C2 axis of symmetry and showed mirror-image CD spectra with a bisignate couplet in the Soret band region (centered around 420 nm and 452 nm, Figure 57a) and very high molar ellipticities ($[\theta]_{425} = 2.7 \times 10^6$ deg cm^2 $dmol^{-1}$).

Using NMR spectroscopy, the authors showed that the chiral porphyrin dimer displayed high affinities for the histidine methyl and benzyl esters (calculated e.e. up to 80–86%) and lysine benzyl ester (calculated e.e. up to 48%) in organic solvents. The CD spectra of the bisporphyrin-amino acid complexes were not reported. Enantioselective binding of (+)-**PD1** and (–)-**PD1** bisporphyrins with

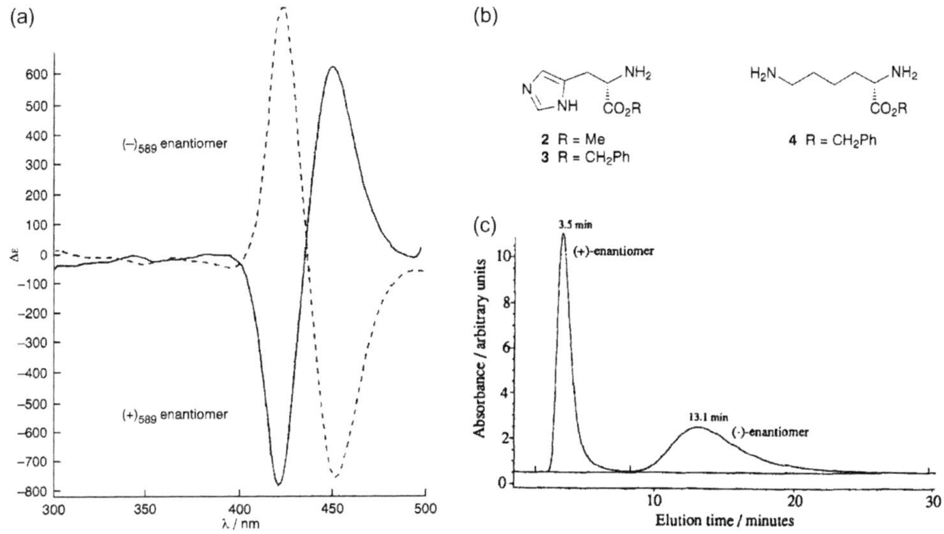

Figure 57. (a) Soret band CD spectra of the enantiomers of Zn(II) bisporphyrin **PD1** in $CHCl_3$. Adapted from Ref. 29 with permission of the Royal Society of Chemistry. (b) Structures of histidine methyl ester and histidine and lysine benzyl esters. (c) Trace of the HPLC separation of (±)-bis-porphyrin **PD1** on a silica gel column pre-saturated with L-histidine benzyl ester (9:1 pentane:chloroform; flow rate 1.0 mL/min, detection: 430 nm). Adapted from Ref. 30 with permission of Elsevier.

L-histidine benzyl ester (K = 1.1×10^8 dm^3/mol *vs.* 1.2×10^7 dm^3/mol) has successfully been used for resolution of **PD1** enantiomers by flash column chromatography on silica pre-saturated with L-histidine benzyl ester (Figure 57c).[30] CD data of the porphyrin-amino acid complexes have not been reported.

A flexible chiral Tröger's base cobalt-bis-porphyrin derivative **PD2** (Figure 30a) was explored by Dolensky and coworkers in 2014 as a potential enantioselective receptor of amino acid methyl esters.[31] Contrary to the bis-porphyrin **PD1**, the bis-porphyrin **PD2** displayed the highest enantioselectivity for lysine methyl ester (1.1×10^6 M^{-1}) rather than histidine methyl ester as determined by UV-vis and CD titration experiments. Higher conformation freedom of bis-porphyrin **PD2** over structurally similar bis-porphyrin **PD1** resulted in lower enantioselectivity but allowed the correlation between the sign of the bisignate couplet in the Soret region with the chirality of the amino acid guest, as observed for Lys-OMe enantiomers (Figure 58b).

In 2005, Ema, Sakai and coworkers reported a C2 symmetry chiral bis-porphyrin **PD26** (Figure 59a) as a receptor and chiral NMR shift reagent for chiral diamines.[125] Chiral bis-porphyrin **PD26** was synthesized as a pure enantiomer from a chiral hydroxy-porphyrin prepared by kinetic resolution with lipase (lipase

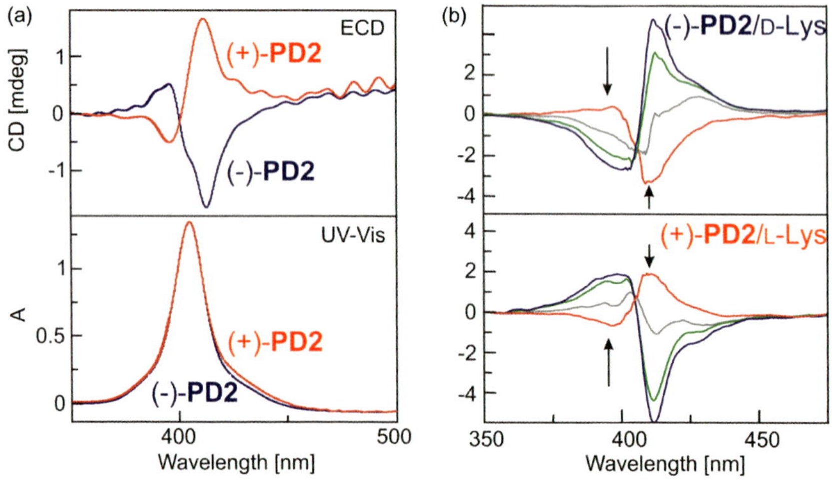

Figure 58. (a) CD and UV-vis spectra of (+)-**PD2** and (–)-**PD2** in CH_2Cl_2. (b) Titration CD spectra of (–)- and (+)-**PD2** with increasing amount of lysine enantiomers. Adapted from Ref. 31 with permission of Wiley-VCH.

PS, *Burkholderia cepacia* lipase immobilized on Celite). The CD spectra of (*R,R*)- and (*S,S*)-**PD26** showed nearly mirror-image (–/+) and (+/–) bisignate couplets, respectively, in the Soret region with CD bands at ~428 nm and ~420 nm (Figure 59b). The bisignate couplet was attributed to the chiral exciton coupling between the two porphyrin chromophores. (*R,R*)- and (*S,S*)-**PD26** successfully functioned as highly sensitive chiral shift reagents for various bifunctional compounds including diamine, aziridines and isooxazoline (Figure 59c).

Addition of achiral as well as chiral diamines amplified and red-shifted the CD couplet by fixing the conformation of the two porphyrin units whereas the signs of the bisignate CD couplets were governed by the chirality of the diamines including 1,2-diphenyl-1,2-ethylenediamine (Figure 59d).

Takeuchi and coworkers have used chiral double-decker porphyrins **PD12a** having one chiral substituent per double-decker[107] and **PD12b** bearing two chiral substituents per double decker[108] for highly enantioselective recognition of dicarboxylic acid (*i.e.*, cyclohexane-(1*R*,2*R*)-dicarboxylic acid) by the control of non-linear responses (Figure 60). **PD12a** exhibited positive homotropic allosterism on binding cyclohexane-1,2-dicarboxylic acids and formed 1:3 complexes with both the (1*R*,2*R*) enantiomer *via* seven H-bonds and the (1*S*,2*S*) enantiomer with six H-bonds as suggested by computational methods. On the other hand, **PD12b** formed 1:2 complex with (1*R*,2*R*) and 1:3 complex with (1*S*,2*S*) cyclohexane-1,2-dicarboxylic acids. The CD spectra of both **PD12a** and **PD12b** displayed a bisignate signal with negative CD bands around 400 nm and positive bands around

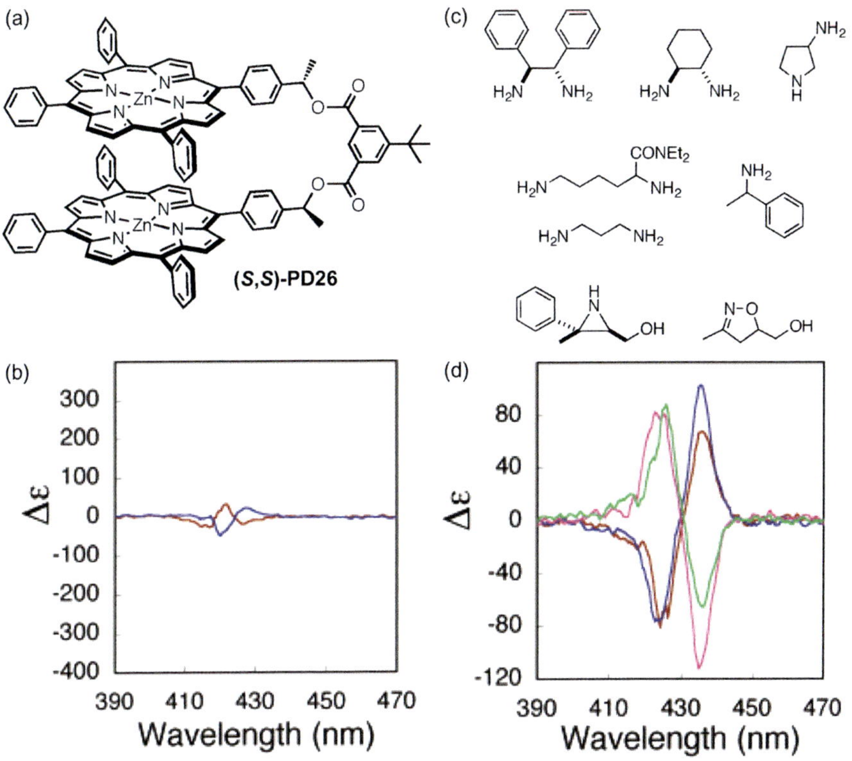

Figure 59. (a) Structure of chiral bis-porphyrin **PD26** and diamines. (b) CD spectra of (*R,R*)- and (*S,S*)-**PD26**. (c) Structures of chiral guest diamines, aziridines and isooxazoline. (d) CD spectra of (*R,R*)-**PD26** the presence of (*R*)- (red curve) and (*S*)- diphenylethylenediamine (magenta curve) and CD spectra of (*S,S*)-**PD26** in presence of (*R*)- (blue curve) and (*S*)- diphenylethylenediamine (green curve). Adapted from Ref. 125 with permission of the American Chemical Society.

310 nm (Figure 60a, 60c). The addition of excess triethylamine to **PD12a** or **PD12b** caused disruption of the intermolecular H-bonds and disappearance of the positive and significant decrease of the negative CD band (Figure 60a, 60c). Both double-decker porphyrins exhibited strong enantioselectivity towards cyclohexane-(1*R*,2*R*)-dicarboxylic acid as was confirmed by CD titrations of **PD12a** (Figure 60b) and **PD12b** (Figure 60d) with racemic cyclohexane-(1,2)-dicarboxylic acid. The CD spectra also suggested suitable preorganization of **PD12** for the chiral dicarboxylic acid guest.

In 2011 and 2012, Ema, Sakai and coworkers reported a chiral free-base porphyrin dimer (*R*)- **PD9** (Figure 38a) possessing a macrocyclic cavity for intercalation of achiral and chiral aromatic guests.[103,104] The porphyrin dimer (*R*)-**PD9** CD spectrum displayed a complex pattern in the 250–600 nm region (Figure 61a).

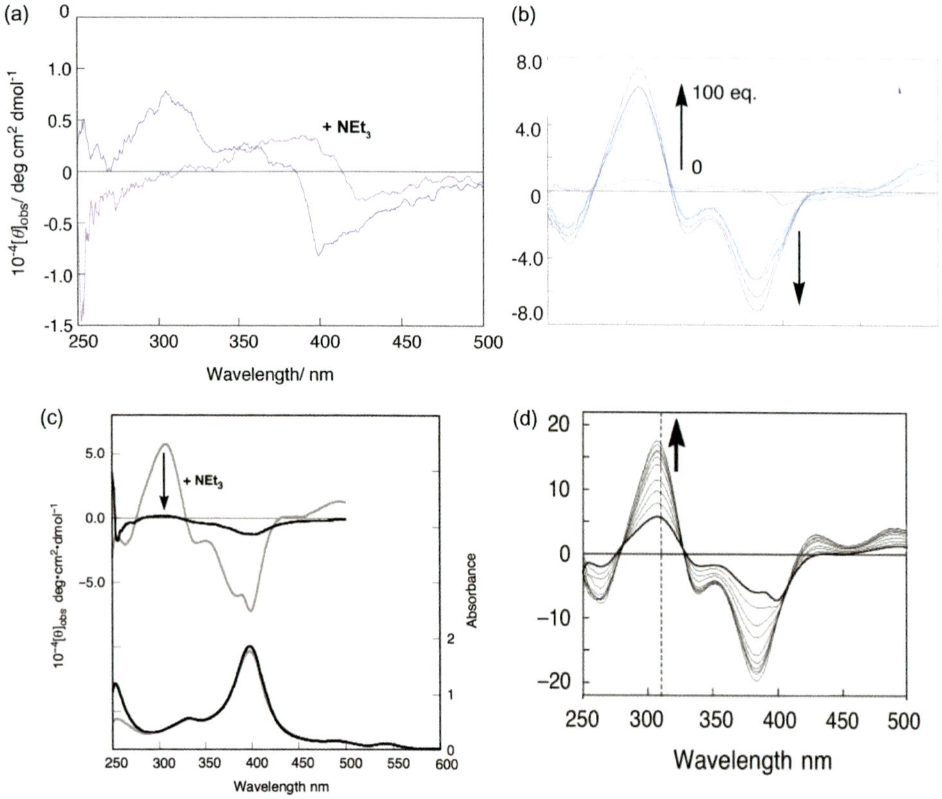

Figure 60. (a) CD spectra of **PD12a** in tetrachloroethane-tetrahydrofuran 30:1 (v/v) without (blue curve) and with (purple curve) triethylamine. (b) Titration CD spectra of **PD12a** with increasing amounts of cyclohexane-(1*R*,2*R*)-dicarboxylic acid. Adapted from Ref. 107 with permission of the American Chemical Society. (c) CD and UV spectra of **PD12b** in tetrachloroethane-tetrahydrofuran 30:1 (v/v) without and with triethylamine. (d) Titration CD spectra of **PD12b** with increasing amount of cyclohexane-(1*R*,2*R*)-dicarboxylic acid. Adapted from Ref. 108 with permission of Taylor & Francis.

Porphyrin dimer (*R*)-**PD9** exhibited excellent recognition and discrimination ability of achiral as well as chiral electron-deficient aromatic molecules using the π–π stacking as the driving force of complexation. Importantly, (*R*)-**PD9** displayed good ability to discriminate the enantiomers of chiral esters bearing a dinitrophenyl group by NMR spectroscopy (Figure 61b).

In 2017, Bian, Jiang and coworkers reported chiral binaphthalene-bridged bis-porphyrins (*R*)-**PD27** and (*S*)-**PD27**[101] with C2 symmetry (Figure 62a) as receptors for chiral supramolecular discrimination of diamines.[126] The bis-porphyrins (*R*)-**PD27** and (*S*)-**PD27** have been prepared as pure enantiomers using (*R*)- or (*S*)-BINOL. The CD spectrum of (*R*)-**PD27** displayed a (–/+) bisignate couplet in the porphyrin Soret region with a negative band at ~430 nm and a positive band at ~420 nm corresponding to a counterclockwise orientation of the porphyrin units,

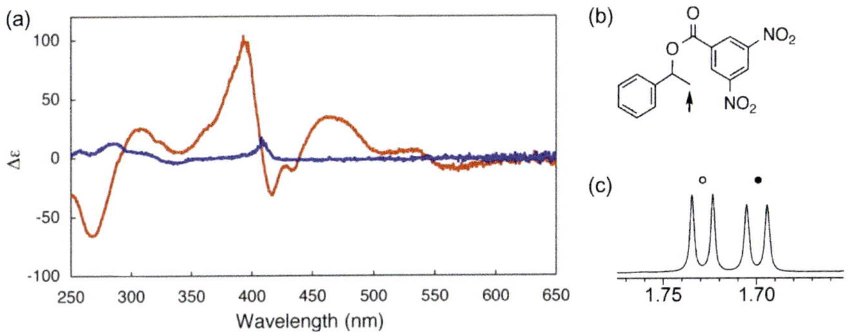

Figure 61. (a) CD spectrum of (*R*)-**PD9** (red curve) and **PD9a** (blue curve) in $CHCl_3$. (b) 600 MHz ^{1}H NMR of chiral but racemic ester of dinitrobenzoic acid in the presence of (*R*)-**PD9** (10 mM, 1 equiv.) in $CDCl_3$. The signals for the arrow marked protons of (*R*)- and (*S*)-enantiomer of the ester are labeled by filled and open circles, respectively. Adapted from Refs. 103 and 104 with permission of the Royal Society of Chemistry.

whereas the CD spectrum of (*S*)-**PD27** displayed a mirror image CD couplet (Figure 62b). Both (*R*)-**PD27** and (*S*)-**PD27** showed enantioselectivity towards chiral 1,2-diamines as determined by UV-vis spectroscopy with the highest value of 5.07 detected for (1*R*,2*R*)/(1*S*,2*S*)-diaminecyclohexane. Importantly, mirror-image CD responses were observed for (*R*)-**PD27** and (*S*)-**PD27** in the presence of (1*R*,2*R*)- as well as (1*S*,2*S*)-diaminecyclohexane (Figure 62c).

B. Supramolecular Recognition and Optical Resolution of Chiral Carbon Nanotubes and Fullerenes by Chiral Porphyrins and Bis-Porphyrins

The electronic, optical and structural properties of carbon nanomaterials such as single-walled carbon nanotubes (SWNTs) and fullerenes are determined by their structure, *i.e.*, 3D orientation of carbon atoms. The majority of SWNTs are chiral (except armchair and zigzag tubes) and several higher fullerenes such as $\mathbf{C}_{76}$, $\mathbf{C}_{78}$ and $\mathbf{C}_{84}$ are chiral and exist as pairs of right-handed and left-handed enantiomers with asymmetrically curved surfaces. Since synthesis of SWNTs yielded racemic mixtures of both helicities much effort has been dedicated to the separation of the non-superimposable mirror image forms. However, chiral carbon nanomaterials lack any appropriate functional groups for point recognition making their chiral discrimination a most challenging task. Chiral porphyrin monomers and dimers have been shown to be among the most efficient molecular hosts for optical resolution of SWNT and fullerenes.

Chiral porphyrin dimers (*i.e.*, chiral diporphyrin nano-tweezers) have played a crucial role in accessing optically pure SWNTs. The first successful separation of the optically active SWNTs (commercially available CoMoCAT) was reported by Komatsu and coworkers in 2007.[7] Their approach utilized molecular recognition of racemic SWNTs by chiral, optically pure enantiomeric diporphyrin nano-tweezer

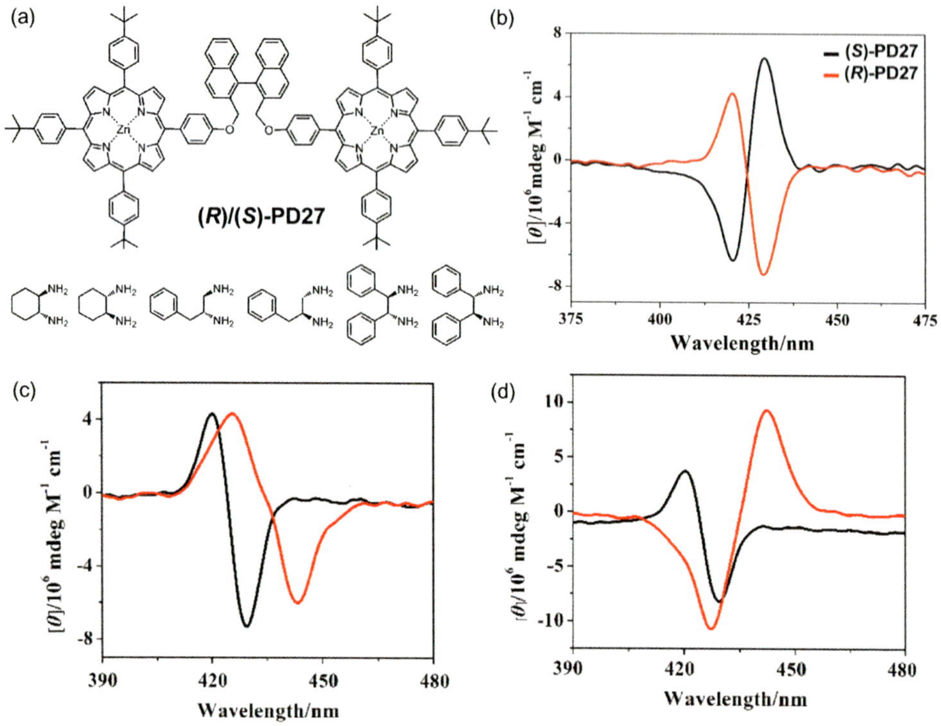

Figure 62. (a) Structures of chiral bis-porphyrins (*R*)-**PD27** and (*S*)-**PD27** and chiral diamines. (b) CD spectra of chiral bis-porphyrins (*R*)-**PD27** (red curve) and (*S*)-**PD27** (black curve). (c) CD spectra of (*R*)-**PD27** before (black curve) and after (red curve) the addition of (1*R*,2*R*)-diaminecyclohexane (20 equiv.). (d) CD spectra of (*R*)-**PD27** before (black curve) and after (red curve) the addition of (1*S*,2*S*)-diaminecyclohexane (20 equiv.). Adapted from Ref. 126 with permission of the American Chemical Society.

hosts (*S*)-**PD28** and (*R*)-**PD28** (Figure 63a). Diporphyrin nano-tweezers (*S*)-**PD28** and (*R*)-**PD28** were synthesized from (*R*)- and (*S*)-phenylalaninal and displayed mirror-image bisignate CD signals in the porphyrin Soret band region with CD bands at 411 nm and 422 nm (Figure 63c). The complexes of nano-tweezer with SWNTs (SWNTs:(*R*)-**PD28** and SWNTs:(*S*)-**PD28**); Figure 63b) displayed red-shifted bisignate CD signals at 415 nm and 430 nm as well as newly induced intense bisignate CD signals at 433 nm and 448 nm (Figure 63c). The authors explained the red shift of the CD by electronic interactions of the nano-tweezer with the SWNTs whereas the new bisignate was rationalized as a result of a locked conformation of the two porphyrin planes. Importantly, exceptional electronic, photophysical and photochemical properties of the porphyrin-SWNTs chiral nano-composites make them promising materials for applications in chiral catalysis or biosensing. Removal of chiral diporphyrin nano-tweezers from SWNTs yielded

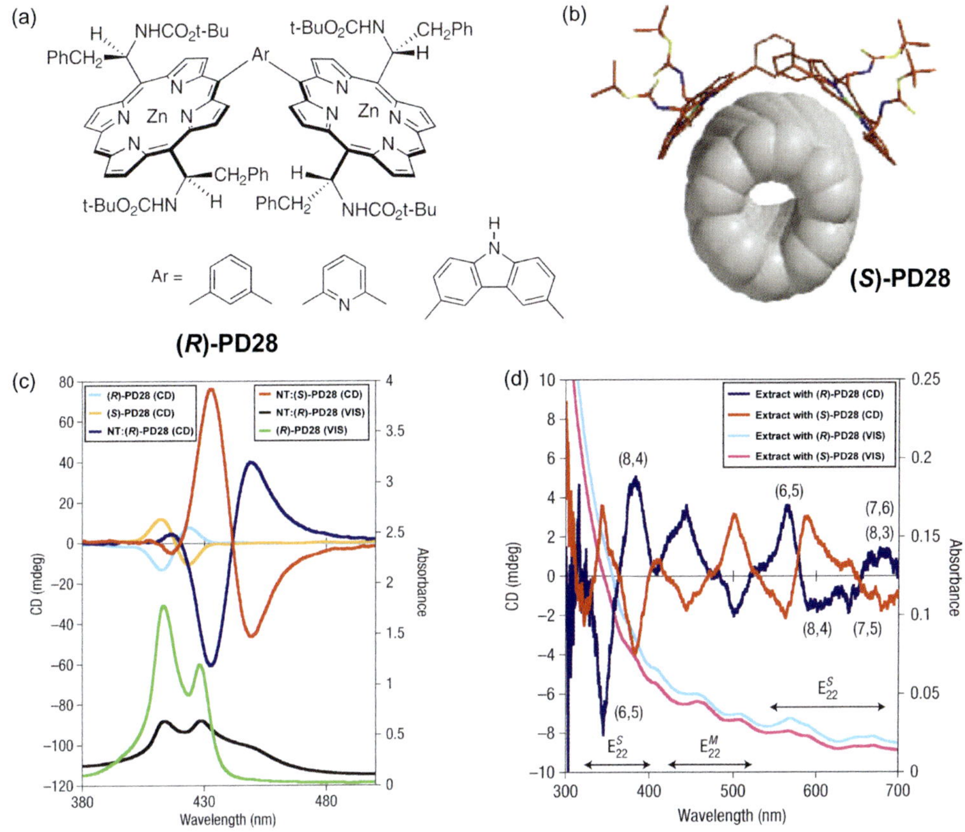

Figure 63. (a) Structure of the chiral nano-tweezers (*R*)-**PD28**. (b) Comprehensive molecular mechanics energy-minimized structure of the left-handed-SWNT:(*S*)-**PD28** complexes. (c) CD (top) and UV-vis spectra of nano-tweezers ((*R*)-**PD28** and (*S*)-**PD28**) and their complexes with CoMaCAT SWNTs: SWNTs:(*R*)-**PD28** and SWNTs:(*S*)-**PD28**). (d) CD and UV-vis spectra of SWNTs extracted with (*R*)-**PD28** and (*S*)-**PD28** after removal of the chiral nano-tweezers. Adapted from Ref. 7 with permission of Nature Publishing Group.

optically active SWNTs with mirror-image CD spectra in the 300–1,100 nm region (Figure 63d). Similar CD spectra (*i.e.*, bisignate CD in the Soret region) and CD behavior (red shift and appearance on intense bisignate CD) were reported by Komatsu and coworkers for 2,6-pyridine-, 3,6-carbazolylene-, *N*-substituted-3,3-carbazolylene and phenanthrene-bridged structural derivative of **PD28**.[127–130]

In 2009, Komatsu and coworkers used chiral monoporphyrin **1** (Figure 64a) prepared in two steps from commercially available (*R*)- or (*S*)-*N*-BOC-phenylalaninal for a host–guest resolution and diameter-based enrichment of SWNTs.[131] (*R*)-**PM34** and (*S*)-**PM34** displayed mirror-image negative and positive CD bands in the Soret band region (406 nm, Figure 64b). The supramolecular complexes of (*R*)-**PM34** and (*S*)-**PM34** with the CoMoCAT SWNTs showed mirror image CD

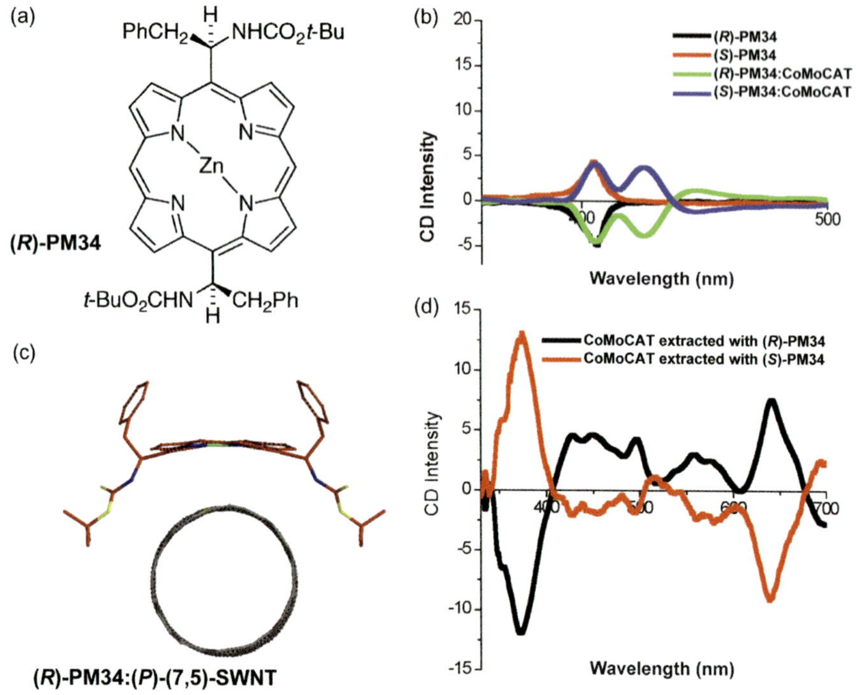

Figure 64. (a) Structure of the chiral porphyrin **PM34**. (b) CD spectra of (*R*)-**PM34**, (*S*)-**PM34** and their complexes with chiral CoMaCAT SWNTs. (c) Computer-generated molecular modeling of the complex structures (view along the SWNT axis) of 1:1 complexes of (*S*)-**PM34**:(*P*)-**(7,5)-SWNT**. (d) CD spectra of the D_2O/SDBS solutions of porphyrin-free CoMoCAT SWNTs extracted with (*R*)-**PM34** (black line) and (*S*)-**PM34** (red line). Adapted from Ref. 131 with permission of the American Chemical Society.

spectra consisting of the CD bands at 406 nm, as well as a newly induced bisignate CD signal (425 nm and 446 nm, Figure 64b). The presence of two CD signals (a single band and a bisignate feature) was explained by the presence of free as well as SWNT-complexed porphyrins co-existing in the solution. The induction of a bisignate CD signal was proposed (based on molecular modeling) to originate from the conformational asymmetry of the porphyrin unit (Figure 64c). Simple synthesis of chiral monoporphyrin **PM34** together with excellent recognition characteristics opened the route for large-scale preparation of optically active SWNTs.[132]

Peng and coworkers used water soluble 5,15-bis-substituted porphyrins functionalized with chiral hydrophilic substituents ((*R*)-**PM35** and (*S*)-**PM35**) to prepare the first example of stable, water-soluble supramolecular porphyrin-SWNT complexes that could be dried and redissolved.[133] The chiral porphyrins (*R*)-**PM35** and (*S*)-**PM35** (Figure 65a) were synthesized from commercially available (*R*)- or (*S*)-*N*-BOC-phenylalaninal and dipyrromethane using the

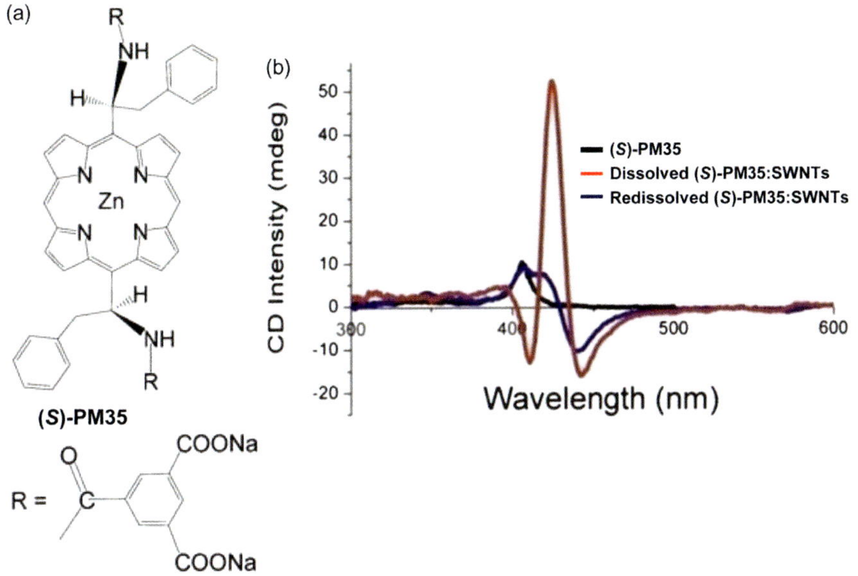

Figure 65. (a) Structure of the chiral porphyrin (*S*)-**PM35** and (b) CD spectra of (*S*)-**PM35** along with the dissolved and re-dissolved (*S*)-**PM35:SWNT** (CoMoCAT) complexes. Adapted from Ref. 133 with permission of the Royal Society of Chemistry.

Lindsey method and displayed negative and positive bands, respectively, in the porphyrin Soret band region (408 nm, Figure 65b). The authors explained these spectra by invoking asymmetrical conformations of porphyrins induced by the chiral substituents. On the other hand, the (*R*)-**PM35:SWNT** complex showed a (+/–/+) trisignate CD feature in the Soret region (443 nm, 426 nm and 412 nm; Figure 65b) rationalized by the SWNT induced locked conformation of the chiral porphyrin and appearance of two sets of overlapping bisignate CD signals (412/426 nm and 426/443 nm). Interestingly, the original CD band of the free porphyrin at 408 nm was absent. Re-dissolution of dried (*R*)-**PM35:SWNT** complex caused a dramatic change of the CD profile in the Soret region: a (+/–; from the long wavelength) bisignate signal of the complex and the negative CD band of the free porphyrin were detected, altogether indicating a different binding geometry (Figure 65b). Importantly, aggregation of porphyrin-coated SWNTs was prevented, likely due to the electrostatic repulsions between the adsorbed charged porphyrins.

In 2013, Komatsu and coworkers reported simultaneous discrimination of chirality, diameter and metallicity of SWNTs with chiral nanoclipper dimers as molecular hosts.[134] In comparison with previously reported nanotweezer dimers,[7,127–130,132] the nanoclipper dimers were designed[134] with (i) a longer (>1.4 nm) yet conformationally restricted spacer, (ii) nearly parallel orientations of the

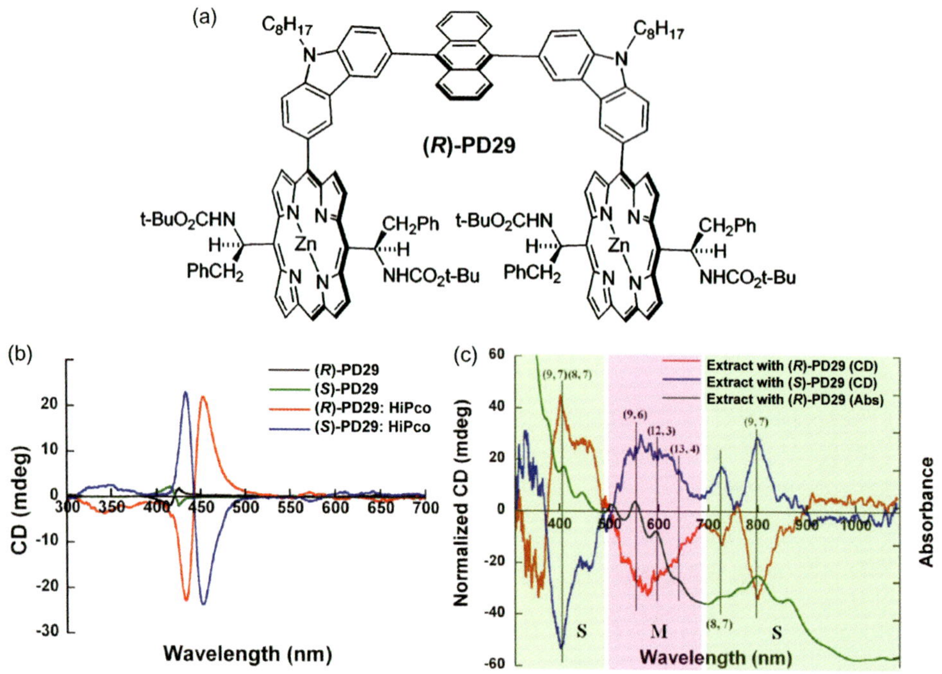

Figure 66. (a) Structure of chiral diporphyrin nanoclipper (*R*)-**PD29**. (b) CD spectra of free chiral nano-clippers (*R*)-**PD29** and (*S*)-**PD29** and their complexes with SWNTs (76-CoMoCAT). (c) CD spectra of porphyrin-free chiral SWNTs extracted with (*R*)-**PD29** and (*S*)-**PD29** and absorption spectra of SWNTs extracted with (*S*)-**PD29**. Green and pink regions indicate semiconducting (S) and metallic (M), respectively. Adapted from Ref. 134 with permission of the American Chemical Society.

porphyrin units, and (iii) stronger π–π stacking interaction of two porphyrins and anthracene with the surface of a SWNT to allow successful recognition of SWNT with larger diameters (>1.0 nm) than with nanotweezers. Nanoclipper dimers (*R*)-**PD29** and (*S*)-**PD29** (Figure 66a) were synthesized as pure enantiomers using (*R*)- or (*S*)-*N*-BOC-phenylalaninal as the chiral starting material. Nanoclippers (*R*)-**PD29** and (*S*)-**PD29** displayed positive and negative bisignate CD signals, respectively, in the porphyrin Soret region (Figure 66b, black and green curves). Complexation of (*R*)-**PD29** or (*S*)-**PD29** with SWNTs (HiPco) gave rise to intense red-shifted mirror-image bisignate CD signals of the nanoclippers:SWNT complex (450 and 436 nm; Figure 66b). The appearance of (+/–) bisignate CD signals in free (*R*)-**PD29** and (*R*)-**PD29**:SWNT complexes (Figure 4b, red curve) was explained by the clockwise orientation of two porphyrin units induced by the stereogenic centers and restricted conformational freedom of the latter. Analogously, a counterclockwise orientation has been proposed for (*S*)-**PD29** and (*S*)-**PD29**:SWNT complex (Figure 66b, blue curve). Due to use of (*R*)-**PD29** and

(*S*)-**PD29** chiral supramolecular hosts, chiral semiconducting SWNTs with diameters >1.0 nm (*i.e.*, (9,7) and (8,7)) and chiral metallic SWNTs (*i.e.*, (9,6), (12,3), and (13,4)) have been successfully isolated and identified for the first time (Figure 66c). Chiral diporphyrin nanoclippers with an even longer spacer (~1.9 nm) consisting of carbazole-pyrene-carbazole have been used for optical resolution of double-walled carbon nanotubes (DWNTs).[135]

VI. Concluding Remarks

This chapter highlighted the studies of some inherently chiral monomeric and dimeric porphyrins, prepared using a wide range of structural modification methods to improve the design of novel structures and their applications. In particular, the chapter focused on the use of ECD spectroscopy for establishing the absolute configurations. However, although significant progress has been made in the analysis of the CD spectra of these types of systems, the results of molecular modeling calculations are extensively utilized by many research groups. It must be emphasized that the CD spectroscopy is highly sensitive to small changes in the conformation of the molecule; this aspect from one side represents a limit of the chiroptical technique, while on the other it can be exploited to study the effect of the interactions with other molecules.

Notably, some examples of supramolecular self-assembly and molecular recognition by chiral porphyrins have also been reported. The design of chiral supramolecular multiporphyrin arrays is a promising strategy for a broad range of potential applications in areas such as chlorophyll mimic systems, sensors, optoelectronics, nanoelectronic devices and also nonlinear optical and chiroptical materials. These systems need an optimal mesoscopic organization of the constituent donor and acceptor chromophores to achieve efficient communication between individual components (*i.e.*, energy or charge-transfer). Therefore, CD spectroscopy is a useful technique to investigate the arrangement adopted by assembled porphyrins. The major application for optically active porphyrins is the supramolecular recognition and optical resolution. Indeed, exploiting the synthetic versatility of porphyrins makes it possible to design selective hosts for separate enantiomeric mixtures of nanomaterials such as SWCNT or fullerenes as well as specific probes for sensing and resolution of chiral biomolecules including aminoacids, peptides and nucleic acid.

VII. References

1. Palyi, G.; Zucchi, C.; Caglioti, L. *Progress in Biological Chirality*; **2004**, Elsevier Ltd.
2. Moji Christianah, A. Use of chiral excipients in formulations containing chiral drugs. *In Chirality in Drug Design and Development*; CRC Press, **2004**.

3. Wang, Z.; Cheng, F.; Winsor, T.; Liu, Y. *Nanotechnology* **2016**, 27:412001, 1–20.
4. Nakanishi, K.; Berova, N.; Woody, R. *Circular Dichroism: Principles and Applications*. VCH, **1994**.
5. Harada, N.; Nakanishi, K. *Circular Dichroic Spectroscopy — Exciton Coupling in Organic Stereochemistry*. University Science Books: Mill Valley, CA, **1983**.
6. Hayashi, T.; Nonoguchi, M.; Aya, T.; Ogoshi, H. *Tetrahedron Lett.* **1997**, *38*, 1603–1606.
7. Peng, X.; Komatsu, N.; Bhattacharya, S.; Shimawaki, T.; Aonuma, S.; Kimura, T.; Osuka, A. *Nat Nano* **2007**, *2*, 361–365.
8. Schwenninger, R.; Ramondenc, Y.; Wurst, K.; Schlögl, J.; Kräutler, B. *Chem. Eur. J.* **2000**, *6*, 1214–1223.
9. Sibrian-Vazquez, M.; Jensen, T. J.; Fronczek, F. R.; Hammer, R. P.; Vicente, M. G. H. *Bioconjugate Chem.* **2005**, *16*, 852–863.
10. Karikis, K.; Georgilis, E.; Charalambidis, G.; Petrou, A.; Vakuliuk, O.; Chatziioannou, T.; Raptaki, I.; Tsovola, S.; Papakyriacou, I.; Mitraki, A.; Gryko, D. T.; Coutsolelos, A. G. *Chem. Eur. J.* **2016**, *22*, 11245–11252.
11. Sibrian-Vazquez, M.; Jensen, T. J.; Hammer, R. P.; Vicente, M. G. H. *J. Med. Chem.* **2006**, *49*, 1364–1372.
12. Czuchajowski, L.; Habdas, J.; Niedbala, H.; Wandrekar, V. *J. Heterocyclic Chem.* **1992**, *29*, 479–486.
13. Balaz, M.; Holmes, A. E.; Benedetti, M.; Proni, G.; Berova, N. *Bioorg. Med. Chem.* **2005**, *13*, 2413–2421.
14. Balaz, M.; Steinkruger, J. D.; Ellestad, G. A.; Berova, N. *Org. Lett.* **2005**, *7*, 5613–5616.
15. Mammana, A.; Asakawa, T.; Bitsch-Jensen, K.; Wolfe, A.; Chaturantabut, S.; Otani, Y.; Li, X. X.; Li, Z. M.; Nakanishi, K.; Balaz, M.; Ellestad, G. A.; Berova, N. *Bioorg. Med. Chem.* **2008**, *16*, 6544–6551.
16. Frolova, E. I.; Ivanova, E. M.; Zarytova, V. F.; Abramova, T. V.; Vlassov, V. V. *FEBS Lett.* **1990**, *269*, 101–104.
17. Kus, P.; Knerr, G.; Czuchajowski, L. *Tetrahedron Lett.* **1990**, *31*, 5133–5134.
18. Bouamaied, I.; Fendt, L. A.; Wiesner, M.; Haussinger, D.; Amiot, N.; Thoni, S.; Stulz, E. *Pure Appl. Chem.* **2006**, *78*, 2003–2014.
19. Oancea, S.; Formaggio, F.; Campestrini, S.; Broxterman, Q. B.; Kaptein, B.; Toniolo, C. *Biopolymers* **2003**, *72*, 105–115.
20. Sakakibara, K.; Nakatsubo, F.; French, A. D.; Rosenau, T. *Chem. Commun.* **2012**, *48*, 7672–7674.
21. Carofiglio, T.; Fornasier, R.; Lucchini, V.; Simonato, L.; Tonellato, U. *J. Org. Chem.* **2000**, *65*, 9013–9021.
22. Matile, S.; Berova, N.; Nakanishi, K.; Fleischhauer, J.; Woody, R. W. *J. Am. Chem. Soc.* **1996**, *118*, 5198–5206.
23. Kuroda, Y.; Kato, Y.; Higashioji, T.; Hasegawa, J.-y.; Kawanami, S.; Takahashi, M.; Shiraishi, N.; Tanabe, K.; Ogoshi, H. *J. Am. Chem. Soc.* **1995**, *117*, 10950–10958.
24. Mizutani, T.; Ema, T.; Tomita, T.; Kuroda, Y.; Ogoshi, H. *J. Am. Chem. Soc.* **1994**, *116*, 4240–4250.
25. Chappaz-Gillot, C.; Canard, G.; Andreoli, F.; Vanthuyne, N.; Giorgi, M.; Naubron, J. V.; Monnier, V.; Rosas, R.; Roussel, C.; Balaban, T. S. *Eur. J. Org. Chem.* **2012**, 6526–6536.
26. Matano, Y.; Shinokura, T.; Matsumoto, K.; Imahori, H.; Nakano, H. *Chem. Asian J.* **2007**, *2*, 1417–1429.
27. Yang, L. G.; Zhou, Y.; Zhu, M. L.; Zhao, L. Y.; Wei, L. Y.; Bian, Y. Z. *J. Org. Chem.* **2013**, *78*, 9949–9955.

28. Peng, X.; Sternberg, E.; Dolphin, D. *Electrophoresis* **2002**, *23*, 93–101.
29. Crossley, M. J.; Mackay, L. G.; Try, A. C. *J. Chem. Soc., Chem. Commun.* **1995**, 1925–1927.
30. Allen, P. R.; Reek, J. N. H.; Try, A. C.; Crossley, M. J. *Tetrahedron: Asymmetry* **1997**, *8*, 1161–1164.
31. Tatar, A.; Valik, M.; Novotna, J.; Havlik, M.; Dolensky, B.; Kral, V.; Urbanova, M. *Chirality* **2014**, *26*, 361–367.
32. Shoji, Y.; Tashiro, K.; Aida, T. *J. Am. Chem. Soc.* **2006**, *128*, 10690–10691.
33. Yoshida, N.; Osuka, A. *Tetrahedron Lett.* **2000**, *41*, 9287–9291.
34. Muranaka, A.; Asano, Y.; Tsuda, A.; Osuka, A.; Kobayashi, N. *ChemPhysChem* **2006**, *7*, 1235–1240.
35. Hamazawa, A.; Yano, T.; Nishioka, T.; Kinoshita, I.; Isobe, K.; Yano, S.; Wright, L. J.; Collins, T. J. *Chem. Lett.* **2003**, 32(1), 20–21.
36. Schulze, M.; Scherer, A.; Hampel, F.; Stryker, J. M.; Tykwinski, R. R. *Chem. Eur. J.* **2016**, *22*, 3378–3386.
37. Littler, B. J.; Ciringh, Y.; Lindsey, J. S. *J. Org. Chem.* **1999**, *64*, 2864–2872.
38. Mullins, O. C.; Sabbah, H.; Eyssautier, J.; Pomerantz, A. E.; Barrõ, L.; Andrews, A. B.; Ruiz-Morales, Y.; Mostowfi, F.; McFarlane, R.; Goual, L.; Lepkowicz, R.; Cooper, T.; Orbulescu, J.; Leblanc, R. M.; Edwards, J.; Zare, R. N. *Energy Fuels* **2012**, *26*, 3986–4003.
39. Gray, M. R.; Tykwinski, R. R.; Stryker, J. M.; Tan, X. *Energy Fuels* **2011**, *25*, 3125–3134.
40. Scherer, A.; Hampel, F.; Gray, M. R.; Stryker, J. M.; Tykwinski, R. R. *J. Phys. Org. Chem.* **2012**, *25*, 597–606.
41. Schulze, M.; Scott, D. E.; Scherer, A.; Hampel, F.; Hamilton, R. J.; Gray, M. R.; Tykwinski, R. R.; Stryker, J. M. *Org. Lett.* **2015**, *17*, 5930–5933.
42. Cantonetti, V.; Monti, D.; Venanzi, M.; Bombelli, C.; Ceccacci, F.; Mancini G. *Tetrahedron: Asymmetry* **2004**, *15*, 1969–1977.
43. Monti, D.; Cantonetti, V.; Venanzi, M.; Ceccacci, F.; Bombelli, C.; Mancini, G. *Chem. Commun.*, **2004**, 972–973.
44. Jintoku, H.; Shimoda, S.; Takafuji, M.; Sagawa, T.; Ihara, H. *Mol. Cryst. Liq. Cryst.* **2011**, *539*, 63/[403]–67/[407].
45. Shu-Jun, W.; Wen-Juan, R.; Xiao-Jing, Z.; Dai-Bing, L.; Zhi-Ang, Z. *Chin. J. Chem.* **2005**, *23*, 44–49.
46. Wytko, J. A.; Graf, E.; Weiss, J. *J. Org. Chem.* **1992**, *57*, 1015.
47. Takeuchi, M.; Fujikoshi, C.; Kubo, Y.; Kaneko, K.; Shinkai, S. *Angew. Chem., Int. Ed.* **2006**, *45*, 5494.
48. Gehrold, A. C.; Bruhn, T.; Schneider, H.; Radius, U.; Bringmann, G. *Org. Lett.* **2015**, *17*, 210–213.
49. Gehrold, A. C.; Bruhn, T.; Schneider, H.; Radius, U.; Bringmann, G. *J. Org. Chem.* **2015**, *80*, 12359–12378.
50. Gazeau, S.; Pècaut, J.; Marchon, J. C. *Chem. Commun.* **2001**, 1644–1645.
51. Gazeau, S.; Pécaut, J.; Haddad, R.; Shelnutt, J. A.; Marchon, J. C. *Eur. J. Inorg. Chem.* **2002**, 2956–2960.
52. Maheut, G.; Castaings, A.; Pècaut, J.; Lawson Daku, L. M.; Pescitelli, G.; Di Bari, L.; Marchon, J.-C. *J. Am. Chem. Soc.* **2006**, *128*, 6347–6356.
53. Fateeva, A.; Pécaut, J.; Bayle, P.-A.; Maldivi, P.; Dubois, L. *Eur. J. Org. Chem.* **2009**, 3845–3848.
54. Di Bari, L.; Pescitelli, G.; Reginato, G.; Salvadori, P. *Chirality* **2001**, *13*, 548–555.
55. Whitesell, J. K. *Chem. Rev.* **1989**, *89*, 1581–1590.
56. Rubio, M.; Merchán, M.; Ortí, E.; Roos, B. O. *Chem. Phys.* **1994**, *179*, 395–409.

57. Mason, S. F.; Seal, R. H.; Roberts D. R. *Tetrahedron* **1974**, *30*, 1671–1682.
58. Gouterman, M. In: Dolphin D, editor. The porphyrins, Vol. 3. *In Physical Chemistry, Part A*. New York: Academic Press; **1962**.
59. Lu, J.; Meng, Q.; Zhang, H.; Wang, H.; Ma, Z.; Sun, L. *Inorg. Chim. Acta* **2014**, *423*, 250–255.
60. Lu, J.; Wu, L.; Jiang, J.; Zhang, X. *Eur. J. Inorg. Chem.* **2010**, 4000–4008.
61. Lu, J.; Deng, Y.; Zhang, X.; Kobayashi, N.; Jiang, J. *Inorg. Chem.* **2011**, *50*, 2562–2567.
62. Lo, W.-C.; Che, C.-M.; Cheng, K.-F.; Mak, T. C. W. *Chem. Commun.* **1997**, 1205–1206.
63. Hirasawa, K.; Yuge, H.; Miyamoto, T. K. *Acta Cryst.* **2008**, *C64*, m97–m100.
64. Lu, J.; Wu, L.; Jing, L.; Xu, X.; Zhang, X. *Dyes and Pigments* **2012**, *94*, 169–174.
65. Tamaru, S.; Takeuchi, M.; Sano, M.; Shinkai, S. *Angew. Chem. Int. Ed.* **2002**, *41*, 853–856.
66. Hameren, R.; Van Buul, A. M.; Castriciano, M. A.; Villari, V.; Micali, N.; Schön, P.; Speller, S.; Scolaro, L. M.; Rowan, A. E.; Elemans, J. A.; Nolte, R. J. *Nano Lett.* **2008**, *8*, 253–259.
67. Kral, V.; Pataridis, S.; Setnicka, V.; Zaruba, K.; Urbanova, M.; Volka, K. *Tetrahedron* **2005**, *61*, 5499–5506.
68. Li, L.-L.; Diau, E.W.-G. *Chem. Soc. Rev.* **2013**, *42*, 291–304.
69. Ding, Y. B.; Tang, Y. Y.; Zhu, W. H.; Xie, Y. S. *Chem. Soc. Rev.* **2015**, *44*, 1101–1112.
70. Andernach, R. E.; Rossbauer, S.; Ashraf, R. S.; Faber, H.; Anthopoulus, T. D.; McCulloch, I.; Heeney, M.; Bronstein, H. A. *ChemPhysChem* **2015**, *16*, 1223–1230.
71. Liu, H.; Xu, J.; Li, Y.; Li, Y. *Acc. Chem. Res.* **2010**, *43*, 1496–1508.
72. Senge, M. O.; Fazekas, M.; Notaras, E. G. A.; Blau, W. J.; Zawadzka, M.; Locos, O. B.; Ni Mhuircheartaigh, E. M. *Adv. Mater.* **2007**, *19*, 2737–2774.
73. Balaban, T. S.; Bhise, A. D.; Fischer, M.; Linke-Schaetzel, M.; Roussel, C.; Vanthuyne, N. *Angew. Chem. Int. Ed.* **2003**, *42*, 2140–2144.
74. Hoeben, F. J. M.; Wolffs, M.; Zhang, J.; De Feyter, S.; Leclère, P.; Schenning, A. P. H. J.; Meijer, E. W. *J. Am. Chem. Soc.* **2007**, *129*, 9819–9828.
75. Oliveras-González, C.; Di Meo, F.; González-Campo, A.; Beljonne, D.; Norman, P.; Simón-Sorbed, M.; Linares, M.; Amabilino, D. B. *J. Am. Chem. Soc.* **2015**, *137*, 15795–15808.
76. Monti, D.; Venanzi, M.; Mancini, G.; Di Natale, C.; Paolesse, R. *Chem. Commun.* **2005**, 2471–2473.
77. Monti, D.; Venanzi, M.; Stefanelli, M.; Sorrenti, A.; Mancini, G.; Di Natale, C.; Paolesse, R. *J. Am. Chem. Soc.* **2007**, *129*, 6688–6689.
78. Jintoku, H.; Sagawa, T.; Takafuji, M.; Ihara, H. *Org. Biomol. Chem.* **2009**, *7*, 2430–2434.
79. Symonyi, M.; Bikadi, Z.; Zsila, F.; Deli, J. *Chirality* **2003**, *15*, 680–698.
80. Wang, Q.; Chen, Y.; Ma, P.; Lu, J.; Zhang, X.; Jiang, J. *J. Mater. Chem.* **2011**, *21*, 8057–8065.
81. Berova, N.; Nakanishi K.; Woody, R. *Circular Dichroism: Princples and Applications*, 2nd ed., Wiley-VCH: New York, **2000**: pp. 337–382;
82. Pattus, F.; Heitz, F.; Martinez, C.; Provencher S. W.; Lazdunski, C. *Eur. J. Biochem.* **1985**, *152*, 681–689.
83. Villa, A.; Mark, A. E.; Saracino, G. A. A.; Cosentino, U.; Pitea, D.; Moro G.; Salmona, M. *J. Phys. Chem. B* **2006**, *110*, 1423–1428.
84. Nieto-Ortega, B.; Ramirez, F. J.; Amabilino, D. B.; Linares, M.; Beljonne, D.; Lopez Navarrete, J. T.; Casado J. *Chem. Commun.*, **2012**, *48*, 9147–9149.
85. Castriciano, M. A.; Zagami, R.; Trapani, M.; Romeo, A.; Patanè, S.; Monsù Scolaro, L. *Chirality* **2005**, *27*, 900–906.
86. McRae, E. G.; Kasha, M. *J. Chem. Phys.* **1958**, *28*, 721–722.
87. Helmich, F.; Lee, C. C.; Schenning, A. P. H. J.; Meijer, E. W. *J. Am. Chem. Soc.* **2010**, *132*, 16753–16755.

88. Helmich, F.; Lee, C. C.; Nieuwenhuizen, M. M. L.; Gielen, J. C.; Christianen, P. C. M.; Larsen, A.; Fytas, G.; Leclere, P.; Schenning, A. P. H. J.; Meijer, E. W. *Angew. Chem., Int. Ed.* **2010**, *49*, 3939–3942.
89. Charalambidis, G.; Georgilis, E.; Panda, M. K.; Anson, C. E.; Powell, A. K.; Doyle, S.; Moss, D.; Jochum, T.; Horton, P. N.; Coles, S. J.; Linares, M.; Beljonne, D.; Naubron, J.-V.; Conradt, J.; Kalt, H.; Mitraki, A.; Coutsolelos, A. G.; Balaban, T. S. *Nat. Commun.* **2016**, *7*, 12657–12668.
90. Harada, N.; Chen, S. L.; Nakanishi, K. *J. Am. Chem. Soc.* **1975**, *97*, 5345–5352.
91. Berova, N.; Nakanishi, K., Exciton Chirality Method: Principles and Applications. In *Circular Dichroism, Principles and Applications*, 2nd ed.; Berova, N.; Nakanishi, K.; Woody, R. W., Eds. Wiley-VCH: New York, **2000**: pp. 337–382.
92. Ema, T.; Nemugaki, S.; Tsuboi, S.; Utaka, M. *Tetrahedron Lett.* **1995**, *36*, 5905–5908.
93. Matile, S.; Berova, N.; Nakanishi, K.; Novkova, S.; Philipova, I.; Blagoev, B. *J. Am. Chem. Soc.* **1995**, *117*, 7021–7022.
94. Matile, S.; Berova, N.; Nakanishi, K.; Fleischhauer, J.; Woody, R. W. *J. Am. Chem. Soc.* **1996**, *118*, 5198–5206.
95. Pescitelli, G.; Gabriel, S.; Wang, Y.; Fleischhauer, J.; Woody, R. W.; Berova, N. *J. Am. Chem. Soc.* **2003**, *125*, 7613–7628.
96. Tanaka, K.; Itagaki, Y.; Satake, M.; Naoki, H.; Yasumoto, T.; Nakanishi, K.; Berova, N. *J. Am. Chem. Soc.* **2005**, *127*, 9561–9570.
97. Tsubaki, K.; Takaishi, K.; Tanaka, H.; Miura, M.; Kawabata, T. *Org. Lett.* **2006**, *8*, 2587–2590.
98. Ema, T.; Misawa, S.; Nemugaki, S.; Sakai, T.; Utaka, M. *Chem. Lett.* **1997**, *26*, 487–488.
99. Flores, V.; Nguyen, C. K.; Sindelar, C. A.; Vasquez, L. D.; Shachter, A. M. *Tetrahedron Lett.* **1996**, *37*, 8633–8636.
100. Hayashi, T.; Aya, T.; Nonoguchi, M.; Mizutani, T.; Hisaeda, Y.; Kitagawa, S.; Ogoshi, H. *Tetrahedron* **2002**, *58*, 2803–2811.
101. Lu, W.; Li, L.; Yang, H.; Zhao, L.; Qi, D.; Bian, Y.; Jiang, J. *Dyes Pigm.* **2017**, *137*, 608–614.
102. Lu, W. X.; Wang, C. M.; Li, X. Y.; Zhan, X. P.; Qi, D. D.; Bian, Y. Z. *Inorg. Chem. Commun.* **2017**, *81*, 18–21.
103. Ema, T.; Ura, N.; Eguchi, K.; Ise, Y.; Sakai, T. *Chem. Commun.* **2011**, *47*, 6090–6092.
104. Ema, T.; Ura, N.; Eguchi, K.; Sakai, T. *Bull. Chem. Soc. Jpn.* **2012**, *85*, 101–109.
105. MacMillan, J. B.; Molinski, T. F. *J. Am. Chem. Soc.* **2004**, *126*, 9944–9945.
106. Arai, T.; Takei, K.; Ogawa, Y.; Nishino, N. *Eur. J. Org. Chem.* **2005**, *2005*, 5067–5076.
107. Ikeda, T.; Hirata, O.; Takeuchi, M.; Shinkai, S. *J. Am. Chem. Soc.* **2006**, *128*, 16008–16009.
108. Ikeda, T.; Sada, K.; Shinkai, S.; Takeuchi, M. *Supramol. Chem.* **2011**, *23*, 59–64.
109. Holmes, A. E.; Das, D.; Canary, J. W. *J. Am. Chem. Soc.* **2007**, *129*, 1506–1507.
110. Gehrold, A. C.; Bruhn, T.; Bringmann, G. *J. Org. Chem.* **2016**, *81*, 1075–1088.
111. Yoshida, N.; Ishizuka, T.; Osuka, A.; Jeong, D. H.; Cho, H. S.; Kim, D.; Matsuzaki, Y.; Nogami, A.; Tanaka, K. *Chem. Eur. J.* **2003**, *9*, 58–75.
112. Ouyang, Q.; Zhu, Y. Z.; Li, Y. C.; Wei, H. B.; Zheng, J. Y. *J. Org. Chem.* **2009**, *74*, 3164–3167.
113. Bringmann, G.; Gotz, D. C. G.; Gulder, T. A.; Gehrke, T. H.; Bruhn, T.; Kupfer, T.; Radacki, K.; Braunschweig, H.; Heckmann, A.; Lambertt, C. *J. Am. Chem. Soc.* **2008**, *130*, 17812–17825.
114. Bringmann, G.; Rüdenauer, S.; Götz, D. C. G.; Gulder, T. A. M.; Reichert, M. *Org. Lett.* **2006**, *8*, 4743–4746.
115. Tsuda, A.; Furuta, H.; Osuka, A. *J. Am. Chem. Soc.* **2001**, *123*, 10304–10321.
116. Gotz, D. C. G.; Bruhn, T.; Senge, M. O.; Bringmann, G. *J. Org. Chem.* **2009**, *74*, 8005–8020.

117. Harada, N.; Nakanishi, K.; Berova, N., Electronic CD Exciton Chirality Method: Principles and Applications. In *Comprehensive Chiroptical Spectroscopy*, John Wiley & Sons, Inc.: 2012; pp. 115–166.
118. Shoji, Y.; Tashiro, K.; Aida, T. *Chirality* **2008**, *20*, 420–424.
119. Kuroda, Y.; Kato, Y.; Higashioji, T.; Ogoshi, H. *Angew. Chem Int. Ed. Engl.* **1993**, *32*, 723–724.
120. Mizutani, T.; Ema, T.; Yoshida, T.; Kuroda, Y.; Ogoshi, H. *Inorg. Chem.* **1993**, *32*, 2072–2077.
121. Wang, C. Z.; Zhu, Z. A.; Li, Y.; Chen, Y. T.; Wen, X.; Miao, F. M.; Chan, W. L.; Chan, A. S. C. *New J. Chem.* **2001**, *25*, 801–806.
122. Imai, H.; Munakata, H.; Uemori, Y.; Sakura, N. *Inorg. Chem.* **2004**, *43*, 1211–1213.
123. Schwenninger, R.; Schlögl, J.; Maynollo, J.; Gruber, K.; Ochsenbein, P.; Bürgi, H.-B.; Konrat, R.; Kräutler, B. *Chem. Eur. J.* **2001**, *7*, 2676–2686.
124. Fa, H.-B.; Zhao, L.; Wang, X.-Q.; Yu, J.-H.; Huang, Y.-B.; Yang, M.; Wang, D.-J. *Eur. J. Inorg. Chem.* **2006**, 2006, 4355–4361.
125. Ema, T.; Ouchi, N.; Doi, T.; Korenaga, T.; Sakai, T. *Org. Lett.* **2005**, *7*, 3985–3988.
126. Lu, W.; Yang, H.; Li, X.; Wang, C.; Zhan, X.; Qi, D.; Bian, Y.; Jiang, J. *Inorg. Chem.* **2017**, *56*, 8223–8231.
127. Peng, X.; Komatsu, N.; Kimura, T.; Osuka, A. *J. Am. Chem. Soc.* **2007**, *129*, 15947–15953.
128. Peng, X. B.; Komatsu, N.; Kimura, T.; Osuka, A. *ACS Nano* **2008**, *2*, 2045–2050.
129. Wang, F.; Matsuda, K.; Rahman, A.; Peng, X. B.; Kimura, T.; Komatsu, N. *J. Am. Chem. Soc.* **2010**, *132*, 10876–10881.
130. Wang, F.; Matsuda, K.; Rahman, A.; Kimura, T.; Komatsu, N. *Nanoscale* **2011**, *3*, 4117–4124.
131. Peng, X. B.; Wang, F.; Kimura, T.; Komatsu, N.; Osuka, A. *J. Phys. Chem. C* **2009**, *113*, 9108–9113.
132. Peng, X. B.; Wang, F.; Bauri, A. K.; Rahman, A.; Komatsu, N. *Chem. Lett.* **2010**, *39*, 1022–1027.
133. Peng, X. B.; Qin, H. M.; Li, L. S.; Huang, Y. Y.; Peng, J. B.; Cao, Y.; Komatsu, N. *J. Mater. Chem.* **2012**, *22*, 5764–5769.
134. Liu, G.; Wang, F.; Chaunchaiyakul, S.; Saito, Y.; Bauri, A. K.; Kimura, T.; Kuwahara, Y.; Komatsu, N. *J. Am. Chem. Soc.* **2013**, *135*, 4805–4814.
135. Liu, G.; Saito, Y.; Nishio-Hamane, D.; Bauri, A. K.; Flahaut, E.; Kimura, T.; Komatsu, N. *J. Mater. Chem. A* **2014**, *2*, 19067–19074.

Index